AF411775

Flock of sheep grazing on field stubble near Gordion. The flat Citadel mound is visible midground to the far left, and the Küş Tepe is the conical mound in the center, through the trees. (Early July, 1992)

GORDION SPECIAL STUDIES

I: *The Nonverbal Graffiti, Dipinti, and Stamps* by Lynn E. Roller, 1987

II: *The Terracotta Figurines and Related Vessels* by Irene Bald Romano, 1995

III: *Gordion Seals and Sealings: Individuals and Society* by Elspeth R. M. Dusinberre, 2005

IV: *The Incised Drawings from Early Phrygian Gordion* by Lynn E. Roller, 2009

GORDION EXCAVATIONS FINAL REPORTS

I: *Three Great Early Tumuli* by Rodney S. Young, 1982

II: *The Lesser Phrygian Tumuli. Part 1: The Inhumations* by Ellen L. Kohler, 1995

III: *The Bronze Age* by Ann Gunter, 1991

IV: *The Early Phrygian Pottery* by G. Kenneth Sams, 1994

MUSEUM MONOGRAPH 131

GORDION SPECIAL STUDIES V

Botanical Aspects of Environment and Economy at Gordion, Turkey

Naomi F. Miller

UNIVERSITY OF PENNSYLVANIA MUSEUM OF ARCHAEOLOGY AND ANTHROPOLOGY

PHILADELPHIA

The publication of this volume was made possible by a generous grant from an anonymous donor.

LIBRARY OF CONGRESS CATALOGING-IN-PUBLICATION DATA

Miller, Naomi Frances.
   Botanical aspects of environment and economy at Gordion, Turkey / Naomi F. Miller.
      p. cm. -- (Gordion special studies ; 5) (Museum monograph ; 131)
   Includes bibliographical references and index.
   ISBN 978-1-934536-15-5 (hardcover : alk. paper)
   1. Gordion (Extinct city)--Antiquities. 2. Gordion (Extinct city)--Environmental conditions. 3. Excavations
(Archaeology)--Turkey--Gordion (Extinct city) 4. Plant remains (Archaeology)--Turkey--Gordion (Extinct
city) 5. Land use--Turkey--Gordion (Extinct city) 6. Agriculture--Turkey--Gordion (Extinct city) 7. Landscape
changes--Turkey--Gordion (Extinct city) 8. Turkey--Antiquities. I. Title.
   DS156.G6M55 2010
   939'.26--dc22
                        2010024007

ISBN-13: 978-1-934536-15-5 (cloth)
ISBN-10: 1-934536-15-6 (cloth)

Published for the University of Pennsylvania Museum of Archaeology and Anthropology by
the University of Pennsylvania Press.

Printed in the United States of America on acid-free paper.

Contents

CD Contents vi

List of Illustrations vii

List of Tables ix

Preface xi

1. Archaeological Background 1
   Stratigraphy and Chronology 2
   Archaeobotanical Questions 6

2. Environment, Vegetation, and Land Use 9
   Topography, Soils, and Water 9
   Climate near Gordion 11
   Modern Vegetation Overview 12
   Recent Land Use—Agriculture, Animal Husbandry, Fuel 12
   Ancient Climate and Vegetation 17

3. Field to Laboratory: Collection and Processing of Wood Charcoal and Flotation Samples 21
   Nature of the Deposits—Burnt Buildings vs. Ordinary Occupation Debris 21
   Field Collection of Wood Charcoal 21
   Field Sampling for Flotation 21
   Representativeness 22
   Laboratory Procedures—Samples, Sorting, Recording, and Quantification 22

4. Analysis of the Wood Charcoal Sample 25
   Archaeological Context 25
   Methodological and Analytical Assumptions 25
   The Taxa: Ecological Significance 27
   Distribution of the Charcoal in Time and Space 29
   Results of the Charcoal Analysis 34

5. Analysis of the Flotation Samples 37
   Methodological and Analytical Assumptions 37
   Quantification of the Remains from Occupation Debris 37
   The Taxa: Economic and Ecological Significance 39
   Distribution of the Taxa in Time and Space 49
   Flotation Samples from Burned Buildings 59

6.   Interpretation—Summary and Conclusions   63

     Vegetation Cover and Changes over Time   63

     Integrated Economies and Archaeobiological Data   67

     Cultural Affiliation   69

     Summary of Results   70

Appendix A. Flotation Samples: Laboratory Protocol for Gordion   73

Appendix B. Wood Charcoal Identification Criteria   76

Appendix C. Vegetation Survey   80

Appendix D. Wild and Weedy Taxa: Seed Identification and Ecological Information   114

Appendix E. Charcoal Samples   141

Appendix F. Flotation Samples   168

Bibliography   265

Index   271

Author Note   273

CD CONTENTS

Appendix C: Vegetation survey
     YH C1   Localities
     YH C2   Plant list
Appendix D: Seed measures
     YH D1   Barley measurements
     YH D2   Barley twistedness
     YH D3   Free-threshing wheat measurements
     YH D4   Einkorn measurements
     YH D5   Emmer measurements
     YH D6   Other Poaceae measurements
     YH D7   Lentil measurements
     YH D8   Flax measurements
Appendix E: Charcoal samples
     YH E1   Inventory of charcoal samples
     YH E2   Data from charcoal samples
Appendix F: Flotation samples
     YH F1   Inventory of flotation samples
     YH F2   Data from flotation samples
     YH F3   Contents of heavy fractions

Illustrations

Sheep grazing near Gordion with the Citadel mound in the background — *Frontispiece*

1.1 Excavation units 1988/1989, three burnt buildings — 3

2.1 Sakarya Valley and Gordion region — 10

2.2 Polatlı climate diagram — 11

2.3 Polatlı precipitation, by growing season (July–June), 1929 to 2007 — 11

2.4 Vegetation in the immediate environs of Gordion — 13

2.5 Woodland vegetation in the Gordion region — 14

3.1 Density distribution of charred material — 24

4.1 Major charcoal types (weighted percent) — 28

4.2 Ubiquity (percent) of major charcoal types — 29

4.3 Comparison of percent by weight, count, and ubiquity for the major types — 30

5.1 Median density according to deposit type — 39

5.2 Median charred material density — 41

5.3 Triticum rachis fragments — 41

5.4 Einkorn (*Triticum monococcum)* and emmer (*Triticum dicoccum)* percent of total wheat grain and rachis segments — 41

5.5 Rachis fragments — 45

5.6 Hordeum, *Triticum aestivum*/durum, *Vicia ervilia*, Lens, insects that infest Lens — 46

5.7 Proportions of wheat and barley — 47

5.8 *Setaria italica* relative to *Triticum aestivum/durum* and *Hordeum vulgare* — 47

5.9 Pistachio (*Pistacia*) — 48

5.10 Flax (*Linum usitatissimum*) — 49

5.11 Ubiquity of crops — 52

5.12 Ubiquity of plants of steppe — 53

5.13 Ubiquity of plants of moist areas — 55

5.14 Ubiquity of plants of disturbance — 58

5.15 Ubiquity of other common taxa — 58

5.16 Comparison of measures of seeds to wood charcoal — 59

5.17 Distribution of values of wild:cereal — 60

5.18 Wild:cereal (% of samples by value) — 60

5.19 Mean and median wild:cereal — 61

5.20 Common types (*Trigonella* and *Galium*) (percent of total number of seeds per period) — 61

5.21 Plants of overgrazed steppe (percent of total number of seeds per period) — 62

5.22   Ruderal plants (percent of total number of seeds per period)   62

5.23   *Galium*, ruderal, overgrazed, combined (percent of total number of seeds per period)   62

5.24   Floodplain types (percent of total number of seeds per period)   62

5.25   Indicators of irrigation and streamsides (percent of total number of seeds per period)   62

6.1   Major food mammals   68

6.2   Caprids, herder animals, and wild:cereal   70

A.1   Flotation sample data sheet example   75

D.1   Apiaceae   116

D.2   Apiaceae   117

D.3   Asteraceae   118

D.4   Asteraceae   120

D.5   Brassicaceae, Caryophyllaceae   121

D.6   Chenopodiaceae, Cyperaceae, Polygonaceae   122

D.7   Polygonum, Thlaspi, Brassicaceae, Eremopyrum

D.8   Dipsacaceae, Euphorbiaceae, Fabaceae   126

D.9   Geraniaceae, Lamiaceae, Papaveraceae   128

D.10   Poaceae   130

D.11   Poaceae   132

D.12   Frequency of widths of *Stipa* seeds according to shape   140

D.13   Poaceae, Primulaceae, Ranunculaceae, *Salsola*   134

D.14   Rubiaceae, Scrophulariaceae, Solanaceae, Verbenaceae, Zygophyllaceae, Unknowns   136

D.15   Unknowns, *Beta*   138

Tables

1.1	Yassıhöyük stratigraphic sequence	4
2.1	Ekrem Bekler's yield estimates	15
2.2	Seasonal round according to Remzi Yılmaz	15
3.1	Number of distinct taxa from hand-picked charcoal samples from occupation debris	23
4.1	Charcoal from occupation debris (% by weight, count, and ubiquity)	26
4.2	Charcoal from burnt buildings (% by weight and ubiquity)	31
5.1	Distribution of flotation samples across time	38
5.2	Density of charred material from flotation samples by deposit type	38
5.3	Density of charred material from flotation samples by date	40
5.4	*Triticum aestivum/durum* measurements from debris and grain concentrations and by grain shape	42, 43
5.5	*Triticum monococcum* measurements	43
5.6	*Hordeum vulgare* var. *distichum* and *H. vulgare* var. *hexastichum* indicators	44
5.7	*Hordeum* measurements for selected samples	44
5.8	Ecological grouping for common or diagnostic types	50
5.9	Ubiquity (%) of cultigen taxa appearing in 25% or more of the samples	51
5.10	Ubiquity (%) of wild and weedy taxa appearing in 50% or more samples (and *Peganum*)	54
5.11	Summary chart based on averages by sample	56
5.12	Comparison of average seed:charcoal ratios, southeast Turkey and northwest Syria	56
5.13	Summary chart based on amounts of cultigens and summed percents of wild and weedy types	57
5.14	Sample distribution of the wild:cereal ratio	58
6.1	Bone counts	67
6.2	Percent of food animals	67
6.3	Economic indicators	68
6.4	Other indicators of plant use	69
C1	Locality descriptions	81
C2	Plants collected or seen near Gordion	82
D1	YH-Apiaceae 2 measurements	115
D2	YH-Apiaceae 4/8 measurements	115
D3	YH-Apiaceae 10/Unknown 13 measurements	115

D4	Archaeological *Stipa* measurements	131
E1a, b	Inventory of charcoal samples	144
E2a, b	Weight and count of charcoal from occupation debris	151
E3a, b	Weight and count of charcoal from burned buildings	165
F1	Inventory of analyzed flotation samples with provenience descriptions	172
F2	Flotation sample data	181

Preface

Ifirst visited Gordion as a tourist in 1983, little expecting I would return five years later to participate in the renewed excavations led by Mary Voigt under the auspices of the University of Pennsylvania Museum. The Penn Museum–sponsored excavations had been suspended after the 1974 death of Professor Rodney S. Young, but scholars returned every summer for study seasons. In the summer of 1987, a small team led by Robert H. Dyson, Jr., Director of the University of Pennsylvania Museum, went to Gordion to assess its potential for a renewed, focused archaeological investigation. Dyson, Voigt, and William Sumner realized that modern excavation methods and analyses of a deep sounding adjacent to or within the older excavated area of the Citadel Mound could provide chronological control and archaeobiological evidence that was not available at the time of the original work.

My field research at Gordion has been supported by the project through grants awarded to Mary Voigt from the National Endowment for the Humanities and the National Geographic Society. In addition, I received travel grants for field work from the American Philosophical Society, Yener Yılmaz, and the University of Pennsylvania Museum. The University Research Foundation (University of Pennsylvania) funded a trip to London and Edinburgh for herbarium work; to this day I do not know if that award was granted in the category of social sciences, natural sciences, or humanities, but I am grateful that Keith DeVries was willing to be my faculty collaborator in that enterprise. Keith also inspired several expeditions to investigate the Gordion landscape. I am grateful to Bob Dyson and the University of Pennsylvania Museum for supporting this and other archaeobotanical research by employing me for more than 20 years.

This volume is a study of the archaeobotanical remains recovered from the excavations of 1988 and 1989, but much of the botanical and ethnobotanical work that informs the discussion was carried out in subsequent years. I would particularly like to acknowledge with fond memories Remzi Yılmaz, Ekrem Bekler, and Muamer Bektöre, who provided practical help with many aspects of the field work, helped me understand Turkish and Turkish culture, and answered my questions about local agricultural practices. I also thank other team members for their help and insights, especially John "Mac" Marston, Ayşe Gürsan-Salzmann, Richard Liebhart, Melinda Zeder, and Bob Henrickson, along with Hüseyin Fırıncıoglu (retired from the Field Crop Research Institute) and Mecit Vural (Gazi University). I am grateful to botanists at the herbaria of the Royal Botanical Gardens of Edinburgh and Kew, who helped me identify most of the voucher specimens I collected between 1988 and 1992. I also appreciated free access to the herbarium established and organized by Gordon Hillman and Mark Nesbitt that is housed at the British Institute in Ankara.

Over the years I have conferred with several colleagues about archaeological seed and plant part identifications; I particularly would like to credit Gordon Hillman for identifying the flax seeds from the Destruction Level. Although I personally sorted and identified most of the samples reported here, over the years I have supervised several laboratory assistants who sorted some of them: Wies van den Brink, Christine Hide, Shannon Palus, Bridget Crowell, and Nancy Mahoney.

David French loaned the British Institute in Ankara flotation set-up to the project in 1988 and 1989, and Mark Nesbitt not only patiently explained how to run it, but provided detailed instructions on how to have our own built. Our government representatives, sent to us from the Museum of Anatolian Civilizations or the General Directorate of Museums and Monuments, have been very supportive of the archaeobotanical research, especially Nurhan Ülgen, Halil Demirdelen, and Vahap Kaya. The Meteoroloji Bakanlığı and the Polatlı Meteoroloji İstasyonu kindly provided the weather data.

My thinking has benefited tremendously from conversations with more colleagues than I can name, but I have learned the most from Mary Voigt, Ben Marsh, Ayşe Gürsan-Salzmann, and Keith DeVries. My former supervisor, Stuart Fleming, always supported my work and that of others on the Gordion project, assigning assistants, reading manuscripts, and acting as a sounding board for issues surrounding the project. Lindsay Shafer worked on digitizing my original pencil drawings of the seeds, and Shannon Palus and Nina Johnson prepared the seed photographs. The two reviewers, Mark Nesbitt and Alexia Smith, gave copious constructive comments, which I have done my best to address.

Finally, I would like to thank Project Director G. Kenneth Sams, Excavation Director Mary M. Voigt, and the entire team, foreign and local, for their enthusiastic, if sometimes bemused, cooperation in setting up and running the flotation machine; taking charcoal and soil samples; and for humoring me as I try to explain the difference between wheat and barley, or teach them to identify *üzerlik* (also known as wild rue or *Peganum harmala*) that, for your information, is the most obvious plant that grows all around the dig house!

1

Archaeological Background

The archaeological site of Gordion is most famous as the home of the Phrygian king Midas and as the place where Alexander the Great cut the Gordian knot on his way to conquer Asia. Located in central Anatolia near the confluence of the Porsuk and Sakarya rivers, Gordion also lies on historic trade routes between east and west, as well as north to the Black Sea. Very favorably situated for long-distance trade, Gordion's setting is marginal for cultivation, but well-suited to pastoral production. It is therefore not surprising that with the exception of a single Chalcolithic site (Kealhofer 2005), the earliest settlements in the region are fairly late—they date to the Early Bronze Age (late 3rd millennium BC). The earliest known levels of Gordion, too, date to the Early Bronze Age, and occupation of at least some part of the site was nearly continuous through at least Roman times (second half of the 1st century BC); a Medieval settlement is also attested (Voigt 2005). Pre-Chalcolithic occupation in this part of the Sakarya valley is evidenced by abraded Late Paleolithic flint tools that erode out of Pleistocene conglomerates and occasionally turn up in flotation samples and other excavated sediments.

The most prominent sites in the archaeological region in which Gordion lies are Gordion itself and at least 100 Phrygian-period burial mounds. Archaeological surveys have recorded sites mostly dating between the Early Bronze Age and the modern era (Kealhofer 2005). Gordion is comprised of the 13-ha Yassıhöyük (literally, "flat mound"), also referred to as the Citadel Mound (and in some earlier project publications as the City Mound), which is surrounded by an inner town and fortification system (Küçük Höyük and Kuş Tepe) that encloses an area of 51 ha. By the mid-1st millennium BC, settlement had expanded to an extensive outer town with an estimated total settlement area of about 1.5 square km (Ben Marsh, e-mail 9/16/08). The plant remains discussed in this report all come from excavations in the eastern part of the Citadel Mound.

Gordion is known through both history and archaeology. The best-known ancient references to Phrygian Gordion and its king Midas are found in Herodotus's *Histories*. Other ancient references, mostly Greek, occur in the works of Xenophon, Arrian, Plutarch, and Livy. Modern archaeological interest in Gordion came through Classicists' knowledge of ancient Greek contact with the Phrygian world. The ancient settlement mound was identified as Gordion and excavated in 1901 by Gustav and Alfred Körte (Körte and Körte 1904; Sams 2005:10). A University of Pennsylvania team led by Rodney S. Young, a professor of Classical Archaeology, began excavations in 1950.

Young's excavations (1950–1974) focused on the Early Phrygian levels at Gordion and Middle Phrygian burial mounds. This work established a rough chronological framework for the region. Analysis and conservation continued under the direction of Keith DeVries after Young's death in 1974. Fieldwork, however, was suspended until 1987, when a small team from the University of Pennsylvania Museum assessed the possibilities for a new project. In cooperation with the University of North Carolina, Chapel Hill, the Penn Museum renewed excavation under the direction of Mary M. Voigt in 1988. (For the history of the excavations, see DeVries et al. n.d.; Sams 2005; Voigt 2005.) Voigt established a stratigraphic sequence for the site based on the excavations of 1988 and 1989. Paleoethnobotanical research is an integral part of the renewed program of excavation and surface survey at Gordion. Since the 1990s, extensive excavation of Phrygian and later deposits has been carried out.

Analysis of those archaeobotanical remains has begun (Marston 2003, 2010; Miller 2007).

Charred plant remains from Gordion provide the best evidence for tracing long-term changes in vegetation and plant use that in turn reflect many aspects of ancient economy and society in the Sakarya basin over several millennia. That the remains originate from a single site is an important limitation for a study that seeks to understand regional trends. Nevertheless, many specific questions can be addressed with these data concerning the nature of the original vegetation, the relationship between agriculture and pastoral production, irrigation, and ethnic markers. This report deals with archaeobotanical remains dating from the Late Bronze Age to the Medieval period that were excavated during the 1988 and 1989 seasons at Gordion. The assemblage consists of charcoal hand-picked during excavation and charred seed and wood remains obtained by the flotation of systematically collected soil samples. In subsequent years, I conducted informal botanical surveys in the region and collected voucher specimens and comparative material that are currently housed in my laboratory at the University of Pennsylvania Museum, Philadelphia. This work has informed both the identifications and interpretations presented here.

Stratigraphy and Chronology

The excavations of 1988/89 were limited to the Yassıhöyük Citadel Mound. Young's work had exposed the royal precinct—or at least elite quarter—of the Early Phrygian period, about 5 m below the modern surface; the excavated area covers about 2.5 ha (Voigt and Henrickson 2000a:39). To minimize the excavation area needed to obtain a stratigraphic sequence, Voigt set the upper excavation units (Operations 1, 2, and 7) at the edge of Young's main excavation area. The uppermost level included one of Young's sherd dumps and then descended from Medieval deposits down to the Early Phrygian royal precinct. Physically but not stratigraphically discontinuous, the lower units (Operations 3–6; 8–11; 14 [below 3–6]) were placed in an Early Phrygian courtyard area and the architectural remains from that level are preserved for touristic purposes (Voigt 2005; Fig. 1.1). The excavation of the lower trenches extended to a small area of Middle Bronze Age date. The project used a lot and locus system for excavation, recording, and analysis. In particular, a lot represents a contiguous unit of excavated earth, ideally from a single depositional stratum; it is the basic unit of excavation. A locus is comprised of one or more contiguous lots that ideally represent a "significant stratigraphic unit." Lots and loci may also be arbitrarily defined (for example, in exploratory trenches).

Voigt has discussed the stratigraphy and the cultural and historical associations in detail (1994). She developed the YHSS numbering system—a shorthand representation of the stratigraphic analysis—to aid in the recording and sorting of the various data classes generated by the project. Re-analysis of artifacts and stratigraphy, along with new radiocarbon dates of short-lived seeds, led to a major revision of the chronology of the first millennium BC levels (DeVries et al. 2003; Voigt 2005) that is used in this volume.

The Yassıhöyük Stratigraphic Sequence and Characteristics of Deposits Sampled for Botanical Remains

The Yassıhöyük Stratigraphic Sequence (YHSS) assigns the excavated deposits to broad chronostratigraphic units that roughly correspond to more traditional archaeological periods (Voigt 1996). Numbered one to ten from top to bottom (Table 1.1), each of these large units is divided into a series of stratigraphic contexts defined with a minimum three-digit code (thus, deposits within YHSS 7 are assigned a number between 700 and 799). Decimal places are added as the complexity and understanding of the deposits warrant (thus, 725 is a floor deposit of a burned building in YHSS 7, and 725.04 is an oven within that building). The Early Phrygian Destruction level (YHSS 6A) is at the base of the upper trenches and the top of the lower ones. The discussion here emphasizes the time periods for which there is substantial archaeobotanical data.

MIDDLE BRONZE AGE (YHSS 10). These deposits pre-date 1500 BC. A single deposit of less than 1 cubic m volume was sampled; two samples from an erosion surface were analyzed.

LATE BRONZE AGE (YHSS 8/9), ca. 1500–12th century BC. Initially YHSS 9 was assigned to the Early Hittite Empire period; the excavated area (and flotation samples taken) consisted primarily of lensed trash

Op. 2, Abandoned Village
Structure (YHSS 3)

Op. 1, Terrace Building 2A (YHSS 6)

Op. 9, Burnt Reed House (BRH) (YHSS 7)

Fig. 1.1   1988/1989 excavation units. Top: Early Phrygian Destruction Level (plan) (source: Gordion archive). Bottom: Three burnt buildings discussed in this volume.

Table 1.1. Yassıhöyük stratigraphic sequence, approximate dates (source: Voigt 2005:27)

YHSS 1	Medieval	13–14th century AD
YHSS 2	Roman [not in these samples]	early 1st–5th century AD
YHSS 3	Hellenistic	330–mid-2nd century BC
YHSS 4	Late Phrygian	540–330 BC
YHSS 5	Middle Phrygian	800–540 BC
YHSS 6A	Early Phrygian ("Destruction Level")	900–800 BC
YHSS 6B	Early Phrygian (courtyards)	950–900 BC
YHSS 7	Early Iron Age	12th century–950 BC
YHSS 8/9	Late Bronze Age	1500–12th century BC
YHSS 10	Middle Bronze Age	2000–1500 BC
—	Early Bronze Age [not in these samples]	2500–2000 BC

and some exterior surfaces; there were no structures. YHSS 8 was assigned to the Late Hittite Empire. The only structure was single-room CBH, a stone-lined cellar with no internal features. Samples analyzed from this phase are mainly from pits, a hearth, and floor deposits. According to Voigt (1996), samples from YHSS 8 and 9 can be grouped for comparisons with the Early Iron Age and later deposits, since there is no break in the cultural sequence at this time.

EARLY IRON AGE (YHSS 7), ca. 12th century–950 BC. The Early Iron Age (EIA) deposits are treated as one chronological unit, though they can be put in three stratigraphic groups. Samples from the earliest EIA, YHSS stratum numbers 730 and higher, come from various features (ovens, pits) associated with domestic structures and activities. A burnt reed structure (BRH, stratum number 725) is roughly in the middle of the deposits assigned to YHSS 7. Due to the *in situ* charring, the floated material is not comparable to ordinary occupation debris and so is listed and treated separately in this report. The most recent samples from YHSS 7 are mostly from wash and later Early Iron Age pits (705).

EARLY PHRYGIAN PERIOD COURTYARDS (YHSS 6B), 950–900 BC. The distinct stratigraphic break between YHSS 7 and 6 signals a change in function from ordinary domestic to elite quarters. YHSS 6B, which consists primarily of clay fills and construction debris, yielded very few botanical remains.

EARLY PHRYGIAN DESTRUCTION LEVEL (YHSS 6A), ca. 900–800 BC. On a grander scale, the buildings of the Destruction Level suffered the fate of the burnt reed structure 725. Similarly, the charred construction debris and *in situ* room contents are not comparable to ordinary occupation debris and are treated separately in this analysis. The deposits analyzed here come from the anteroom of Terrace Building 2. Broadly, there is clear stratigraphic continuity between YHSS 6B and 6A, but the YHSS 6B deposits excavated in 1988/1989 are in the center of the old excavation, and the YHSS 6A deposits are at its edge.

MIDDLE PHRYGIAN (YHSS 5), ca. 800–540 BC. Soon after the fire in YHSS 6A, the center of the Citadel Mound was leveled and covered with a thick (4 to 6 m) layer of clay (Voigt 2007). This phase is poorly represented in the stratigraphic sounding, so only a few samples were taken, mostly from post-occupation deposits within the cellar of Middle Phrygian building I:2 and a few later pits. This makes generalizations difficult.

LATE PHRYGIAN (YHSS 4), ca. 540–330 BC. Thanks to a large number of trash-filled pits in the excavated area, many flotation samples yielding quite a bit of material were taken. There are also a few samples from hearths. However, remains of structures were fragmentary due to Hellenistic stone-robbing. Despite the fairly small exposure the neighborhood can be characterized as an "industrial" area (Voigt 1996).

HELLENISTIC (YHSS 3), ca. 330–mid-to-late 2nd century BC. Two phases have been distinguished, YHSS 3B, ca. 330–mid-3rd century BC, and YHSS 3A, mid-3rd–mid-2nd century BC. The industrial nature of the excavated area continues in the lower part of this stratum (YHSS 3B), and most of the samples come from a series of hearths. During YHSS 3A, Galations (European Celts) arrived at Gordion. A burned structure, part of the Galatian "Abandoned Village" of YHSS 3A, lies above 3B. Functionally, these flotation samples are most usefully compared to those of the YHSS 7 BRH structure and Terrace Building 2A of the YHSS 6 Early Phrygian Destruction Level. A few Late Hellenistic (YHSS 3A) pits and wall fragments lie above.

MEDIEVAL (YHSS 1), 13th–14th century AD. Voigt (1994) reserved YHSS 2 for Roman period deposits; in the 1988/1989 excavation area, however, there is a stratigraphic gap. Roman material has been excavated in the northwestern zone of the Citadel Mound elsewhere on the site recently (see Goldman 2005; Marston 2010; Miller 2007a, 2007b). The few Medieval samples in the YHSS sounding come primarily from a few pits and an oven.

Yassıhöyük Stratigraphic Sequence in Cultural Context

All archaeological periods were important for the people living in them, but some stand out thanks to the breadth and depth of present-day knowledge of the time. Texts—those that were never lost as well as those known only from excavation—are an independent source of information against which one can compare the archaeological materials. Also, members of the Gordion team, working with excavation, survey, archival, and other data, continue to refine our understanding of the sequence.

BRONZE AGE SETTLEMENT. In addition to Gordion, there are a few Early Bronze Age sites within a 10-km radius of the site. During the Middle and especially the Late Bronze Age in the region, Gordion was in the orbit of the Hittite empire (Voigt 1994:276). Despite the uncertain environment for farming, there were many settlements within 15 km of the site (Kealhofer 2005). Perhaps integration into the Hittite economy allowed people simply to move away or to trade for foodstuffs in bad years, or the local adaptation would have encouraged putting more effort into herding to see people through hard times.

THE PHRYGIAN QUESTION. In line with Herodotus's and Strabo's writings, the Phrygians are thought to have originated in southeastern Europe (Sams 1988; Voigt and Henrickson 2000a, 2000b). Keith Devries (2000:18) has mapped the plausible extent of Phrygia (at least 7th to 4th centuries BC) in west central Anatolia through rock inscriptions in the Phrygian language and other epigraphic finds. Sometime after Hittite (YHSS 9–8) ceramic evidence of connections with the Sakarya valley ceases and before the establishment of the royal precinct (YHSS 6), Phrygians had settled at Gordion. Voigt (1994:277) sees a stratigraphic break between YHSS 8/9 and 7, along with a suite of cultural changes. Voigt and Henrickson (2000:46) have argued the changes reflect the arrival of a new group of people, the Phrygians. For example, a possible ceramic marker is the Early Iron Age handmade pottery characteristic of early YHSS 7, which replaced the wheel-made Hittite ceramics; among other possibilities, at the very least this would indicate a change in ceramic production and distribution (Henrickson 1993). Despite the apparent continuity in settlement between the earlier and later YHSS 7 Early Iron Age deposits, the pottery at the end of the phase is again wheel-made, and indeed, is indistinguishable from that of Early Phrygian YHSS 6B.

THE EARLY PHRYGIAN DESTRUCTION LEVEL (YHSS 6B). Rodney Young's major investigation of the Yassıhöyük mound stopped at the Early Phrygian royal precinct. The area exposed by his excavation included presumed royal residences in the center and at the edge a series of attached megarons (Terrace Buildings 1–8), the back walls of which presented a single face to the central area. These buildings appear to have functioned as service buildings for the elite quarter. The buildings had been destroyed in a catastrophic fire, now dated to about 800 BC. Though no skeletons were found, the fire was so intense it melted pottery and vitrified the silicates in some of the wood and seeds. Young and others associated the fire with the Kimmerian invasion mentioned by ancient authors, but even before the current re-dating to 800 BC, that view was not tenable (Voigt 2007). Rather, the Early Phrygian rulers had begun a major revamping of the fortification system at the time of the catastrophic, but accidental conflagration. A clear stratigraphic break, the Destruction Level is culturally continuous with what lies above.

MIDDLE PHRYGIAN REBUILDING (YHSS 5). One of the most mysterious aspects of Gordion is the clay layer that seals the Destruction Level. Over much of the excavated area, the buildings built into the clay layer follow the general lines of the earlier, now buried, structures. It is therefore not surprising that "The YHSS 5 (Middle Phrygian) ceramic assemblage is clearly derived from that of YHSS 6 (Early Phrygian) both typologically and technologically" (Henrickson 1993:132). Henrickson remarks that this assemblage is restricted to local types. It is during the Middle Phrygian period that the settlement expanded considerably. Excavation and surface survey suggest relatively dense occupation over an area of approximately 160 ha (Voigt and Henrickson 2000a); the Citadel Mound lay in the middle of a settlement whose maximum extent was about 1 km north-south and 2 km east-west. This archaeological evidence for expanded settlement, the massive earth-moving and reconstruction of the palace quarter, continued tumulus building, and a plethora of imported wares suggest it was a very prosperous time (DeVries 2005; Henrickson 1993:140; Voigt 2005). Regional survey, too, suggests the Middle Phrygian was a time of prosperity and agricultural expansion (Kealhofer 2005). This conclusion is fully consistent with what one might expect, given the textual evidence for the power and expansion of Gordion during the 8th century.

The outstanding feature of the Middle Phrygian and subsequent landscapes was the burial tumuli that dot the countryside, especially Tumulus MM ("Midas Mound") and the cluster nearby. About a hundred tumuli have been mapped. They are distributed within a 10-km radius of the Citadel. Tumulus building in the region ended by the 2nd century BC.

LATE PHRYGIAN ECONOMIC EXPANSION (YHSS 4). The Late Phrygian phase at Gordion is the time of the Persian/Achaemenid conquest. Gordion's political importance probably had waned, but it appears to have been a prosperous economic center; most of the Greek pottery comes from these deposits, demonstrating contact with the west, as well (Henrickson 1993; Voigt 1994).

HELLENIZATION AND THE MEETING OF PEOPLES. In 333 BC, Alexander the Great arrived at Gordion and incorporated the area into his emerging empire. In the Early Hellenistic (YHSS 3B) ceramic assemblage, "the adoption of Greek forms becomes even more pervasive, affecting even basic types like cooking pots" (Henrickson 1993:155). Around 250 BC, finds, both spectacular and quotidian, confirm the Celtic (Galatian) occupation at Gordion attested by ancient texts (Dandoy et al. 2002; Voigt 2003).

MEDIEVAL (YHSS 1). During the Medieval period new cultural interactions might have had some affect on land use. In the case of Gordion, there is enough pig bone to suggest the presence of a resident non-Muslim population. We might expect that the influx of Central Asian Turkic tribes and political unification of new regions under Islam to have influenced trade networks and the material, including plants, that traveled along the routes.

Archaeobotanical Questions

The previous sections give some general archaeological and cultural background. Samples from the stratigraphic excavation contain a record of close to 2000 years. Data from plant macroremains, charred wood, seeds, and other plant parts can address a number of issues concerning ancient plant use, land use, and landscape. The long sequence allows us to trace vegetation history in the region and evaluate the extent and nature of human impact. Charred wood indirectly provides evidence of forest composition, and the remains themselves come from fuel and construction. From the seeds of cultigens and wild plants we can infer the relative importance of agriculture and pastoralism over time. Somewhat more directly, the charred remains leave evidence of crop choice. The intensity of land use for agricultural and pastoral pursuits would have varied, too. In conjunction with the other archaeological interpretations, the botanical data can enrich our understanding of agriculture and economy in the Sakarya valley.

Original Vegetation, Climate, and Changes in Land Use Intensity

Even today, and certainly in antiquity, climate is one of the major determinants of vegetation. For central Anatolia, Aytuğ (1970) proposes a landscape of anthropogenic steppe, certainly around Ankara but even around Gordion. As Walter points out, "Die Grenze zwischen Wald und Steppe wird in Zentralanatolien noch dadurch kompliziert, daß dieses Land keine Hochebene im eigentlichen Sinne darstellt. Vielmehr wechseln weite Beckenlandschaften (als 'ova' bezeichnet) mit Gebirgsrücken ab. Auf den höheren Erhebungen findet man noch Waldreste, während die tiefer liegenden Teile baumlos sind" (1956:97). [The boundary between forest and steppe in central Anatolia is complex, as this land is not a plateau in the proper sense. Basin landscapes, called 'ova,' alternate with mountain ridges. On the higher slopes one finds relict woodland, while the low-lying parts are treeless.] He uses an analogy between Ankara and Salt Lake City to conclude that the natural vegetation would be grassy steppe. At least in the United States, comparable *Artemisia* steppe occurs in Nevada, with less than 300 mm (winter) rainfall. Around Ankara, in a fenced area, Walter saw perennial grasses, including various *Stipa, Bromus tomentellus, B. erectus, Festuca sulcata, Phleum* sp., *Melica* sp., and other plants. He therefore suggests, at least for Ankara, an original *Stipa–Bromus tomentellus* steppe, and similar vegetation along the route to Eskişehir. *Artemisia fragrans* grows at the same elevation range.

Two types of natural vegetation characterize the central Anatolian steppe: perennial grassland and *Artemisia* steppe. Botanists have argued about whether the *Artemisia* steppe is disturbed grassland or original vegetation cover (Walter 1956:98). I think it likely that around Gordion, whose elevation is so close to the steppe-forest boundary, relatively favorable conditions prevailed, allowing a dense grass cover that could have supported grazing animals, presumably wild in the distant past, but herds of domestic sheep and goat by the Middle Bronze Age. Note that Marsh (2005:168) found "typical grassland soils" in the Sakarya valley below later erosion deposits.

Catchment

Regional surveys and excavation at the site of Gordion give some evidence of population and land use in the Sakarya valley over the archaeological sequence. In different periods, the area from which food supplies were drawn would have expanded or contracted according to population levels and exchange relations (political, social, or economic) with people beyond the valley. With few exceptions, most of the food plants could easily have been produced locally; what cannot be determined from the remains is whether they actually were.

Irrigation

A variety of evidence can potentially bear on the question of whether or not crops were irrigated. The first thing to consider is whether it would have been desirable and possible to irrigate. Given the erratic nature of the climate, any technique that would even out harvests from year to year would be a good thing, especially in those time periods, such as the Middle Phrygian, when there was a relatively high population density. Since the late 1950s when the Sakarya was straightened, the river has been down-cutting the plain, and irrigation requires the use of pumps. Aerial photographs from the 1950s show a very different meandering river regime, but the annual flooding of the first half of the 20th century may itself be a relatively recent phenomenon, post-dating the archaeological deposits (Marsh 2005).

Several types of botanical evidence can address the question, but not all are relevant to the data currently available from Gordion.

1. Weed seeds of irrigated and unirrigated fields. Due to the unfortunate (for the archaeobotanist) practice of suppressing weed growth in the fields, I am unable to evaluate the ancient seed assemblage through comparison with the modern field weed composition. The evidence of the sedges, however, supports the view that habitats available for grazing were not constant, and that irrigation was most significant during the Middle Phrygian and Medieval periods.

2. Crop choice. Some crops would have been irrigated because they are summer-grown (millets, and in the Medieval samples, cotton and rice). The samples from the 1988/1989 excavation have few millets and show a suggestive association with the sedge seeds in the Medieval period, but not earlier. If wheat and barley were irrigated, one might expect some association with seeds of wet areas. Namely, in a situation (including the present) where both are cultivated, wheat is more likely than barley to be irrigated because it is less drought resistant and, favored as food, is the more valuable crop. Similarly, six-row barley is more likely to be irrigated than the two-row type. The notable stability in the proportion of wheat to barley reveals no identifiable change in irrigation practices of the two major cereals (wheat or barley).

3. Measurements of cereal grains. As discussed above, there is a similar lack of positive evidence

for changes in irrigation practices based on the plumpness of the wheat and barley grains.

Ben Marsh (in Voigt and Young [1999:n. 6]) has suggested that clean clay used to cap the Early Phrygian level on the Citadel Mound may have consisted of sediments originating from "hydraulic work at the time of the reconstruction (e.g., digging irrigation canals or drainage ditches)"; both activities, especially the former, support an interpretation that land use for agriculture intensified. It may be no accident, then, that two indications of a relative shift toward the agricultural side of the agropastoral continuum date to this period: a dip in the proportion of sheep and goat and an increase in the wild seed to cereal ratio (see discussion in concluding chapter).

Population Movement

Several questions specific to the culture history of Gordion will also be addressed. For example, do changes in the agropastoral economy reflect changing ties to the world beyond the Sakarya valley? Anatolia has long been a crossroads between east and west, and north and south. Based on both ancient texts and modern archaeology, Gordion has attracted scholarly attention concerning several ancient episodes of migrations, or at least of population movement. One group of questions for Young, Sams, Voigt, and others is: When and under what circumstances did Phrygians arrive in Anatolia, and can they be identified by non-linguistic material remains? The same questions can be asked of the Celtic (Galatian) arrival and presence. Voigt and Henrickson's stratigraphy-based analyses of changes in material culture have generated several hypotheses in this regard. Social, political, and ethnic environment all may affect the agropastoral economy; assigning changes in the archaeobotanical record exclusively to these specific factors would be unwise.

By phrasing these questions somewhat less specifically, however, the archaeobotanical remains could provide some illumination as well. We all know that pots do not equal people, and archaeological cultures (typically recognized by pottery) do not equal ethnic groups. It is hard to think that plant remains could unequivocally distinguish Gordion's place in the orbit of the Hittites (YHSS 8/9) from its independence during the heyday of the Phrygians (YHSS 7, 6, 5), or mark the Persian conquest (YHSS 4), the arrival of the Celts (Galatians) (YHSS 3), or contacts with the wider Islamic world (YHSS 1). Voigt (1994:276) suggests that certain kinds of domestic, relatively private, habits can help identify cultural markers. Examples include hearth and fireplace form, which could relate to food preparation customs; one archaeobotanical contribution to the discussion is food remains.

Cultural Affiliation

One of the results of the Gordion archaeobotanical study is that much of the evidence for environment and land use in the Sakarya valley shows incremental change that is not correlated in any obvious way with the apparent changes in the population or its cultural affiliation. As will be demonstrated in this study, the most significant change in agricultural strategy occurred during Early and Middle Phrygian times, when cultural continuity prevailed. Despite the dramatic history of population movement and replacement in the Sakarya valley, agricultural strategies appear to have been remarkably stable. I suggest that at a given level of technology within the Near Eastern agricultural tradition, the harsh environment of the Sakarya valley strongly constrains the agricultural possibilities, and that when any newcomers arrived, it behooved them to learn how to be successful farmers from the local population, if they did not already know. This is not to deny any agricultural innovation at all, but that of necessity it was cautiously applied. In conjunction with data and interpretations generated by other researchers, however, two possible expressions of Phrygian identity may be suggested (see discussion in concluding chapter): the consumption of einkorn and a possible "heirloom" artifact made of alder.

2

Environment, Vegetation, and Land Use

Preliminary archaeobotanical work (Miller 1999), geomorphological studies (Marsh 2005), archaeological survey (Kealhofer 2005), and ethnoarchaeological studies (Gürsan-Salzmann 2005) all show that the 20th-century landscape of the Sakarya valley is quite different from that of 3000, 300, or even 30 years ago. Even so, the present-day climate and vegetation provide a baseline against which one can assess the macrobotanical remains. Palynological studies from neighboring regions give independent information with some time depth.

Strong Mediterranean influence on the climate gives much of Turkey cool or cold wet winters and hot dry summers. Elevation, local topography, and distance from the coast create great variation—the climate becomes more continental in the interior, and there is some rain in the summer. Thanks to adequate rainfall, the natural vegetation of the coastal regions of Turkey is forest. Oak and pine dominate the Mediterranean forests of the west and south, and mixed hardwoods are characteristic of the Pontic (Black Sea coast) forests to the north (Zohary 1973:Map 7). Going inland past the coastal mountains ranges, overall precipitation declines; in general, lower elevations experience less rainfall. The lower boundary of the central Anatolian true steppe is approximately 700 m, depending on local conditions. Gordion straddles that elevation boundary, so short-term climate anomalies could have a disproportionately strong effect on the land and the agricultural economy. Even in the absence of climate shifts, normal interannual rainfall variability or human actions that alter such factors as the water table or drainage could affect the natural vegetation cover and moisture available for crops.

Topography, Soils, and Water

Some "natural" processes that might affect plant life occur regardless of human intervention, such as long-term and short-term climate shifts. More locally, down-cutting of the Sakarya River or its opposite, a shifting bed of aggrading streams, would alter the land. At the time scale considered here, the archaeobotanical record reflects predominantly human manipulation of the landscape—intentional earth movement as well as erosion that results from deforestation and overgrazing.

The Sakarya River originates in the western highlands of Anatolia; it flows north through Gordion toward its outlet in the Black Sea. The Porsuk river, which flows through Eskişehir, meets the Sakarya about 4 km north of the site. Over time, the bed of the Sakarya has shifted; with canalization in the late 1950s, it is now down-cutting, but through the first part of the 20th century it meandered and flooded annually. Ben Marsh's geomorphological studies (2005) show several major shifts in the river over the occupation of the site. In fact, until sometime after 600 BC, the river flowed to the east of the Citadel Mound (Marsh 1999).

Gordion is situated in a fertile alluvial valley (Fig. 2.1). Within about 5 km of Gordion, the soils and geological substrate as mapped by Marsh (2000, 2005) occur in several different zones. Today, a narrow riparian strip supports an assortment of woody and herbaceous vegetation. The east side of the valley bottom, annually flooded before the river was straightened in the 1950s, consists of a strip of deep soils eroded from the eastern hillsides, at most 2 km in width but usually narrower. Just east of the floodplain are some marls and

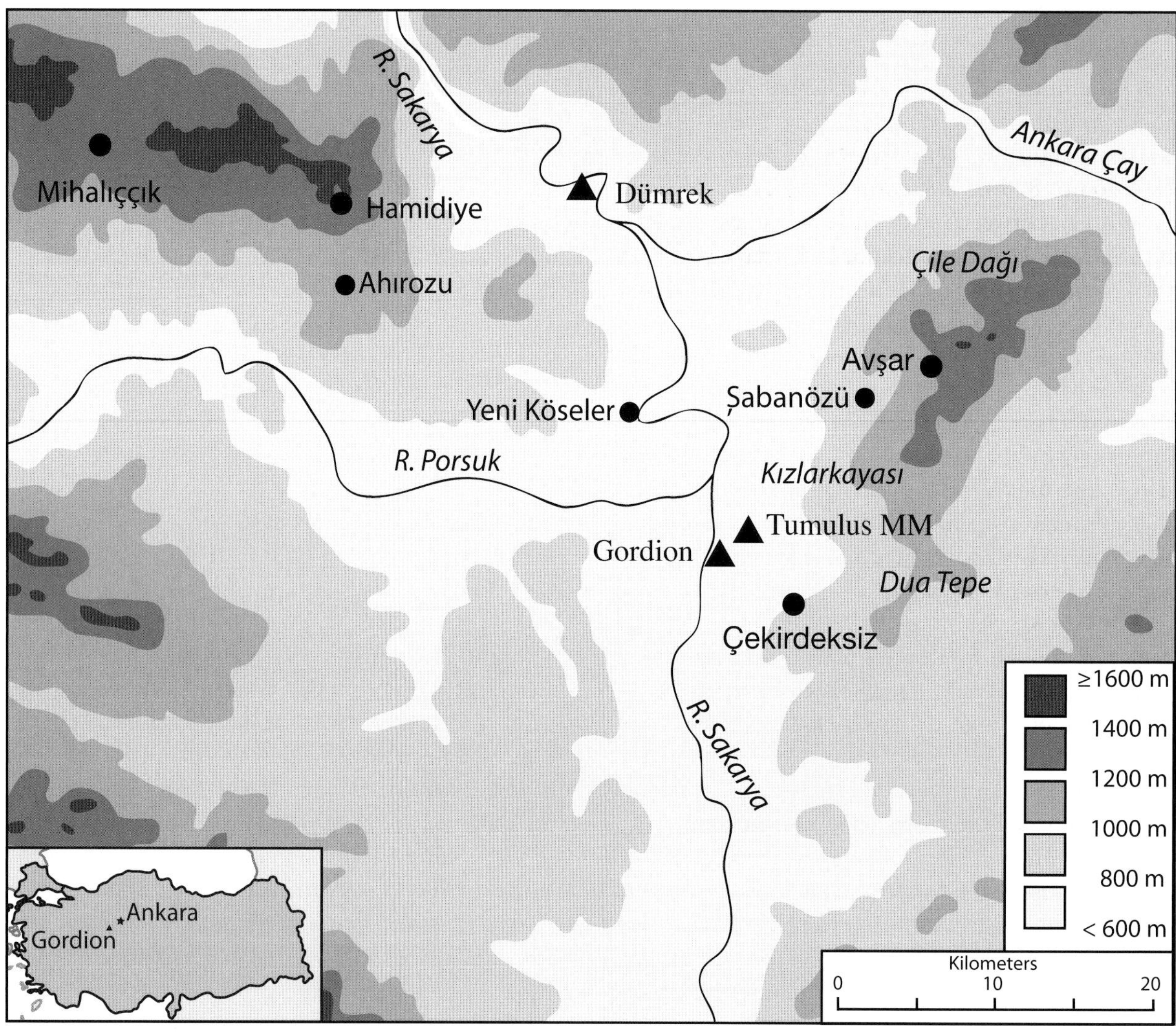

Fig. 2.1 Sakarya Valley and Gordion region (digitized by Nina Johnson).

gypsum outcrops; further east is siltstone pediment with basalt intrusions (Marsh 2000, 2005). To the west of the river are marls and conglomerate plateaus. The arable soils of today include a relatively small area of alluvial marls that are not suited to dry-farming. To the east, marls and upland basaltic soils eroded from the hills above fill the valley (Marsh 2000, 2005). Alternate-year fallow allows the lighter soils to store moisture; the basalt-derived soils are "highly porous and permeable and holding and releasing groundwater throughout the year" (Marsh 2005:164). Most of the soils within 5 km of Gordion fit into this category, and prior to mechanized pump-driven irrigation, the major land use was dry-farmed cereal production and grazing (Gürsan-Salzmann 2005).

Groundwater availability in antiquity would have been greater than it is under the eroded, devegetated conditions of today. According to Marsh, "The streams are shallower and they flow less in the dry (summer) season. Springs also flow much less through the year and they have also been buried if they were close to the streams" (2000); he also points out that recent massive irrigation is lowering the water table. Traces of ancient settlement tend to be located near springs and streams throughout the sequence (Kealhofer 2005).

Climate near Gordion

The nearest town for which meteorological information is available is the district center of Polatlı, which is about 20 km northeast of Gordion at an elevation of 875 m (Meteoroloji 1974). For the forty-one years between 1930 and 1970, the average temperature was 11.9°C, with about 65.5 days/year with the lowest temperature below freezing. Average yearly precipitation was 346.6 mm, with a moisture deficit from June to October (Figs. 2.2, 2.3). The average number of days with snow was 12. These data suggest that Polatlı is within the territory of reasonably secure rainfall agriculture (allowing for some variation hidden by the use of averages, 250 mm/yr is considered the minimum for dry-farmed cereals in the Middle East). The 61-year precipitation average for the July to June agricultural year is 347 mm, with a standard deviation of 62. This suggests fairly erratic rainfall, but generally enough for dry farming. Summers can be cool, and in contrast to much of the Near East, summer downpours are a normal, if occasional, aspect of the climate.

In inner Anatolia, precipitation tends to decline as elevation decreases. Available moisture for natural vegetation as well as for rain-fed crops would be somewhat less in the Sakarya valley near Yassıhöyük, because it is nearly 200 m lower than Polatlı in elevation. The relatively benign variability in Polatlı, therefore, might indicate a high proportion of serious drought years at Gordion. Indeed, from the balcony of the Gordion excavation house, it is common to see summer rainclouds skirt the edge of the valley with-

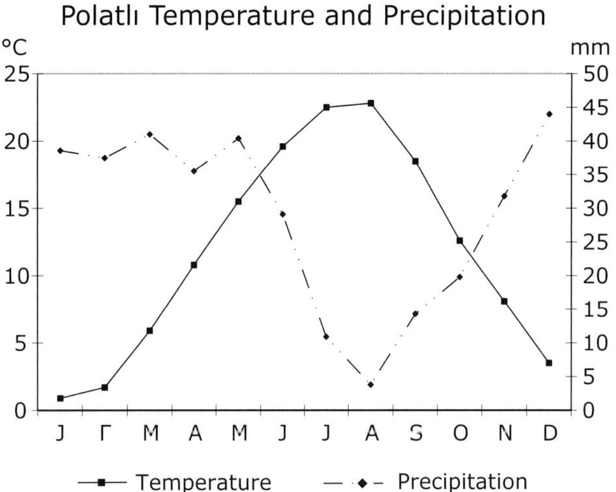

Fig. 2.2 Polatlı climate diagram (39°35′N 32°08′E). Based on 41–yr average. Average monthly minimum always >0°C; absolute minimum <0°C January and February (source: Meteoroloji Bülteni 1974).

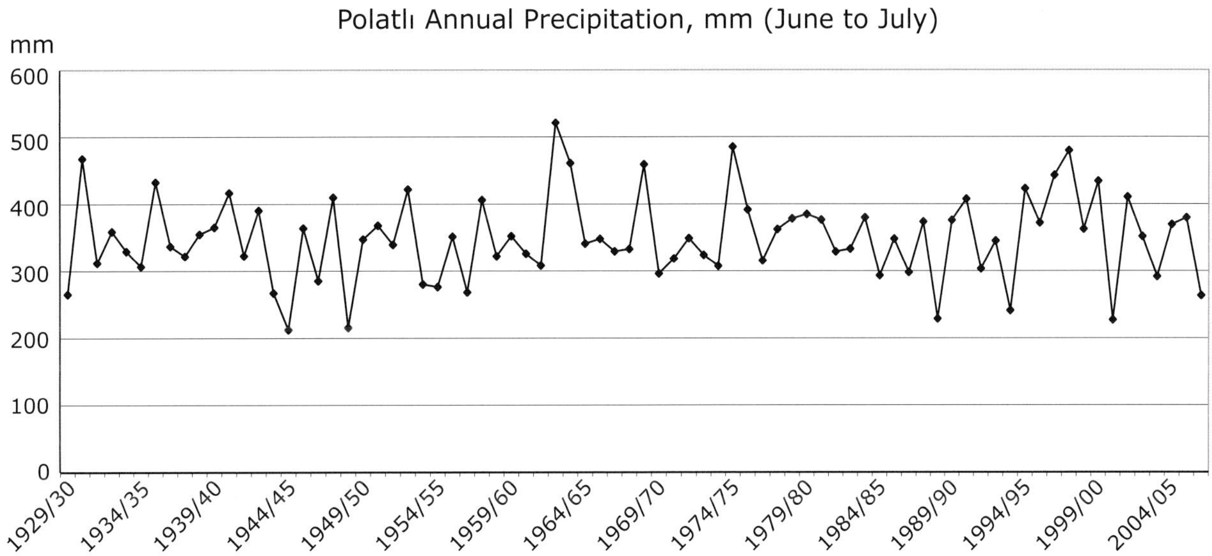

Fig. 2.3 Polatlı precipitation, by growing season (July–June), 1929 to 2007. Seventy-eight-year mean: 349.4 mm, S.D. 64.0 mm (source: Meteoroloji Bakanlık and Polatlı Meteoroloji Istasyonu).

out dropping any moisture. Even with pump irrigation, farming in Yassıhöyük seems risky; the bumper crop of a very wet year, 1988, gave way to nearly total crop failure in 1989. In those years (July to June), precipitation reported in Polatlı was 376.0 mm and 228.9 mm.[1]

Modern Vegetation Overview

Michael Zohary (1973:579) describes the natural vegetation of the Anatolian plateau between 700 and 2000 m as "steppe forest," commenting that the term forest is "not always appropriate to a formation in which the arboreal elements are sometimes so remotely scattered, that one can hardly catch two trees at one glance." This description certainly fits the modern landscape. One should think of this vegetation type as "a steppe sprinkled with solitary trees which under certain conditions may become condensed and turn into a forest-like formation" (ibid.). At an elevation of just under 700 m, Gordion itself would be at the upper boundary of the treeless Anatolian steppe, though terrain at 700 m elevation lies as close as 2 km.

Today, the land around Yassıhöyük is largely devoted to farming and grazing. Any land that can be irrigated is, but all irrigation is carried out with motor-driven pumps. Since the mid-1990s, a government water project has brought irrigation to the slopes, greatly expanding the area of irrigable and irrigated land. In and near the village of Yassıhöyük itself, trees grow primarily in protected gardens and along the banks of the Sakarya River. Isolated trees (*Elaeagnus angustifolia, Ulmus glabra, Prunus amygdalus, Salix* sp.) grow near the edges of some fields.

Between Şabanözü and Avşar, oak (*Quercus pubescens*) grows as close as 15 km from Gordion. To the northwest, the closest stands of junipers (*Juniperus excelsa* and *J. oxycedrus*) mixed with oak en route to Hamidiye are near Ahırozu, about 30 km by road from Gordion (elev. c. 1000 m). About 40 km from

Gordion, the soil changes and oak becomes more common. Continuing on to Hamidiye (Yağ Arslan), about 50 km from Gordion, oak (*Q. pubescens, Q. cerris*) and pine (*Pinus nigra*) grow. Just past Hamidiye, larger trees, mainly pine with an understory of oak and juniper (*J. oxycedrus*), grow in the forest (Figs. 2.4, 2.5). The extent to which the poor aspect of the vegetation is due to climate or human interference (fuel gathering, grazing, and, in antiquity, construction projects) is not entirely clear, but the analyses of archaeological woods from Gordion illuminate this question.

Since 1988, I have conducted informal vegetation surveys in the region, most intensively within 2 km of Gordion. Uncultivated habitats lying within this radius include the riverside, former floodplain, and degraded steppe on a marly siltstone substrate in which *Artemisia* sp. and wild thyme (*Thymus* sp.) dominate. A small patch of grassy steppe vegetation that was relatively undisturbed until the mid-1990s straddles the boundary between Yassıhöyük's fields and those of a neighboring village, Şabanözü (about 13 km to the northeast). Perennial grasses mixed with a variety of other plants covered the slope in 1992, but after irrigation came to the adjacent fields, annual grasses began to replace the native steppe vegetation.

Recent Land Use—Agriculture, Animal Husbandry, Fuel

Even since 1988, land-use patterns in the Sakarya valley near Yassıhöyük have changed. Most obvious to the occasional visitor are the expansion of irrigation to previously dry-farmed fields and the increase in weekend day-trippers from Polatlı and Ankara. At a scale of centuries and millennia, climate fluctuations, shifting river channels, periods of erosion, and many other human and natural factors have affected the landscape, so arguably there is no "ethnographic present." Gürsan-Salzmann (2005) is conducting a comprehensive historical and ethnographic study of the region; here I present a general description based on her work, other published sources, my own observations, and conversations and discussions with some of the villagers who worked for the project (mainly Ekrem Bekler and Remzi Yılmaz) and team members Ayşe Gürsan-Salzmann and Ben Marsh.

---

1. Note that in the calendar years 1988 and 1989, precipitation reported in Polatlı was 407.2 mm and 245.5 mm; for the growing seasons, the figures for the crop of 1988 (July 1987–June 1988) was 373.7 mm and for 1989 (July 1988–June 1989) was 228.9. Particularly good years (precipitation >450 mm) outnumber bad years (precipitation <250 mm) 5 to 3.

a

b

c

d

Fig. 2.4 Immediate environs of Gordion: a. Sakarya vegetation with Kuş Tepe in background; b. overgrazed pasture with Tumulus MM in right midground and Kızlarkayası outcrop in background; c. "grassy steppe" in 1992, before irrigation wrought land-use changes; d. former "grassy steppe" in 2007, after irrigation led to intensified agricultural activity in adjacent fields (Tumulus MM visible in center background).

Crops

Agriculture is still the main occupation of the Yassıhöyük villagers, at least during the growing season. The most important field crops are macaroni wheat, two-row barley, sugar beet, onion, sunflower, and melon. The last two of these are also grown in smaller gardens, along with tomato, eggplant, peppers, okra, and other vegetables for home consumption and market sale. Lentils and chickpeas are also grown. Several crops that were common in recent memory are no longer grown: rye, which is still a common weed of wheat fields, and cumin. One retired farmer (pers. comm., July 8, 1994) mentioned

three kinds of barley that were once grown: *beyaz arpa* (white barley), *siyah arpa* (black barley), and *peygambar arpa* (pilgrim barley; common oat?). Several crops were grown for oil: *keten* (linseed, flax; *Linum usitatissimum*), *konjit/susam* (sesame; *Sesamum indicum*), and a plant he called *zıra* (possible mishearing or variant of *zeyrek* [flax], Ertuğ 2000). An older farmer remembers growing *burçak* (bitter vetch; *Vicia ervilia*). Bitter vetch is harder to harvest than other fodder crops, so its culture declined after mechanization (H. Fırıncıoğlu, pers. comm., July 12, 2001). These discontinued crops were not irrigated, as the villagers did not have pumps then. Rye was and barley is grown primarily for fodder.

Fig. 2.5 Woodland vegetation in the Gordion region: a. oak (*Quercus pubescens*) below Avsar, ca. 900m; b. juniper (*J. oxyce-drus*) and oak (*Quercus pubescens*) en route to the pine forest at Hamidiye, over 1000 m; c. pine (*Pinus nigra*), juniper (*J. excelsa* and *J. oxycedrus*) and oak (*Q. pubescens*) grow together in pine forest zone, ca. 1400 m; d. juniper (*J. excelsa*) en route to pine forest, ca. 1200 m; e. pine (*Pinus nigra*) in clearing in pine forest.

Grain yields depend on moisture availability, the crop rotation, and the application of fertilizer (see Gürsan-Salzmann 2005). One farmer (E. Bekler, July 18, 1994) said that barley yields can be relatively low because wheat is more likely to be planted after a fallow year, when the soil is more fertile. He used to sow unirrigated wheat at a rate of 20 kg/*dunam* (ca. 20 kg/ha), for an expected yield of about 10 *teneke* (130–150 kg). In a dry year, a field would yield 7–10 teneke; the best years' yields are about 15–20 teneke. Irrigated wheat, which takes a lot of fertilizer and water, will typically yield 25–30 teneke; yields in a dry year would be 13–15 teneke, and in the best years could be as high as 35 teneke. Twenty-six kg of unir-rigated barley planted after a fallow year ordinarily yield 20 teneke. The yield in a dry year would be only 5 or 6 teneke, and in a wet year would be 20. For ir-rigated barley, if you plant two teneke, you can ex-pect a return of 30–35 teneke; in a dry year the yield would be 7 or 8 teneke, and 40 in the best year (Table 2.1).

Farmers' yields mentioned to Gürsan-Salzmann averaged about 200–250 kg/dunam for unirrigated, and 450–500 kg/dunam for irrigated wheat (about the same or slightly higher than E. Bekler's estimates of about 150 up to 300 kg/dunam for unirrigated wheat in a good year, and 325–450, up to 500 kg/dunam for irrigated wheat in a good year).

Table 2.1. Ekrem Bekler's yield estimates (pers. comm., July 18, 1994)

	Amount sown per *dunam* or hectare	Yield in *teneke** under different conditions		
		Drought	Normal	Wet
Unirrigated wheat	20 kg	7–10	10	15–20
Irrigated wheat	20 kg	13–15	25–30	35
Unirrigated barley	20 kg	5–6	20	20
Irrigated barley	2 *teneke* (26–30 kg)	7–8	30–35	40

*1 teneke ≈13–15 kg

Table 2.2. Seasonal round according to Remzi Yılmaz (pers. comm., 1992)

Month	Activity
September/October	Plant winter wheat and barley
Mid-November to mid-March	Rains come
December	First frost
Mid-December	Ground freezes for up to a month, 5–20 cm
March–May	Irrigate wheat (and barley)
April	Plant summer crops: cumin, sugarbeet
May	Plant summer crops: lentil, chickpea, melon, watermelon, garden crops
Mid-June to mid-July	Barley harvest, followed by wheat harvest; note that harvest is a few weeks earlier in Sakarya valley than between Polatlı and An-kara; in a dry year, harvest may be several weeks earlier
July	Harvest cumin, chickpea, lentil
July, August	Harvest melons, garden crops; weed sugar beet
September	Harvest sugar beet

Planting Year

The agricultural year begins in the fall, before the winter rains, when winter cereals are planted (Table 2.2). In addition to the additional labor input for irrigation, nowadays farmers use commercial fertilizer and weed-killer (the grain fields, at least, do not have broad-leaf weeds; sugar beet and onion are intensively hand-weeded).

Although irrigation is not necessary for wheat and barley cultivation in this part of Turkey, under irrigation the cereals are watered three times (March, April, May). Sugar beet takes seven waterings; it also must be thinned and weeded. Cumin and the pulses do not have to be irrigated.

Water Sources: Irrigation and the Sakarya

Prior to the deepening and subsequent down-cutting of the river channel and the introduction of pumps, the villagers grew irrigated rice on the floodplain (A. Gürsan-Salzmann, e-mail, January 18, 2007). The Sakarya would begin to rise in February. The waters would be highest in March, but the valley would be flooded through April (E. Bekler, pers. comm., July 25, 1993). Along the river, willow, poplar, wild pear and apple, and elm grew in dense thickets (*bük*) where wild pigs resided in great numbers. Most of the plain was grazed rather than farmed. According to Remzi Yılmaz (pers. comm., July 9, 1993), there used to be more mosquitoes, pasture plants, and *kamış* (reeds and cattails—*Phragmites* and *Typha*). Field irrigation was limited to low-lying areas near the Sakarya, and rice was grown near the river.

With gasoline-fueled pumps, fields can be irrigated as far as 1500 m away from the river, but more commonly no more than 500–700 m. Piping is assembled as needed, so there is no need to dig irrigation ditches. A government-sponsored water project that was in operation by 1995 has brought water to areas never before irrigated. In the past, wheat was more likely to be irrigated than barley. Today, even sunflower is watered, even though it used to be dry-farmed. It is probably no coincidence that traditional dry-farmed crops like flax, bitter vetch, and cumin have fallen out of favor. The plateau west of the Sakarya is still farmed without irrigation, however; in 1996, the main crop grown there was barley, along with a little wheat.

Some Aspects of Animal Husbandry

A variety of animals are kept in the village: cows, sheep, a few goats, fowl (geese, turkeys, chickens). Given the vagaries of the weather and the market, mixed farming is an important strategy in the valley. A recent study in the Polatlı region has shown that goat husbandry, and reliance on pastoralism in general, is more important in the hills than on the plain (H. Fırıncıoğlu, pers. comm., July 12, 2001); those who live in the mountains earn a lower proportion of their income from field crops, and they raise more goats and fewer sheep than people who live on the plain. This area depended much more heavily on pastoralism in the 19th and early part of the 20th century than it does today (Gürsan-Salzmann 2005).

Overgrazing, a problem in much of Turkey, is certainly occurring at Yassıhöyük. Fields are owned by individuals, but pasture is owned by the village. In the Polatlı area, the greatest stress on the pasture occurs in the spring, just when the perennial grasses and many other wild plants are flowering and fruiting: 50% of the fodder comes from pasture in April and May, and 100% in June (H. Fırıncıoğlu, pers. comm., July 12, 2001). Animals are stall-fed at least part of the year, so fodder must be grown or purchased. During the winter, the animals are taken out of their stalls to be watered, but the ground is too muddy for them to graze. With overgrazing, plants such as *üzerlik* and *tiken* (wild rue and camelthorn; *Peganum harmala* and *Alhagi pseudalhagi*) increase. When fresh, they are avoided by the herds—*Peganum* does not taste good and *Alhagi* is spiny—but in the winter, when they have dried, sheep and goat will eat both (E. Bekler, pers. comm., July 27, 1993). Their archaeological presence is therefore an indicator of poor pasture.

Farming and herding have different seasonal labor and land requirements, some of which are mutually exclusive. A mixed strategy can enhance food security and provide products suitable for exchange in a broader system. The changing agropastoral economy at Gordion had to balance the goals of achieving security and surplus in an agriculturally marginal but commercially central environment.

Fuel

Fuel is necessary for domestic cooking and heating. Nowadays, bottled gas (*töp*) and coal (*kömür*) are readily available for purchase. Other fuel-consuming activities known ethnographically or attested in the archaeological deposits include gypsum plaster (*tatlı kireç*) production, ceramic firing, and metalworking. The traditional fuels for these activities are wood, charcoal, and dung.

Before the river was deepened, men used to cut wood in the woods along the river (*bük* [thicket]), and in the dry months, women would make dung cakes. In the old days, shepherds would go out to the hills a few kilometers from Yassıhöyük for days; family members would bring them food. People would sweep up the dung and bring it back as the main fuel. Dung was also collected from the animal pens, as it is to this day.

One use of dung cake fuel was to make gypsum plaster (E. Bekler, pers. comm., July 29, 1996). Gypsum from Kızlarkayası would be collected and burned for three or four days in a big pile (several meters high), until it got soft and powdery. The resulting gypsum plaster could then be applied to walls when mixed with water. It is not as good as the store-bought kind (*acı kireç* [lime plaster]) because it is powdery and comes off on your clothes. People would apply it three or four times a year, at holidays.

Dung is used for fuel in several forms, and its quality varies. Sheep, goat, and cow dung are all used for fuel, even today. Fuel from sheep and goat pellets is better than cow dung because it is inherently more compact, and after it has accumulated in the stalls over the winter, it is even denser.

Not surprisingly, there are several Turkish words for the different types of dung and dung fuel. Seona Anderson (1994/1995) gives a detailed description of the various forms used near Aksaray; there appears to be some difference in usage between the people she spoke to and Ekrem Bekler, a retired farmer. In Yassıhöyük, the two most commonly used terms are *tezek* and *kerme.*

> *tezek:* general (and common) word for dried dung used as fuel (same in Aksaray)
> *kerme:* sheep dung slabs dug out from stalls, also called *kemre.* In Aksaray, this term is used for winter cow dung mixed with straw and water, unshaped
> *kiğ:* old word for sheep dung used as fuel (also used in Aksaray)
> *kaba tezeği:* dry cow pats (*yaban tezeği* in Aksaray)

> *el yapması:* cow dung shaped by hand (*yapma* in Aksaray)
> *mayıs:* cow or horse dung
> *davar mayısı:* sheep or goat dung

Several words reported by Anderson's consultants were unfamiliar to E. Bekler:

> *sarma:* sheep dung dug out from byres
> *kön:* soil-like by-product of cow dung—bedding or fertilizer
> *kareli:* made from bits of *kerme* and *sarma*
> *kerpiç/kasnak:* molded cow dung with water, straw: E. Bekler knows the word as molded cow and/or horse dung; *kerpiç* is also the Turkish word for mudbrick

Ancient Climate and Vegetation

Climate reconstructions are based primarily on proxy data, as there are few direct indicators of past climate conditions. Geomorphological, botanical (pollen, phytolith, and macroremains), and soil studies commonly reveal more about vegetation cover—episodes or erosion or deforestation—than they do about climate. This is particularly true for the more recent (post-Bronze Age) periods, when human impact on the vegetation is so great that it masks the natural fluctuations of climate (Miller 1997a). And of course, dating non-archaeological deposits at a sufficiently fine scale to be useful is also problematic.

In any case, the climate record of central Anatolia for the past 3000 years is thin. In southwestern Turkey, both the Beyşehir and Söğüt pollen diagrams show an expansion of pine (apparent increase in moisture) at about 1000 BC, with pine remaining important until the top of the core; the modern deforested state of the vegetation post-dates the core (van Zeist and Bottema 1991:81). About 120 km to the northeast, in the Yeniçağa core (inconclusive dates for the past 3000 years), Bottema et al. (1993/1994: 33) consider vegetation changes to be the result of human activities rather than climate. Regardless of any minor shifts in climate, what would seem to be a constant for agropastoral strategies at Gordion, however, is a high interannual variability.

Earlier Thought on the Vegetation around Gordion and Archaeobotanical Finds

At least since the discovery of the great tumuli, with their incredibly well-preserved timbers and wooden furnishings, questions arose concerning the forests of the present virtually treeless landscape around Gordion. Rodney Young, director of the excavations from 1950 to 1974, wrote,

> Although the modern landscape is as bare as bare can be and the few trees that grow now—poplar, willow, and wild pear—are limited to the margins of the river and the irrigation ditches, the hillsides were certainly once clothed in woods which have since disappeared. The profusion with which wood was used in Phrygian construction and the size of the timbers preclude importation from very far away. (Young 1960:3)

The wood and charcoal assemblage from Young's excavations provided significant information about the state of the forest and trade during Phrygian times. Pine timbers used for construction on the Citadel Mound were commonly re-used (Kuniholm 1977:48). These large timbers would have been valued because they would have been difficult to transport, even over short distances.

Most of the wood from the excavations of 1950–1973 comes from the Early Phrygian Destruction Level on the Citadel Mound and Middle Phrygian period tombs. The charred wooden beams and construction material found on the Citadel Mound in the Early Phrygian Terrace Buildings and elsewhere are pine (*Pinus*) (Kuniholm and Tarter 1989; Kuniholm 1990; this report). Construction materials identified from Middle Phrygian Tumulus MM include pine (*Pinus*—wall, ceiling, beam), juniper (*Juniperus*—exterior), Lebanon cedar (*Cedrus libani*—floor) (Kayacık and Aytuğ 1968).[2] The tables, screen, and

coffin from Tumulus MM were made of boxwood (*Buxus sempervirens*), juniper (*Juniperus*), walnut (*Juglans regia*) (Aytuğ 1988). The coffin woods have recently been determined to be pine and Lebanon cedar (Blanchette and Simpson 1992). The chamber in Tumulus P was made of black pine (*Pinus nigra* subsp. *pallasiana*) logs, with some internal planks of juniper (Aytuğ and Pehlivan 1989); furnishings are boxwood, juniper, walnut, and poplar, according to Aytuğ and Pehlivan (1989).

Although modern sources for some of the woods found on the Yassıhöyük mound and in the tumuli are fairly close to Gordion, other types, for example, *Cedrus libani,* would have come from moister, higher (>1000 m) regions in Turkey well over 100 km away (Davis 1965; Kuniholm 1977: Pl.12). Proximity is therefore a relative concept, and it is not possible to specify the exact distance between Gordion and its sources of wood. Young seems to have envisioned fairly dense woodland during Phrygian times at the edge of the Sakarya valley (i.e., within 2 km of Gordion) (K. DeVries, pers. comm., 1991). The current research does suggest a somewhat more wooded landscape than can be seen today. By the 1st millennium BC, timber was being transported far and wide in the Near East. If indeed large cedars came from over 100 km away, Phrygian technology was clearly adequate to transport more local timbers like pine and juniper. As Richard Liebhart (pers. comm., 1999) points out, some of the logs in the chamber in Tumulus MM have "a flattened channel with a hole cut near the large end...[which] shows how the Phrygians transported logs: the cutting with its hole was placed on an axle between two wheels (a pin in the axle fit into the hole), and the smaller and lighter end of the log was lifted up, turning the log into a make-shift wagon for easier transport" (see Young 1981:86). In contrast, ethnographic analogy suggests that fuel-gathering would probably not have been economical over distances greater than about 50 to 75 km.

Other work on woods from Gordion and its tumuli has been carried out primarily by H. Kayacık and B. Aytuğ (1968) and Aytuğ (1988) on the construction and furnishings of the MM Tumulus, by Blanch-

---

2. An earlier identification by Kayacık and Aytuğ (1968), Aytuğ (1988), Aytuğ and Pehlivan (1989) has been revised; the yew (*Taxus baccata*) reported in the structure and furnishings of the Tumulus MM is now recognized to be pine (Blanchette and Simpson 1992). Some of the specific determinations in the earlier works (especially *Pinus silvestris* rather than *P. nigra,* and *Juniperus foetidissima* rather than *J. excelsa*) should probably be revised on phytogeographical grounds. Similarly, some

of R. Young's observations may not be valid (see Gordion I appendices).

ette and Simpson (1992) on "Midas's " coffin, and by Aytuǧ and Görcelioǧlu (1988) and Aytuǧ and Pehlivan (1989) on Tumulus P tomb and furnishings.

The results of the 1988 and 1989 excavations reported in Chapter 4 add considerably to the interpretations based on the materials excavated in the 1950–1973 seasons. First, the new samples greatly extend the time range of documented wood use: Middle Bronze Age to the Medieval period. Second, much of the charcoal is the residue of incompletely burned fuel, a better indicator of the state of the local woodland contemporary with a given deposit than are valuable timbers and possibly rare or exotic products of the cabinetmaker's art.

3

Field to Laboratory: Collection and Processing of Wood Charcoal and Flotation Samples

Nature of the Deposits – Burnt Buildings vs. Ordinary Occupation Debris

Two basic types of deposits were encountered in the 1988/1989 seasons—burnt buildings and ordinary occupation debris. Because the former is likely to include a substantial amount of construction debris and even some food stores, and the latter is likely to include a substantial amount of spent fuel, there is no reason to sample and analyze them in the same way. The three burnt structures represent different occupation phases and types of houses: the Burnt Reed House is an Iron Age wattle-and-daub domestic structure; Terrace Building 2 probably housed support staff for the Early Phrygian Destruction Level elite quarter; and a Hellenistic structure from the so-called Abandoned Village occupation appears to be domestic. In contrast to the more ordinary occupation debris, the plant material from these buildings is a mixture of construction debris and whatever seeds and wooden objects were left behind after the fire.

Field Collection of Wood Charcoal

Supervisors and workers were told to collect wood charcoal visible in the course of excavation and from screened deposits. If it was obvious that the fragments came from a single large piece, they were asked to label the bag as such. It was not practical to even try to collect all the charcoal from the burned buildings. Nevertheless, relatively large samples of fractured beams and other construction materials were collected. Most charcoal from ordinary occupation debris came from pieces scattered in the excavated 'lot' of soil.

Some large pieces were wrapped in string and sent to the Cornell Tree Ring Laboratory in Ithaca, New York, as possibly useful for dendrochronological study.

Field Sampling for Flotation

Excavation supervisors were told to take samples for flotation of approximately 10–15 liters of an archaeological deposit, which they put into heavy-duty plastic bags. Some excavators were more conscientious than others, but the guidelines were to take samples from: all hearths and pits; places with a high density of charred material (e.g., trash deposits); just above ancient floors and surfaces (from ca. 5 cm above down to the upper edge of the surface); sediments associated with hearths, pits, and other sampled features ("control samples"); any deposit about which an excavator was curious. Along with wood charcoal samples, flotation samples were taken from the burned buildings, too.

In 1988 and 1989, flotation was accomplished with the aid of a Siraf-like machine (French 1971) built by Mark Nesbitt and loaned to the project by the British Institute of Archaeology in Ankara; Nesbitt also provided detailed instructions on its use. Rather than a stiff inset lined with metal screening, the heavy fraction of the samples was caught in synthetic window-screen mesh (ca. 1 mm squares, variable). The light fractions flowed into polyester cloth set in an agricultural sieve through which only dust could pass. The dried samples were transferred to plastic bags and sent to the University of Pennsylvania Museum with the permission of the Museum of Anatolian Civilizations in Ankara.

Representativeness

Most of the botanical macroremains from Gordion are preserved in charred form. Flotation of sediment samples concentrates remains that are dispersed in the site matrix. Most such material is assumed to represent incompletely burned fuel remnants redeposited as trash, intentionally burned trash, or accidentally burned material (see, for example, Hillman 1984; Miller and Smart 1984; Minnis 1981). We also floated samples from burned buildings, partly in order to be able to make quantitative comparisons with the dispersed material, and partly to pick up small items mixed in with or part of the charred construction debris.

In an ideal world, large pieces of charcoal would be recovered at the same rate as ceramics and bone, so questions about representativeness within excavation units would be irrelevant; one would, of course, still have to worry about how the excavation units were chosen. In reality, however, not all charcoal was recovered, because it tends to be smaller and of less obvious interest to the archaeologists and workmen than artifacts. In this context, relative amounts of the different types are more significant than absolute quantities, so the hand-picked charcoal from occupation debris is treated as though it is a fair representation of what theoretically could have been collected.

The goal of sampling for flotation was to get a collection of charred seeds and wood representative of the remains in the excavated deposits. At Gordion, as in most archaeological sites, excavation units were not chosen randomly, but rather in relation to the archaeological, historical, and chronological questions outlined in Chapter 1. Therefore, interpretations presented here are not based on formal statistical significance of the quantified remains. Rather, the sampling for macroremains aimed at obtaining an assemblage that would reflect what was in the excavated deposits and that would include enough material to analyze.

Although I cannot provide any statistical certainty, it is likely that the source of charred remains in settlement debris (excluding burnt buildings) is redeposited hearth sweepings. Partial justification for this conclusion is that for any given time period, different types of deposits tend to share taxa (i.e., hearths yield the same range of taxa as pits or trash).

Laboratory Procedures—Samples, Sorting, Recording, and Quantification

Wood Charcoal

Excavators were encouraged to collect all charcoal chunks seen in the field (except for charcoal in flotation soil samples). This ideal was not met for either the burnt buildings or ordinary occupation debris. And in the case of the three burnt buildings (Burnt Reed House, Terrace Building 2A, and the so-called Abandoned Village occupation, Fig. 1.1), there was too much charcoal for all to have been collected. Therefore, density of wood charcoal at Gordion cannot even be approximated.

The samples were mixed with varying amounts of dirt. As the tiniest pieces of charcoal are not readily identified or quantified, samples were sieved through 2 mm mesh, and only the pieces caught in the 2 mm mesh were measured and weighed. An attempt was made to identify pieces with at least one complete growth ring, to avoid over-representing easily identified taxa such as oak.

Two microscopes were used to make determinations. A stereozoom microscope, magnification 7.5–75x, was used for initial determinations, and an incident-light compound microscope was used at magnifications up to 400x, but usually 100x or 200x for smaller features. See Chapter 4 for details of charcoal analysis.

For purposes of analysis, the total weight of wood charcoal per sample is reported, as is the weight of the identified pieces. The amount for each identified taxon is calculated proportional to the total amount of charcoal in the sample.

Sampling and Its Influence on the Interpretation of Diversity

"Sample" refers to charcoal included under one YH#. Consequently, "sample" is a totally arbitrary unit in terms of archaeological context, and there is no set size or excavated volume of deposit from which it comes. A sample may be as small as a single piece of charcoal 2 mm in diameter or enough large pieces to fill a shoebox or two. As noted in the text above, it seemed most reasonable to make major comparisons between

Table 3.1. Number of distinct taxa (excluding "unidentified") from hand-picked charcoal samples from occupation debris

YHSS phase	No. samples	No. distinct taxa	No. pieces analyzed	Wt. pieces analyzed (grams)
1	12	7	67	20.38
3	36	9	232	118.56
4	71	13	475	235.86
5	8	4	54	53.73
6	16	6	76	28.57
7	52	5	251	183.80
8/9	13	4	70	75.61

time periods. Comparative analysis by archaeological context of fuel remains awaits further archaeological analysis (see, e.g., Marston 2009, 2010, n.d.).

Since it was neither possible nor productive to analyze all pieces of charcoal in every sample, I used several criteria to help determine the number of pieces I would examine. The goal was to get a reasonable view of the variability within a sample (cf. Smart and Hoffman 1988).

1. I identified at least 10 pieces per sample, unless a sample had fewer than 10 pieces or if a sample was clearly (on visual inspection) all the same type (this was most often the case for some of the bags of pine from Terrace Building 2A).
2. If only one or two taxa were seen in the first 10 pieces, no more were examined. If more were seen, up to 10 additional pieces were looked at.

The laboratory sampling strategy enhanced the chances that the number of pieces analyzed would be directly associated with the number of types. Indeed, variety is generally associated with the number of pieces analyzed per level, but it is not associated with the weight actually analyzed, on which the interpretation rests (Table 3.1). This makes statements about variety problematic. In particular, as one might expect, the Late Phrygian period with the greatest variety (13 distinct types) had the most pieces analyzed. Even so, the possibility cannot be excluded that changes in variety are due to long-term vegetation change or functional differences between, say, a preponderance

of industrial trash from pits (Late Phrygian, YHSS 4) and household trash (Early Iron, YHSS 7). The later periods do seem to have higher variety; for example, there are almost twice as many types (at least 9) in the Hellenistic deposits as in the Early Iron deposits (at least 5), though both are similarly domestic in character with comparable numbers of charcoal pieces examined (232 and 251). Diversity indices and measures of evenness would have been calculated, but the sample sizes were too small (Popper 1988).

Flotation Samples

Charred material is very well preserved at Gordion, and it was not possible to analyze all samples taken. The goal for analysis was to get a broad functional representation of deposits from different time periods; if more than one sample from a particular archaeological feature was taken, a judgment was made based on amount of material and complexity of the deposit. For example, small samples from a single deposit might be combined for analysis, and several samples from an extensive trashy deposit might be examined to see if the deposit is homogenous or not. Most samples were sorted by me, but occasionally by a student or laboratory assistant. In all cases, I checked the work.

Sorting and analysis instructions that were generally followed for the light fractions appear in Appendix A. The basic procedure was to sieve each sample through graded mesh, partly for ease of sorting. In addition, categories of plant remains are recognizable and identifiable at different sizes, so size-sorting serves an analytical function. Wood charcoal is easy to sort to 2 mm, though pieces smaller than about 5 mm become progressively more difficult to identify. A catch-all category, "charred material >2 mm," was separated out, but not analyzed. It probably includes parenchyma and other plant fragments, but the amounts are very small. Seeds such as grain, pulses, nutshell, and some plant parts may be confidently recognized to 1 mm, and many cultivated and wild seeds easily pass through 1-mm mesh.

Density of charred material (wood, seeds, and other plant parts) from an entire excavated context cannot be calculated for the hand-picked charcoal, but it can be estimated from flotation samples. As not all deposits were sampled for flotation, we cannot assume representativeness for the site, especially because sampling in the field favored deposits thought by the ex-

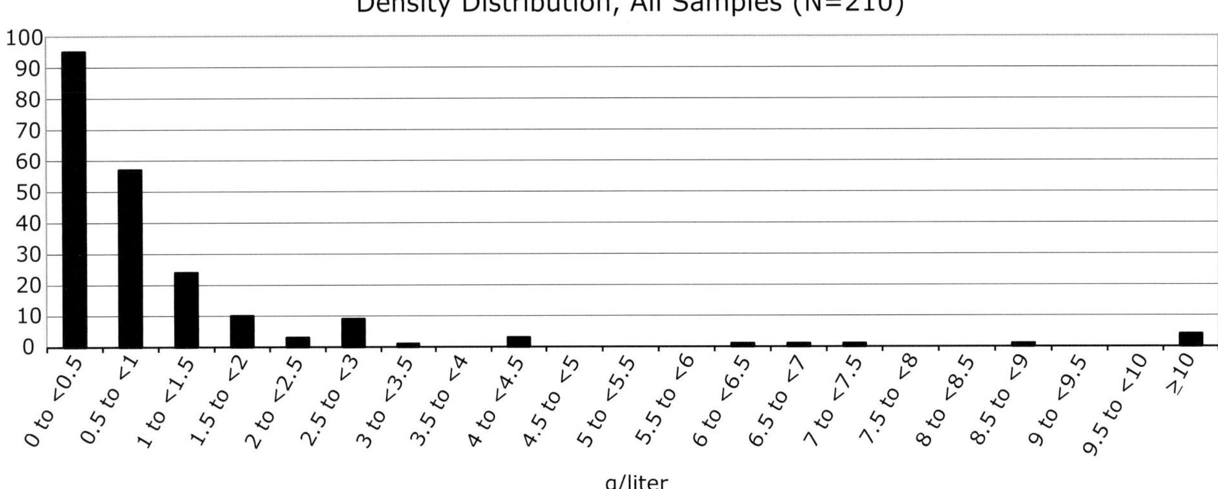

Fig. 3.1 Density distribution of charred material (data in Table 5.3).

cavators to have charred plant remains. Even so, most sample densities (184 out of 224) are below the mean of 1.33 g/liter. As the statistical distribution does not follow a normal curve, the mean density does not describe the population. The median for the samples as a group is only 0.55 g/liter (Fig. 3.1).

Heavy Fractions

At first, the heavy fractions of a small number of samples were examined to 1 mm with the help of a dissecting microscope, but so few seeds were recovered (and those that were included the same types as those found in the light fractions) that further recording would not change the interpretations. Subsequently, the heavy fractions were examined down to 2 mm. Some seeds (especially large rounded ones like bitter vetch and *Galium*) are more likely to sink than others, and one type, wild almond fragments, occurs only in the heavy fraction in the samples reported here. (See Appendix F3 for heavy fraction contents.)

4

Analysis of the Wood Charcoal Sample

Archaeological Context

The stratigraphic sounding undertaken in 1988 and 1989 established a sequence of archaeological phases, and the excavations greatly expanded the amount and variety of plant materials available for study. Excavators were asked to collect all chunks of charcoal seen in the course of excavation; this goal was not reached. The three burnt buildings contained too much charcoal from construction debris, and even from the other kinds of deposits, excavators were somewhat erratic in their zeal to collect plant remains. The new materials come primarily from occupation debris, including pits in residential areas, trash pits, ordinary occupation debris, and occupation debris from an elite quarter. Wood charcoal from the three burnt structures gives evidence of building materials in Early Iron Age, Early Phrygian, and Hellenistic times. Finally, wood from the Tumulus MM tomb chamber and its furnishings identified by the wood anatomists adds another context type.

In short, the archaeobotanical remains come from structures, furnishings, and occupation debris. The structures and furnishings provide material most like traditional archaeological artifact categories and in many respects can be analyzed accordingly, in terms of function, source, and distribution within the site. The most wood and charcoal come from burnt buildings on the Citadel Mound and the wooden tomb at the base of Tumulus MM. The second source of plant materials, tomb furnishings, consists of small but high-status items made of wood. The most widespread material, however, consists of charcoal and seeds from settlement debris.

Methodological and Analytical Assumptions

The Gordion excavation uncovered burnt building levels interspersed with other structures and settlement debris accumulated over time. This means that the charred wood recovered archaeologically came to be on the site for a variety of reasons, the most obvious being as fuel and construction material. With regard to fuel residues, I presume that quantities of the various taxa reflect availability in the local vegetation, in general terms. People are more selective in choosing construction materials. Presumed construction charcoal is therefore tallied and analyzed separately from more ordinary charcoal that is most probably the incompletely burned residue of fuel. In reality, "construction" and "fuel" deposit types are not mutually exclusive. Nevertheless, occasional inaccurate functional designation of charcoal should not mask the overall patterns. The functional assignment of any one sample (material included under one YH#) to fuel or construction may be wrong, but such errors will be insignificant if the number of samples analyzed is large enough.

There are several ways to quantify charcoal remains: weight, count, ubiquity, and volume. Mass is more directly related to ancient fuel use than number of pieces or volume. It is most useful for the analysis of the fuel remains, even though wood density varies between types. For example, oak is very dense, pine is not, and juniper is in between; analysis by weight would therefore tend to over-represent oak, and analysis by volume would over-represent pine. For the fuel charcoals, I report the weight of charcoal larger than 2 mm, as well as the proportion (by weight) of the

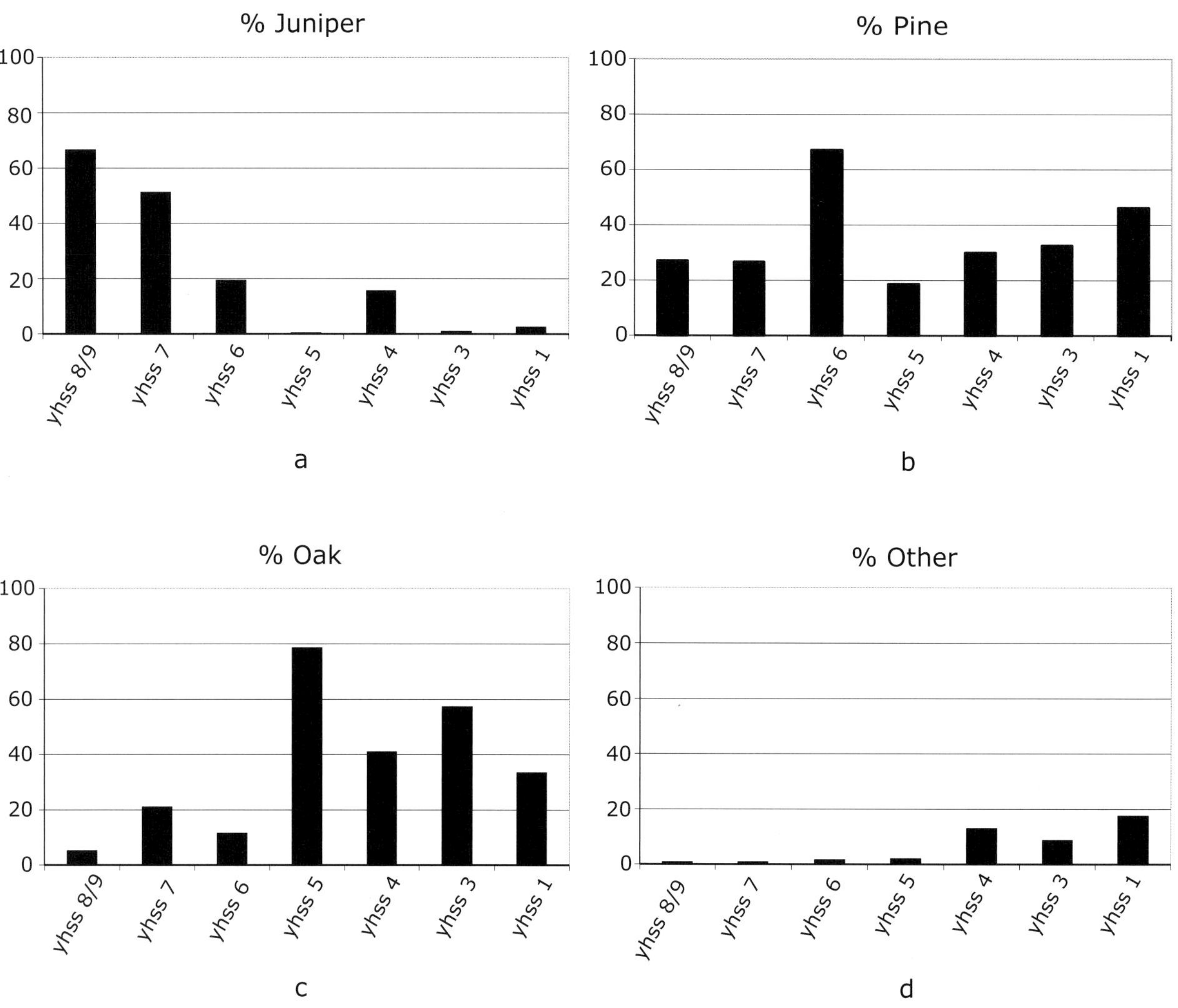

Fig. 4.1 Weighted percent of major charcoal types (by weight; data in Table 4.1a).

to the dredging and straightening of the river, Young saw wild pear there, as well.

Zohary (1973) and Davis (1965–1988) provide information about the habitats and ranges of the remaining wood taxa. Unfortunately, habitats for plant genera must usually be fairly broadly drawn. For example, one type of ash that grows in inner Anatolia, *Fraxinus angustifolia* subsp. *angustifolia,* is found on "dryish, rocky places" (elev. 650–1700 m), and another, *F. angustifolia* subsp. *oxycarpa,* is found "often in wet places, flood plains, by streams in mixed deciduous forest, s.l.–900 m" (Davis 1978:150 ff.). Alder (*Alnus*) would grow along watercourses, though the ancient specimens may not be local. Buckthorn (*Rhamnus*)

is usually "unimportant vegetationally" in the Middle East (Zohary 1973:373).

The charcoals of the minor components of the Gordion assemblage (i.e., anything not pine, oak, or juniper) mostly come from minor components of the steppe-forest vegetation and from along watercourses. Some pieces may have come from trees planted or protected in areas of former steppe-forest. Considering that the non-dominant types are most heavily concentrated in the Late Phrygian levels and later, they may represent secondary succession plants and degraded steppe-forest vegetation (especially pear/hawthorn). That poplar, elm, and tamarisk were more prevalent in later times would also be consistent with this view; other things

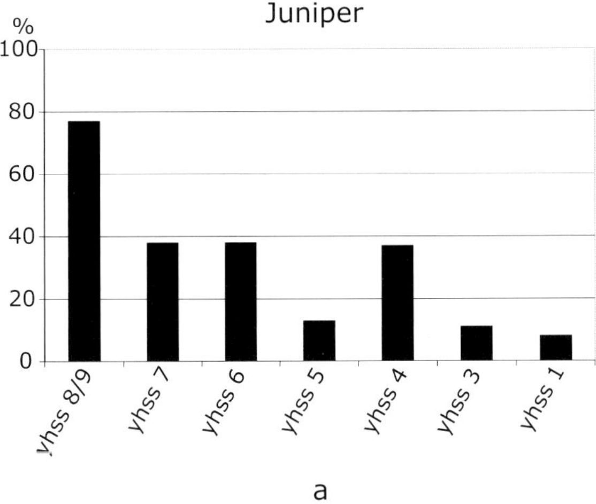

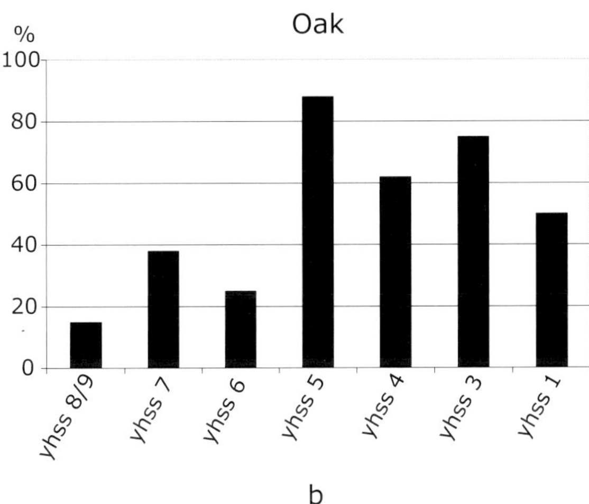

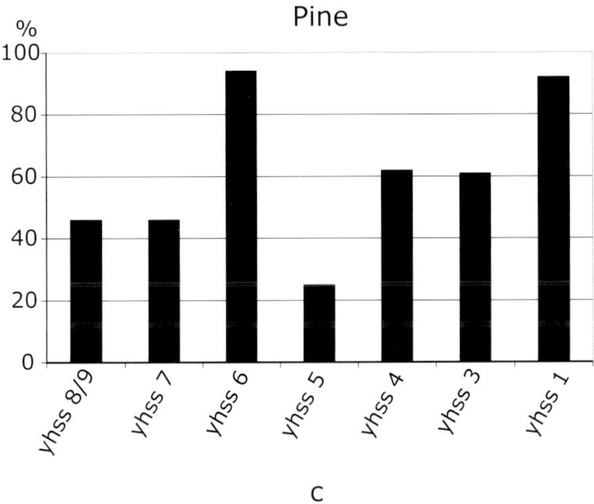

Fig. 4.2 Ubiquity (percent) of major charcoal types (data in Table 4.1c).

being equal, gallery forest can regenerate more easily than dryland types because more moisture is available. Note that the analysis of seeds and other plant parts recovered through flotation sheds additional light on human-induced vegetation change (see below).

Distribution of the Charcoal in Time and Space

Archaeological context provides the key for understanding the distribution of charcoals through the Gordion stratigraphic sequence. Charcoal does not occur "naturally"; simply tallying the different types by time period does not reveal the state of the vegetation. People bring wood onto a site for a variety of purposes—as building materials, tools and furnishings, and fuel. They select woods from among the ones that are most appropriate to a task, taking transport costs and availability into account. For example, a king or a cabinetmaker may be willing to import boxwood from the Black Sea coast to make fine furniture, but it is highly unlikely that such wood would routinely be burned for fuel. In contrast, people are much less discriminating about their choice of fuel wood for cooking, heating, and industry, so availability is a key factor (see Miller 1985).

The Gordion excavations of 1988 and 1989 produced quite a bit of charcoal, but the sequence is fairly long and the excavated area was relatively restricted:

1. Many of the phases are characterized by only a small amount of charcoal, less than 100 g total or fewer than 100 pieces examined (Table 4.1a,b); analysis of even a few additional samples could alter the proportions of the different woods.
2. Different phases are represented by different types of deposits. For example, YHSS 7 (Early Iron Age) and YHSS 3 (Hellenistic) deposits excavated in 1988 and 1989 are characterized by apparently residential architecture, YHSS 6A and 6B (Early Phrygian) deposits have more substantial, "elite" architecture, and most of YHSS 5 (Middle Phrygian) charcoal is from floors and trash pits.
3. Domestic and industrial trash can sometimes be distinguished, or at least inferred from the archaeological context (e.g., Feature 430.04, possible 'metallurgical pit' of YHSS 4), but as of this writing, and aside from the building materials of burnt

By Hellenistic times, it is virtually absent from the assemblage, and presumably absent from the immediate environs of the site. Oak becomes the major fuel wood during the Middle Phrygian period, when Gordion had reached its maximum extent. Oak can be more sustainably harvested than pine or juniper, which could explain its prominence in the assemblage during a time of maximum population. It declines from that early peak, but remains a significant part of the assemblage.

Pine proportions may be partly interpreted with the model of wood exploitation that I have proposed based on the modern vegetation zones and (overland) distance-related transport costs as the factors determining fuel use. Contrary to expectation for the Early Iron Age and Early Phrygian period, pine, which today would have to come from farther away, exceeds or equals oak charcoal by weight and ubiquity. After Middle Phrygian times, the increase in pine follows the model of local depletion of wood sources. Preliminary results of Roman and Medieval samples excavated in 2004 strengthen this impression; pine predominates in these samples (Miller 2007). To explain the distribution of pine and oak in the first part of the sequence, several possibilities come to mind:

1. Pine was mixed with oak at lower elevations closer to Gordion than today. A small stand of relatively short pine grows on Çile Dağı about 25 km by road northeast of Gordion. Remember Bottema and Woldring's (1984) observation that oak tends to replace pine in the pine-oak associations in central Anatolia.
2. The pine peak occurs in the Early Phrygian levels, and the fuel charcoals come from the pre-Destruction Level palace area. Coming from relatively far away, pine may have been a high-status fuel. Perhaps during Early Phrygian times the economics of scale supported specialized charcoal cutters who provided charcoal fuel for the city[6]; during other periods wood, which is heavier and bulkier than charcoal for equivalent heat value, was collected from sources that were a bit closer. Supporting

this view is that clay sources of the Early Phrygian period "must have lain elsewhere in the valley or beyond" (Voigt and Henrickson 2000:51), unlike the earlier sources that came from local Sakarya valley sediments. Against this view is that the wealthiest period of occupation at Gordion is the Middle Phrygian, when household and industrial production presumably put the most stress on the woodlands.

3. Pine grew in the mountains upstream from Gordion, and the fuel transport economy was based on the river, especially before the Middle Phrygian period, making pine a cheaper fuel than oak.
4. Pine was preferred, probably as charcoal, and so it would have been economical to transport it over relatively longer distances. In other words, pine wood has a somewhat lower heat value than oak wood by volume (and maybe by weight), but pine charcoal would have the same or higher heat value by both volume and weight. I suspect this is the case, but do not have directly comparable figures to prove it.
5. The observed trends are simply a result of small sample size, and will not hold up after more charcoal is analyzed.

These explanations are not mutually exclusive, though the river transport option 3 is unlikely. Based on the analysis of the seed remains, the differences between time periods may relate to other aspects of land use (see Chapter 6). In any case, after the Early Iron Age juniper forms a negligible part of the assemblage, and oak and pine predominate from Middle Phrygian times on.

The minor components are more difficult to interpret, primarily because they represent such a small proportion of the total; any patterning and variability are heavily influenced by chance factors that have little to do with ancient plant use: where the excavation units were, the alertness of the excavators, and sampling in the laboratory. Nevertheless, some patterning can be discerned. First, the variety of types recognized is higher from Late Phrygian times on, and those types are not major forest trees. Rather, they are trees characteristic of degraded woodland that one would expect to remain after the steppe-forest is removed (hawthorn and some elms, for example) and types that grow in favorable conditions and might therefore regenerate

---

6. It seems unlikely that lumbering for construction materials could have created a secondary market for the trimmed branches substantial enough to account for the large proportion of pine in this period.

quickly (hydrophilic types like poplar and tamarisk, and in central Anatolia, maybe pear and elm as well).

A few pieces of charcoal may be mulberry (see Appendix B for discussion of taxonomic difficulties); there is a possibility of confusion with elm, but a brief discussion is appropriate. Today, mulberry is planted as a street tree in Polatlı and elsewhere in Turkey; its berries are edible, but people grow it commercially primarily for its leaves, to feed silkworms. Mulberry is not native to the Near East, but would probably have come from central Asia. Historical records place the beginnings of the silk industry in Byzantium to the reign of Justinian (527–565) (Braudel 1979:326), so I am not proposing that silk production had reached Gordion in the Early Iron Age or even by Late Phrygian times. Along with some mulberry wood reported from the well in the Northwest Palace of Assurnasirpal II at Nimrud (Mallowan 1953:25n.), and a single mulberry seed from a Roman deposit at Tell Hadidi (van Zeist and Bakker-Heeres 1985[1988]:308), however, these pieces (if indeed they are mulberry) would constitute the only evidence for that tree from ancient Near Eastern sites west of Iran.

Wood from "planks" from a "tray" on the floor of the Burnt Reed House is tentatively identified as *Alnus viridis* (alder). This species is anatomically quite distinct from two alders native to Turkey and those from Europe. Assuming the geographical distribution of *A. viridis* is the same as in antiquity, the closest source would have been south central Europe. Rather speculatively, one might propose the "tray" was an heirloom, brought by Phrygians who settled at Gordion during YHSS 7B (Voigt and Henrickson 2000).

Charcoal from Burnt Buildings

Several different functional considerations determine wood selected for building—if nothing else, a beam must be long enough, and ideally it should be resistant to decay and not too difficult to work (Table 4.2). Juniper is very resistant to decay, fairly soft, and some species grow tall and straight under favorable conditions (e.g., *J. excelsa*). Juniper would seem well-suited for construction, especially for the pole-and-mud construction of the Burnt Reed House coded as YHSS 7, feature 725. Pine became the timber of choice for the more monumental structures of YHSS 6 (Early Phrygian) times; it would have come from

further away, but the most likely type, *Pinus nigra* var. *pallasiana* grows up to 30 m (compared to 20 m for *Juniperus excelsa*) (Davis 1965; Fig. 2.5d, e). Furthermore, pine grows faster than juniper, so it reaches a wide diameter sooner than juniper. Perhaps the closest oak, *Quercus pubescens*, was avoided in construction because it was hard to work or because the local oaks were too short to span the rooms (see Marston 2009).

YHSS 7 (Early Iron Age) material includes samples from the building collapse and floor deposit of the Burnt Reed House. There is every likelihood that more than just "construction debris" is included on Table 4.2. For example, YH 30419 is a bag of alder charcoal that the excavator labeled "planks." Most of the wood collected from the Burnt Reed House is juniper or pine, but here, too, the samples are mixed with other types: oak, poplar, and elm/mulberry. The proportions of the various fuel charcoals (material from trash deposits) in YHSS 7 deposits do not parallel those from the contemporary BRH (stratum 725). Furthermore, weight and ubiquity, as measures of "importance," do not show an unequivocal trend; by weight, juniper is the predominant fuel and construction wood, but by ubiquity it seems less important than pine and equal to oak in the debris samples.

Early Phrygian construction debris comes from the Destruction Level (YHSS 6A), Terrace Building 2A (YH strata 610, 620). Nearly all the charcoal is pine. The fire was so intense that the bags of pine charcoal were remarkably light for their volume, compared to wood charcoal samples from other phases. The small quantity of oak could have come from room furnishings, and is unlikely to have come from the ceiling beams.

YHSS 3 (Hellenistic) deposits sampled by the Yassıhöyük stratigraphic sounding that yielded charred construction wood include clearly burned roof material from a burnt room in the Abandoned Village (deposits from Operation 2 designated by YH strata 320, 330, 350). Most of the charcoal (by count, weight, and ubiquity) is pine, and there is an admixture of oak and a small amount of ash wood. Due to the fact that one unworked charcoal chunk looks pretty much like another, one cannot exclude the possibility that some of the charcoal comes from burnt furnishings, stored fuel wood, or other wooden items that might have been burned in the room. For example, if excavators had noted particular artifacts or concentrations, it would have been possible to separate construction debris from other functional categories of wooden objects. The fact that the distribution of charcoal types from the burnt room deposits is primarily pine rather than oak supports the view that the charcoal in these deposits is primarily from roof beams.

Building materials in Terrace Building 2A parallel the predominant fuel wood of its time (YHSS 6A and 6B combined)—pine. The situation is quite different in YHSS 3, the Hellenistic deposits. Here, if one looked only at the construction debris, pine would look like the most important wood. By segregating out construction material from ordinary firewood, oak emerges as the main wood in the assemblage. This strongly suggests pine was not as economical for fuel as it had been in Early Phrygian times, though people were still willing to go some distance for construction material. Alternatively, perhaps this was a time when pine was grown or harvested for its timber, with firewood being an incidental use of trimmed branches. Under current conditions near Gordion, cultivated trees must be carefully watered until their roots get deep (K. DeVries, pers. comm., 1991), but such care would at least have had the benefit of reducing transport costs for a bulky item like lumber.

Other Botanical Indicators of Climate and Vegetation Changes: Pollen, Phytoliths, and Seeds

Widespread tree-felling could therefore have shifted the borders between the different vegetation zones, directly (by obliterating the trees) or indirectly (by inducing small changes in available soil moisture). Although the charcoal analysis cannot by itself locate the borders of ancient vegetation zones, a combination of geomorphological and phytolith research might help resolve this issue. For example, in a preliminary geomorphological study, Ben Marsh (1993) suggests that post-Phrygian erosion on a massive scale could be relatively recent; he suggests over-grazing as a cause; sedimentary evidence of massive erosion is Medieval or later (Arlene Rosen (pers. comm., June 12, 1999). The findings are consistent with the botanical evidence, which suggests that forest trees grew relatively close to Gordion as late as the Medieval period.

If erosion has not totally destroyed original soils, or if paleosols can be found, transect sampling of soils across the current theoretical borders of vegetation zones might reveal the vegetation history. Although there are many difficulties with this approach, it is one that has been used in the United States to locate shifts in the prairie and forest edge (Wilding and Drees 1968, Mohlenbrock 1991). This sort of analysis can distinguish leaf phytoliths from grass phytoliths, and conifers could be distinguished by their distinctive cross-field pits in silicified tracheids (S. Mulholland, pers. comm., 1991). Thus, the actual and hypothetical vegetation is well-suited to this kind of analysis.

The vegetation zone least amenable to wood charcoal analysis is, of course, the treeless steppe. The primary steppe vegetation of central Anatolia is probably characterized by a high proportion of grasses (van Zeist et al. 1975:68); a few species of grasses and other plants may serve as indicators of steppe, as opposed to cultivated lands (Appendix C). Over time, even in the absence of cultivation, grazing on the natural steppe might encourage the survival and expansion of such unpalatable types as wild rue (*Peganum harmala*). Fortunately, flotation analysis of seed remains can be informative. The seeds of two anti-pastoral types (wild rue and camelthorn [*Alhagi camelorum*]) do seem to increase in frequency after the Early Phrygian period, and some of the hypothesized steppe plants may decline, as discussed in Chapter 5.

Results of the Charcoal Analysis

1. Oak, pine, and juniper have been major components of the arboreal vegetation within a 50-km radius of Gordion since Late Hittite times. Pine may have been a more important component of the woody vegetation at lower elevations closer to Gordion than it is today.

2. Over the entire sequence, juniper use declined. Oak and pine became the most prominent types in the arboreal vegetation. These shifts in the proportions of juniper, oak, and pine, and the increase in the minor components are evidence of an overall reduction in arboreal vegetation near Gordion.

3. A decline in the three dominant types probably set in by Late Phrygian times.

4. There is no reason to invoke climate change to ex-

plain the inferred changes in wood use. The proposed vegetation shifts are not that dramatic, and mainly involve reduction in the number of trees. These changes are readily explained by human factors. This is not to say that climate did not change at all, or that climate shifts are necessarily irrelevant.

5. Archaeological fuel remains and charred construction debris provide different and complementary information about ancient vegetation and wood use. It is impossible to interpret the charcoal remains adequately without knowing the archaeological context of the finds.

5

Analysis of the Flotation Samples

Methodological and Analytical Assumptions

During the 1988 and 1989 excavations seasons, over 600 flotation samples were taken from about 230 stratigraphically distinct deposits (Table 5.1). A small proportion of these were mixed in antiquity or were not as well excavated as they might have been. In choosing samples to analyze, I tried to get as full a time range as possible, as many *in situ* hearths and pits as there was time for (including multiple samples from complex deposits), and some samples that were from securely dated deposits but otherwise not noteworthy.

It is difficult to generalize about minor differences between the time periods. Ideally, one would see some overall trends. But most comparisons between time periods, though not meaningless, are not easily interpreted. Despite extensive archaeobotanical sampling, patterning is hard to see for a few reasons. First, the samples are very diverse. Even after 200 soil samples had been examined, new plant taxa were occasionally discovered. The number of wild types represented by more than 100 specimens over the entire sequence is about 70 (including unknowns), and those represented by more than 1000 is 3. Second, there are many unknown types or taxa so broadly defined that they cannot be assigned to an ecological niche. For example, I have been able to designate only a few genera that seem to be indicators of steppe. Even so, some patterns have begun to emerge.

Where possible, the summary statistics for each phase give each sample analyzed equal importance (wild:cereal, seed:charcoal). Some small samples have to be excluded from ratio analyses because the denominator would be zero or unmeasurable. In some cases, however, I have decided to add items together as totals for each phase. Statistically, this procedure weights samples by the value of the denominator; for example, the total amount of wheat relative to the total amount of wheat and barley gives more importance to samples with more identified cereal (Miller 1988). The purpose is to give some idea of possible trends, and therefore may serve simply to suggest hypotheses that might be tested using additional data.

Excavators were instructed to take soil samples from the full range of deposits they encountered. In order to maximize the quantity of material recovered without ignoring deposits poor in remains, the sampling procedure emphasized deposits most likely to have charred remains: hearths, pits, and deposits where charred material was obvious. The debris lying on floors was also sampled. Unlike the debris samples, flotation samples from the three burnt buildings more commonly consist of charcoal in overwhelming proportions, with almost no seeds (i.e., fallen roofing and other construction materials), or virtually pure crop seed samples, easily recognized as such by the excavators in the field. For that reason, material from the burnt buildings is analyzed separately.

Quantification of the Remains from Occupation Debris

It is not possible to reconstitute the totality of plants or even of seeds that were brought to the site and burnt. The goal of the procedures for quantifying the archaeobotanical remains is somewhat more modest: it should enable one to discover patterns of deposition and preservation as they vary in time and space. It is also hoped that the presentation of results will be a fair representation of the assemblage. When and if enough samples are analyzed, one would hope that

Table 5.1. Distribution of flotation samples across time (including burnt levels, excluding poorly provenienced YHSS "0")*

YHSS phase	No. samples taken	No. deposits sampled	No. samples analyzed	No. deposits analyzed
1 Medieval	30	13	15	10
3 Hellenistic	83	40	36	29
4 Late Phrygian	131	60	53	48
5 Middle Phrygian	26	17	15	14
6 Early Phrygian	84	14	21	5
7 Iron Age	193	66	78	55
8/9 Late Bronze	65	16	32	13
10 Middle Bronze	3	1	2	1
Totals	615	227	252	175

* The following samples are not included in the summary tables and charts in this chapter because they have virtually no seeds: YH 20127, YH 20303, YH 20785, YH 22022

Table 5.2. Density distribution of charred material from flotation samples [source: Appendix F; file YH F1]

Density (grams/liter)	Collapse (within structures)	Surface (directly over floor, etc.)	Debris (trash, other)	Pit	Pyrotechnic installation
0 to <0.5	8	6	22	40	19
0.5 to <1	7	9	6	26	9
1 to <1.5	3	.	4	16	1
1.5 to <2	.	1	1	6	2
2 to <2.5	.	.	.	3	.
2.5 to <3	.	2	.	7	.
3 to <3.5	.	.	.	.	1
3.5 to <4	.	.	.	.	.
4 to <4.5	.	.	1	2	.
4.5 to <5	.	.	.	.	.
5 to <5.5	.	.	.	.	.
5.5 to <6	.	.	.	.	.
6 to <6.5	.	.	.	1	.
6.5 to <7	.	.	.	1	.
7 to <7.5	.	.	.	1	.
7.5 to <8	.	.	.	.	.
8 to <8.5	.	.	.	.	.
8.5 to <9	.	.	.	.	1
9 to <9.5	.	.	.	.	.
9.5 to <10	.	.	.	.	.
≥10	.	.	1	.	3
No. samples	18	18	35	103	36
Range	0.05 to 1.19	0.21 to 2.97	0.04 to 11.50	0.13 to 7.22	0 to 41.22
Median	0.535	0.61	0.28	0.64	0.45

the observed patterns would be relatively stable, even if one analyzed additional samples.

Density of charred material in the soil samples reflects how intact the material is in a given deposit. The distribution is not normal (see Fig. 3.1), so the mean value is, in fact, meaningless. For that reason, the following discussion compares the median values. The density distribution shows that most samples have little charred material. One might expect the category of deposit would show significant variation, with hearth deposits ("pyro" in Fig. 5.1) having the highest densities of charred material. But even hearths do not necessarily have *in situ* deposits, as they might be emptied of ash daily. This should not be too surprising. In addition, an efficient fire might produce more ash than charcoal. For example, highly combustible straw will burn faster and more completely than wood, and fuel in a fire that is tended will be exposed to oxygen and leave ash rather than charcoal (Table 5.2, Fig. 5.1). With few exceptions, even the material excavated directly from hearths is likely to be from settlement debris rather than from the last fire burned in it. Thus, the charred material in most samples is most likely hearth sweepings redeposited in pits and trash dumps. The distribution of median charred material density by time period is similarly uninterpretable; as with deposit type, most samples have relatively low densities of material (Table 5.3, Fig. 5.2).

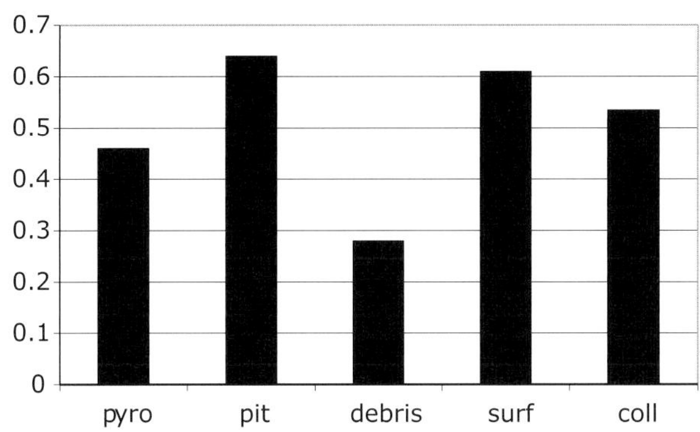

Fig. 5.1 Median density (g/liter) according to deposit type (data in Table 5.2).

Because excavators were asked to collect all the visible wood charcoal they came across, the estimates of charcoal quantities of all samples from a given time period have been added together to generate percentages of the various taxa and changes through time. One can argue with the validity of the results, but had this ideal sampling strategy been followed, the assumption is reasonable. This assumption does not apply to the flotation samples. Not all deposits were sampled, and not all samples were analyzed for this report. Neither the stratigraphic units nor the samples extracted from them come from equivalent soil volume, and they contain different densities of charred material, so in principle it is not appropriate to simply add the taxa, as has been done for the wood charcoal. On the other hand, the density distribution over time and space supports the contention that there is one major source of botanical material: charred residues of fires. Sampling in the field and laboratory favored pits and hearths, and in several cases, more than one sample per stratigraphic unit was analyzed. Nevertheless, based on the provenience information available between 1988 and 1991, samples were chosen for analysis that would represent all time periods and a variety of functional deposits.

A variety of measures can be used to assess the importance of the different plant types for environmental and economic reconstructions. Absolute quantities of remains are less interpretable than densities and relative amounts. Some variables can characterize the samples independent of soil volume or total amount, and will be discussed below: the seed:charcoal ratio as an indicator of fuel choice (dung, wood) and the wild:cereal ratio as an indicator of fodder choice (pasture, fodder crop). To isolate indicators of importance for particular species, percentages and frequencies for some taxa are considered for samples grouped by time period, even though this requires the assumption that the distribution of the remains fairly represents the charred material within the excavated deposits and ultimately for each time period. Insofar as different ways of arranging the data reveal similar or supporting patterns, those patterns will be considered more secure, and are likely to remain stable with further analysis. Some of the patterns are simply a function of sample size; so few samples were taken or so few seeds were recovered from some periods that the full range of taxa cannot be expressed. Note that in many of the graphs, periods YHSS 5, 6B, and 1 have few deposits sampled and fewer than 100 seeds each, and dramatic shifts, especially in those phases, are most safely attributed to chance.

In Appendix F, data are listed individually by sample, so that others may experiment with different assumptions.

The Taxa: Economic and Ecological Significance

The diversity of types in these samples is remarkable; I have already noticed over 100 different genera of seeds and a variety of other plant parts (straw, grain rachis fragments, the heads of two types of composites; for full accounting, see Appendix F, Table F.2). Cultigens include hulled barley (probably two- and six-row types), one-seeded einkorn, emmer, bread or hard wheat, rice, Italian millet, lentil, bitter vetch, chickpea, possibly pea, flax, and grape. There was also some almond and other nutshell. Despite conscientious collecting of voucher specimens and seeds in the area around Gordion, a number of the archaeological types remain unidentified, and many more have been identified only to the level of family. For identification criteria and ecological or economic significance of the wild and weedy taxa, see Appendix D. Appendix C contains details of floristic studies carried out in the Yassıhöyük area. Following common archaeobotanical practice, I refer to the preserved reproductive parts of the plants as seeds, though technically some are fruits (for example, the achenes of the Asteraceae).

Table 5.3. Density distribution of charred material from flotation sample by phase (grams/liter) [data: Appendix F5; file YH F1]

YHSS phase	10	8/9	7	6	5	4	3	1
0 to <0.5	1	14	29	8	.	17	19	8
0.5 to <1	.	9	14	.	5	13	9	6
1 to <1.5	.	4	6	.	2	11	1	.
1.5 to <2	.	.	5	.	1	4	.	.
2 to <2.5	.	.	1	.	1	1	.	.
2.5 to <3	.	2	.	.	3	3	.	1
3 to <3.5	.	1	.	.	.	.	.	.
3.5 to <4	.	.	.	.	.	.	.	.
4 to <4.5	.	1	2	.	.	.	.	.
4.5 to <5	.	.	.	.	.	.	.	.
5 to <5.5	.	.	.	.	.	.	.	.
5.5 to <6	.	.	.	.	.	.	.	.
6 to <6.5	.	.	1	.	.	.	.	.
6.5 to <7	.	.	1	.	.	.	.	.
7 to <7.5	.	.	.	.	1	.	.	.
7.5 to <8	.	.	.	.	.	.	.	.
8 to <8.5	.	.	.	.	.	.	.	.
8.5 to <9	.	.	.	.	.	1	.	.
9 to <9.5	.	.	.	.	.	.	.	.
9.5 to <10	.	.	.	.	.	.	.	.
≥10	.	.	2	.	.	1	1	.
No. of samples	1	31	61	8	13	51	30	15
Range (0–41.22)	0 to .71	0.05 to 4.49	0.09 to 16.02	0 to 0.17	0.55 to 7.22	0.19 to 10.03	0.11 to 41.22	0.06 to 2.65
Median (0.56)	n/a	0.58	0.53	0.105	1.07	0.71	0.365	0.44

Crops

Cereals

Wheat (*Triticum*). At least two broad categories of wheat are present in these samples: free-threshing and hulled. The rachises of bread wheat (*Triticum aestivum*) and macaroni wheat (*T. durum*) sometimes can be distinguished; of 5048 fragments of the free-threshing type, 80% are indeterminable, 18% are from bread wheat, and 2% are from macaroni wheat (Fig. 5.5). A few wheat rachis fragments have a blocky cross-section (Fig. 5.3). The grains of these two taxa are indistinguishable and are designated *T. aestivum/durum* (Fig. 5.6; Table 5.4), including some of the compact type, with length to breadth ratio less than 1.5 (see Hubbard 1992:110). On the data charts the grains are listed as *Triticum aestivum/durum* (Appendix F2). Einkorn (*T. monococcum*) and emmer (*T. dicoccum*), two hulled wheats, also make an appearance as grain (Fig. 5.4a) and spikelet forks (Fig. 5.4b).

Under experimental anoxic conditions that approximate the temperatures of dung- and wood-fueled fires (i.e., over 400°C), the size and shape of bread wheat, macaroni wheat, and emmer are similar to each other (L:B≈1.4–1.6) (Braadbaart and Pine 2005:73). In actual archaeological samples, however, differences have been detected; the L:B measure of an "almost pure" emmer sample from Korucutepe was 2.17, as against "almost pure" free-threshing wheat samples with averages of 1.46, 1.54, 1.60, and 2.05 (van Zeist and Bakker-Heeres 1982 [1985]:243).

Clearly, interpretations of measurements are not straightforward, and the flotation samples contained varying amounts of intact, measurable wheat grains. Even so, grain size and shape can help distinguish different taxa or cultivation practices. For example, among the free-threshing wheats, the compact type is characterized by short broad grains. Within a given variety, irrigation may affect the shape of the grain; measurements of two modern 200-grain samples hinted that irrigation may reduce the L:B ratio (Miller 1982:112). For purposes of the analysis, I treat the grains from each time period as a unit if they come from occupation trash and debris rather than burnt structures. The samples which seem to be cleaned crops are not included. That is, I treat "prime grains" (in the sense of Hillman 1984:23) separately from

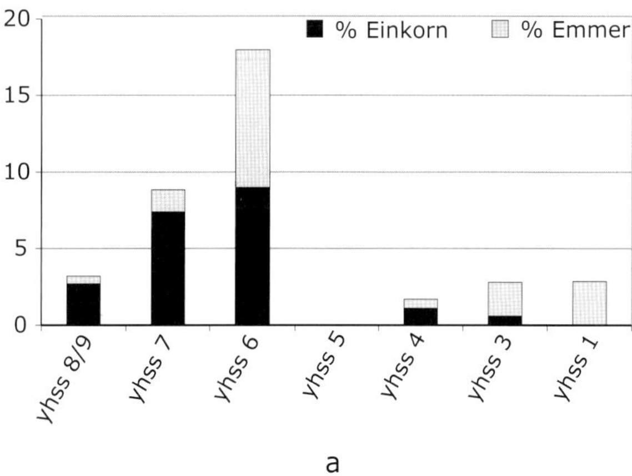

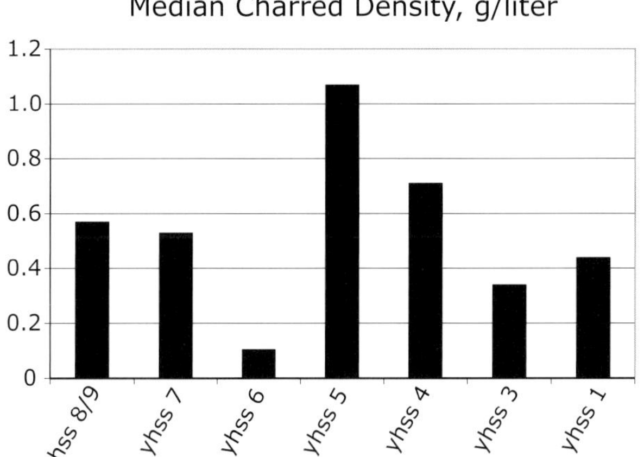

Fig. 5.2 Median charred material density (g/liter) by phase (data in Table 5.3).

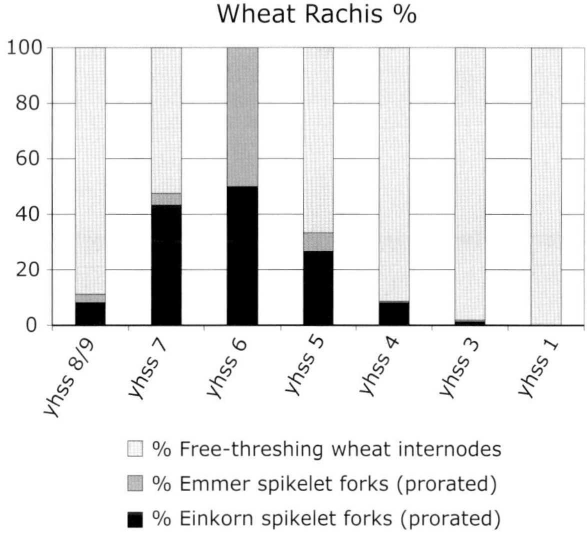

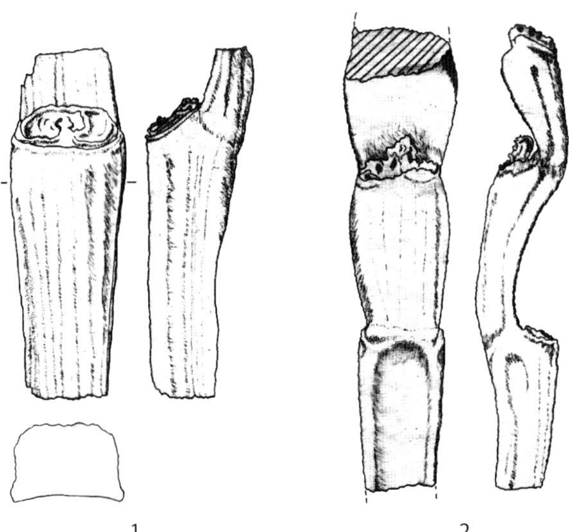

Fig. 5.3 *Triticum* rachis fragments (1, YH 22075; 2, YH 23307); a free-threshing type that has a stem with a square cross-section. It occurs primarily in Late Phrygian (YHSS 4) samples.

Fig. 5.4 Einkorn (*Triticum monococcum*) and emmer (*T. dicoccum*): a. grains as percent of total wheat grain; b. spikelet forks (estimated percent) of emmer and einkorn and internodes of free-threshing wheat (*T. aestivum/durum*) (data in Table 5.2 and file YH F2).

Table 5.4a. *Triticum aestivum/durum* measurements from debris (source: file YH D3)

Phase	N	Length (mm)	Breadth (mm)	Thickness (mm)	L:B	T:B
Medieval (YHSS 1)	17	4.4 (2.8–5.7)	2.8 (1.7–4.0)	2.3 (1.5–3.1)	1.60 (1.20–1.91)	0.82 (0.70–0.94)
Hellenistic (YHSS 3)	126	4.4 (2.3–6.2)	2.7 (1.2–3.8)	2.3 (0.9–3.5)	1.63 (1.15–2.50)	0.84 (0.69–1.05)
Late Phrygian (YHSS 4)	407	4.2 (2.0–5.7)	2.5 (1.2–3.8)	2.1 (1.0–3.5)	1.73 (0.96–2.79)	0.84 (0.55–1.07)
Middle Phrygian (YHSS 5)	14	4.1 (2.8–5.2)	2.5 (1.3–3.4)	2.1 (1.3–2.7)	1.74 (1.29–2.54)	0.87 (0.79–1.00)
Early Phrygian (YHSS 6B)	2	4.7 (4.5–4.9)	3.1 (2.8–3.4)	2.4 (2.3–2.4)	1.52 (1.44–1.61)	0.76 (0.71–0.82)
Early Iron (YHSS 7)	326	4.0 (1.7–5.6)	2.5 (1.1–3.9)	2.1 (0.9–3.1)	1.60 (1.06–2.54)	0.83 (0.57–1.13)
Late Bronze (YHSS 8 & 9)	227	4.1 (2.1–5.9)	2.6 (0.9–3.7)	2.2 (0.8–3.1)	1.61 (1.12–2.64)	0.83 (0.61–1.33)
Middle Bronze (YHSS 10)	2	4.1 (4.0–4.1)	2.2 (2.1–2.2)	2.1 (1.8–2.4)	1.88 (1.86–1.90)	0.98 (0.82–1.14)

Table 5.4b. *Triticum aestivum/durum* measurements from grain concentrations (source: file YH D3)

Phase	YH no. N Stratum	Length (mm)	Breadth (mm)	Thickness (mm)	L:B	T:B
Early Phrygian (YHSS 6A)	33246 N=50 620	3.9 (2.5–5.0)	2.5 (1.6–3.4)	2.0 (1.3–2.7)	1.59 (1.29–1.94)	0.82 (0.67–1.111)
Early Iron (YHSS 7)	33368 N=35 725; barley sample	4.9 (3.4–6.0)	3.1 (2.4–3.9)	2.7 (1.8–3.4)	1.58 (1.38–2.00)	0.88 (0.72–1.07)
Early Iron (YHSS 7)	33382 N=64 725; wheat sample	4.2 (2.9–5.7)	2.6 (1.3–3.5)	2.1 (1.2–3.1)	1.62 (1.21–2.23)	0.82 (0.69–1.04)
Early Iron (YHSS 7)	33402 N=399 725; wheat sample	4.4 (2.5–5.5)	2.8 (1.4–3.7)	2.3 (1.1–3.4)	1.60 (1.07–2.43)	0.83 (0.65–1.19)

the rest, because they are less likely to come from the waste fraction of crop cleaning or burned dung fuel. In general, the "prime grain" (from the BRH, YHSS 725) is larger than the other wheat from the Early Iron Age, but the difference is not great. There is no significant change in the other grain between the Middle Bronze Age and Medieval times, despite the fact that the site underwent major cultural change and that the wheat assemblage may well be heterogeneous (with more than one type of free-threshing wheat). The barley shows a similar homogeneity (see below for discussion).

Table 5.4c *Triticum aestivum/durum* measurements by grain shape (source: file YH D3)

	N	Length (mm)	Breadth (mm)	Thickness (mm)	L:B	T:B
"compact"	763	4.2 (1.7–5.9)	2.9 (1.4–4.0)	2.4 (1.1–3.5)	1.48 (0.96–2.00)	0.84 (0.57–1.33)
"long"	142	3.8 (2.1–5.5)	1.8 (0.9–3.5)	1.5 (0.8–2.9)	2.10 (1.40–2.79)	0.82 (0.55–1.04)
"regular"	549	4.4 (2.3–5.9)	2.6 (1.2–3.6)	2.1 (1.0–3.2)	1.75 (1.24–2.45)	0.83 (0.6.1–1.07)
"compact" YH33402	119	4.2 (2.6–5.1)	2.9 (1.7–3.7)	2.4 (1.7–3.4)	1.46 (1.19–1.71)	0.83 (0.67–1.10)
"regular" YH33402	92	4.5 (2.7–5.4)	2.7 (1.4–3.6)	2.2 (1.1–3.1)	1.69 (1.39–2.43)	0.82 (0.65–1.00)

Some *Triticum aestivum/durum* grains from the TB2A and BRH burnt buildings are most probably cleaned crops (Table 5.4b). The Destruction Level seeds come from a crop sample in a small jar. That wheat is small compared to the other concentrations, probably due to the intense fire. It is similar in shape measures to the wheat sample from the Burnt Reed House. The average lengths and breadths of the measured wheat samples from debris samples and from concentrations are similar to each other (Table 5.4a, b).

Einkorn (*Triticum monococcum*). Although no caches or pure samples were found, it is likely that einkorn was a field crop, at least during the Iron Age (Table 5.5). Einkorn is a minor component of archaeobotanical assemblages in the Near East that date to the 2nd millennium BC and later (Miller 1991). It was known to and grown by the Hittites (Hoffner 1974), but it remained popular in southeastern Europe well after it had become a minor crop plant in the Near East (see Hubbard 1976; Kroll 1991). Never numerous at Gordion, einkorn increases in the Early Iron Age relative to other wheat and barley (Fig. 5.4). It is possible that this anomalous increase was due to Phrygians coming from southeastern Europe as postulated by Voigt and others above. It virtually disappears as a crop after the Early Phrygian period (as a percent of wheat).

Emmer (*Triticum dicoccum*). The presence of low quantities of emmer overall suggests it was either a very minor crop or a minor contaminant of other crops.

Barley (*Hordeum vulgare* var. *distichum* and *Hordeum vulgare* var. *hexastichum*). Both two-row and six-row barley are attested. Of the determinable grains, a large number appear to be twisted (indicative of six-row barley), yet most of the determinable rachis fragments come from the two-row type (Tables 5.6, 5.7). The stems and leaves of both types are good for fodder. The cultivation of two-row barley is totally consistent with what we know about the drinking habits of the Phrygians, not to mention the Hittites before them and every other group that lived at Gordion, up to and including the archaeologists (Sams 1977; Hoffner 1974; pers. obs.). Namely, in non-commercial settings using unimproved varieties, two-row barley might be preferred for beer-making, because it is starchier than the six-row type (see Schwarz and Horsley n.d.). Six-row barley grain is more likely to be fodder, and usually needs more water than the two-row type. In recent times, barley is grown primarily for fodder (grain and leaf), but it may also be eaten.

Table 5.5. *Triticum monococcum* measurements (source: file YH D4)

	N	Length	Breadth	Thickness	L:B	T:B
Whole sequence	109	5.1 (3.2–6.8)	2.3 (1.3–3.4)	2.5 (1.5–3.5)	2.27 (1.53–3.85)	1.08 (0.68–1.57)

Table 5.6 *Hordeum vulgare* var. *distichum* and *H. vulgare* var. *hexastichum* indicators (determinable whole grains and rachis fragments) (source: file YH D2)

YHSS phase	8/9	7	6	5	4	3	1
Whole grains, N	581	872	9	78	1275	474	59
% straight	25	21	44	18	20	28	19
% twisted	35	32	0	10	34	30	20
% indet.	40	46	56	72	46	42	61
Rachis, N	264	486	0	8	815	349	87
% 2-row	73	84	0	63	68	66	74
% 6-row	9	6	0	25	14	15	13
% compact	18	10	0	13	18	19	14

Table 5.7 *Hordeum* measurements for selected samples (str=straight grain; tw=twisted) (source: file YH D1)

YH no. (stratum)	N	Length (mm)	Breadth (mm)	Thickness (mm)	L:B	T:B
29540, str (495.04)	12	6.1 (4.5–7.0)	3.0 (2.1–3.5)	2.3 (1.6–3.1)	2.04 (1.63–2.87)	0.77 (0.65–0.89)
29540, tw (495.04)	19	6.2 (5.0–7.9)	2.9 (2.4–3.7)	2.3 (1.8–3.3)	2.17 (1.72–2.50)	0.81 (0.69–1.00
30664, str (450.10)	46	6.0 (4.8–7.4)	2.7 (2.0–3.6)	2.0 (1.3–2.9)	2.24 (1.90–3.08)	0.76 (0.59–0.87)
30664, tw (450.10)	100	6.0 (4.4–7.7)	2.8 (2.0–3.7)	2.2 (1.3–3.6)	2.12 (1.71–2.85)	0.78 (0.57–1.12)
33573 (620)	127	5.6 (4.0–7.1)	2.9 (1.6–3.7)	2.3 (1.2–3.0)	1.99 (1.50–2.75)	0.80 (0.67–1.00)
33368, str (725)	48	5.8 (4.6–7.5)	2.9 (2.0–3.9)	2.3 (1.2–4.1)	2.03 (1.67–2.60)	0.78 (0.60–1.64)
33368, tw (725)	98	6.1 (4.4–7.5)	3.1 (1.7–4.2)	2.4 (1.5–3.3)	1.95 (1.58–2.82)	0.77 (0.60–1.50)
31603, str (870.03)	19	5.8 (4.3–7.8)	2.9 (2.0–3.5)	2.2 (1.6–2.9)	2.04 (1.62–2.56)	0.78 (0.69–0.86)
31603, tw (870.03)	28	5.6 (4.2–7.2)	2.7 (1.8–3.7)	2.1 (1.0–3.0)	2.10 (1.76–2.83)	0.76 (0.50–0.97)

The proportions of wheat and barley vary. In most periods, barley constitutes more than half of the identified grain (by weight), but less than half by rachis fragments (count) (Fig. 5.7; Fig. 5.6a shows two-row barley).

Millets (*Setaria* sp., *Panicum* sp.). Millets occur in small quantities through most of the sequence, though not always as domesticates. *Setaria italica* shows an increase over time relative to other cereals (the catego-ries bread/hard wheat and barley) (Fig. 5.8). Based on concentrations encountered during R. Young's excavation of the Destruction Level, millets—*Setaria italica* and *Panicum miliaceum*—were undoubtedly crop plants by that time (Nesbitt and Summers 1988). As a summer crop, they would have been irrigated.

Rice (*Oryza sativa*). Six grains from a Medieval period oven (YHSS 150.03) presumably would have

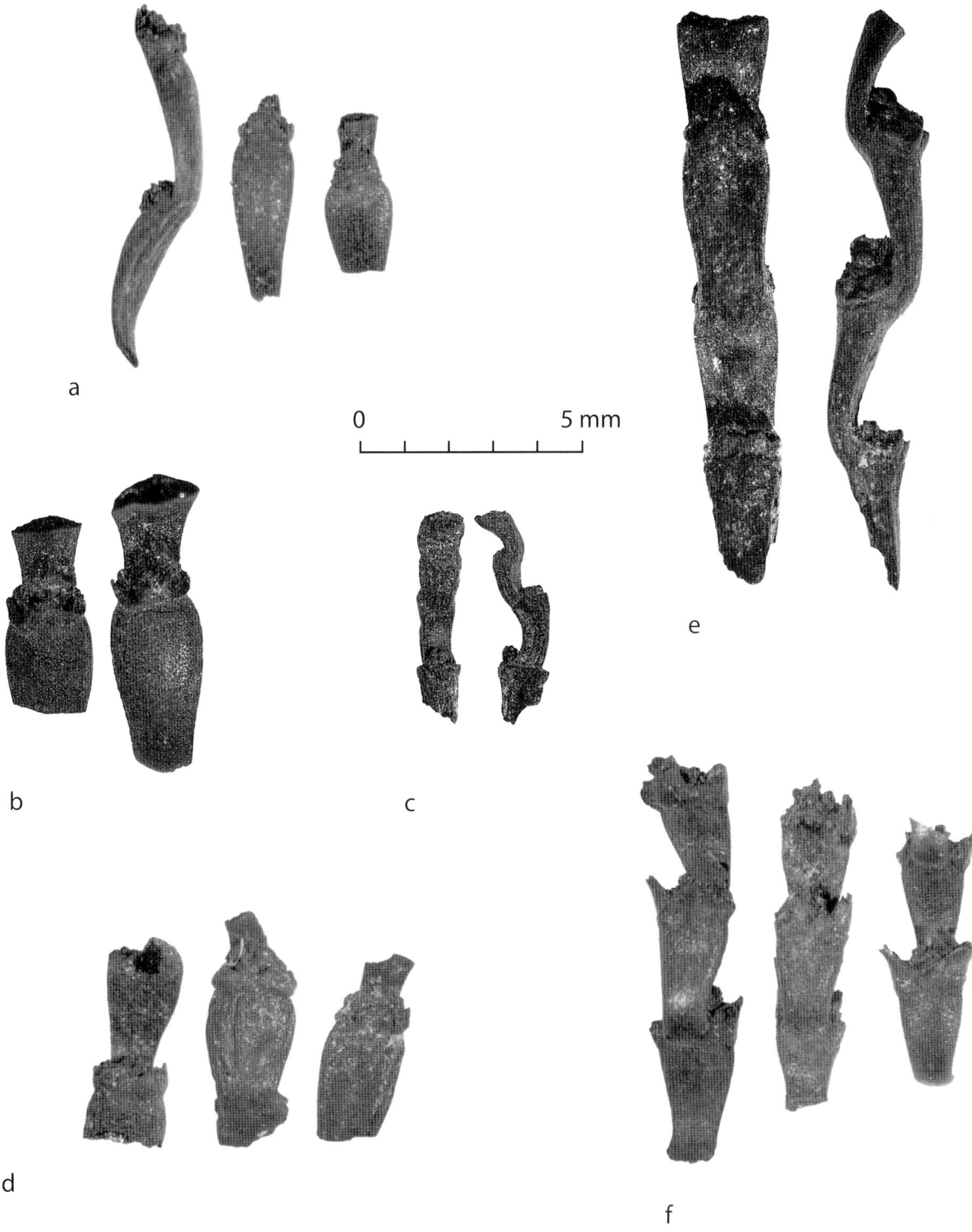

Fig. 5.5 Wheat rachis fragments: a. *Triticum aestivum/durum*, type 1 (YH 26899); b. *T. aestivum/durum*, type 2 (YH 26899); c. *Triticum* or *Hordeum*? (YH 28338); d. *Triticum aestivum* (YH 28338); e. *Triticum aestivum/durum*, type 2 (YH 26899); f. *Triticum durum* (YH 26899).

Fig. 5.6 a. *Hordeum vulgare* (YH 33354); b. *Triticum aestivum/durum* (YH 33379); c. *Vicia ervilia* (YH 33355); d, e. Lens (33243); f. insect remains in Lens (YH 33243) sample: 1–cf. *Sitophilus*, 4–cf. *Bruchus*.

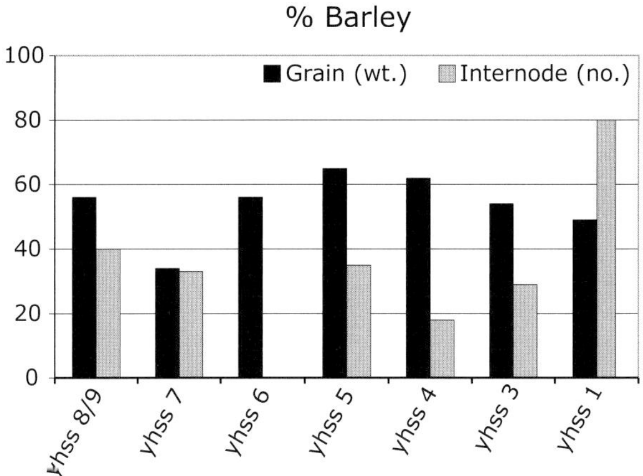

Fig. 5.7 Proportions of barley relative to naked wheat (data in Table 5.13).

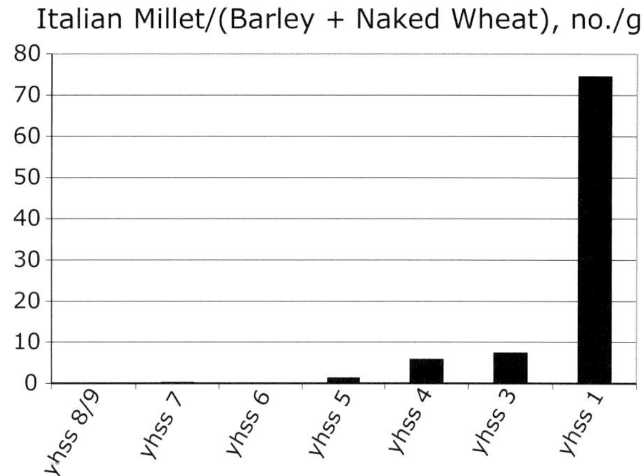

Fig. 5.8 Italian millet (*Setaria italica*) relative to free-threshing wheat (*Triticum aestivum*) and two-row barley (*Hordeum vulgare* var. *distichum*) (no./g) (data in Table 5.13).

been irrigated. One of the rice grains still had a fragment of the hull attached, which suggests it may have been locally grown, and several samples had silicified rice hull fragments. It was grown by the villagers of Yassıhöyük until the late 1950s (Ayşe Gürsan-Salzmann, e-mail January 18, 2007).

Pulses

Bitter Vetch (*Vicia ervilia;* Fig. 5.6c). Bitter vetch is the most common and plentiful cultivated legume in the Gordion assemblage. The finds of a concentration of bitter vetch in the Burnt Reed House (BRH; YHSS 7) and several in the Destruction Level (M. Nesbitt, letter dated 22 January 1989, Gordion archive) show that it was grown as a crop. Although bitter vetch is usually considered a fodder plant, Hans Helbaek identified it from food storage contexts at Late Bronze Age Beycesultan (Helbaek 1961), and at least some of the Gordion remains could represent food. Its toxicity makes it relatively resistant to insects, but special processing is necessary to render the seeds fit for human consumption (Enneking 1995:9). Bitter vetch was an early cultigen in southeastern Europe and Turkey (Zohary and Hopf 1994), but it became a minor crop that is grown primarily for fodder. It occurs in all periods at Gordion; in addition to incidental inclusion in debris samples, a concentration of bitter vetch was found in the Early Iron Age (YHSS 7) burnt structure (YH 33335).

Lentil (*Lens;* Figs. 5.6d, e). Several concentrations of lentil in Terrace Building 2A (YHSS 6A) demonstrate that it, too, was grown. As food for humans, lentils are far superior to bitter vetch because processing is simpler. Yet lentil tends to be less common than bitter vetch in the occupation debris. This suggests that the pulses found in those samples might have originally come from fodder that found its way into dung fuel, or that despite its toxicity, bitter vetch was more commonly eaten in those days than it is today. At issue is a total of 78 lentils and 191 bitter vetch in the assemblage analyzed to date, so generalizations about the relative importance of these pulses are not definitive. One lentil sample from the Terrace Building had an insect infestation. Some lentils have insect damage (perhaps from a *Bruchus* species, suggested by M. Kislev e-mail February 1, 2009). In addition, granary weevil (*Sitophilus* cf. *granarius*, suggested by M. Kislev, A.G. Heiss, and A. Bain), also charred, was encountered (Fig. 5.6f).

Chickpea (*Cicer arietinum*). Its presence in the heavy fractions of two samples (YH 21068, YHSS 3 and YH 23774, YHSS 4) and as incidental inclusions from three Terrace Building concentrations of lentil, flax, and wheat samples is enough to suggest that chickpea may have been grown, but it does not appear to have been a major crop plant. Mark Nesbitt reports a sample with chickpeas from the Rodney Young excavations as well (letter dated 22 January 1989, Gordion archive).

Nuts, Fruits, Oil, and Fiber Plants

Nuts

Almond (*Prunus* spp.), both cultivated and possibly wild, and a thin shelled wild pistachio (*Pistacia*) were found. Wild almond (*Prunus* sp.) was seen in five light fractions and nine heavy fractions, mostly in Early Iron Age (YHSS 7) and Late Phrygian (YHSS 4) contexts. A domesticated almond (*P. amygdalus*) was found in the Destruction Level. Today, a spiny branched wild almond (*acı badem*; *Prunus = Amygdalus orientalis*) grows within about 15 km of the site (near Çekerdeksiz and Dümrek, in both cases on basalt substrate). The domestic type is grown in Yassıhöyük gardens. Pistachio is not cultivated today, and the one identifiable nearly whole nut is of the wild type, similar to *çitlenbik* (*Pistacia* cf. *terebinthus*) that is for sale in the local market and that grows within about 10 km of the site at Dümrek (Fig. 5.9). Mark Nesbitt identified hazelnuts (*Corylus*) from the Destruction Level jars (letter dated 22 January 1989, Gordion archive).

Fruit

Grape (*Vitis vinifera*). Grape occurs only rarely at Gordion, in fragments. Nowadays one sees a few grapevines in gardens, but the vine is not widely grown in the area today. Organic residue analysis identified tartaric acid indicative of wine in vessels from the funerary feast remains of Tumulus MM, but the wine could have been produced elsewhere (McGovern et al. 1999); wine is well-attested in Hittite sources (Hoffner 1974: 39–41), and vineyards were tended in the Ankara region well into the 20th century.

Cherry. A single cherry pit (*Prunus* sp.) occurred in the heavy fraction of YH 29541, a Late Phrygian sample. Mark Nesbitt (letter dated 22 January 1989, Gordion archive) encountered several uncharred, rodent-gnawed cherry pits from Young's excavations.

Hackberry. Although the wood of *Celtis* was not encountered, it is a component of the central Anatolian steppe forest, and *Celtis* cf. *glabrata* was seen growing about 45 km west of Gordion near Yunusemre.

Oil and Fiber Plants

Flax (*Linum usitatissimum;* Fig. 5.10). In addition to two probably wild specimens found in flotation samples, a jar of flax seeds was found in the Destruction Level of Terrace Building 2A (YH 33595). In the same room were similarly placed small jars of obvious food plants (wheat, barley, and lentil). Flax seed size is influenced by irrigation practices (Helbaek 1959), and flax grown for oil tends to have larger seeds than that grown for fiber (Zohary and Hopf 1994: 119). It is likely that the small size of these seeds resulted from loss of mass with exposure to the intense heat of the YHSS 6A destruction, which was high enough to vitrify pottery. The room in which the seeds were found had loomweights and other evidence of weaving (Voigt 1994: 272). The spindle whorls from the Destruction Level include center-weighted whorls suitable for flax spinning (Burke 2005; see Miller et al. 2006). Flax fiber was found in Tumulus MM (Bellinger 1962). Therefore, one might suppose these seeds to be the stock for the fiber plant. It is not possible to ascertain whether the seed was grown for fiber or oil, or whether it was irrigated or not. Note that in an earlier publication I mistakenly reported these seeds to be sesame (Miller 1991:153).[7]

Cotton (*Gossypium*). Six cotton seeds from two Medieval deposits were identified. Like rice and millet, cotton is a summer-irrigated crop.

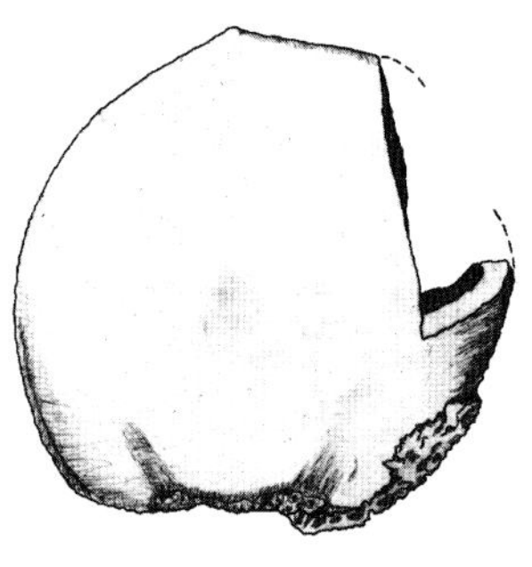

Fig. 5.9 *Pistacia* (YH 28838).

---

7. Gordon Hillman pointed out that the tiny, sesame-sized seeds were asymmetrical; it was the intense fire that reduced their mass.

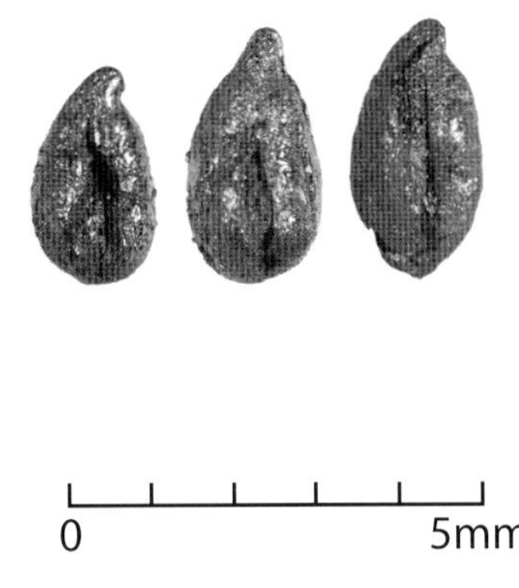

Fig. 5.10 Flax (*Linum usitatissimum*) (YH 33555).

Wild and Weedy Plants

Plant taxa differ in the breadth of their ecological requirements. Some grow in a variety of habitats, and others are quite restricted in their distribution. An entire plant family may characteristically grow in a particular environment (New World cacti in moisture-poor areas, sedges in moist ones), though such tendencies tend to be manifested at lower levels in the taxonomic hierarchy. But even at the level of genus or species, there will always be exceptions. It is clear from the *Flora of Turkey*, as well as personal observation, that very few taxa are restricted to fields (irrigated or unirrigated), gardens, steppe, or streamsides. In an attempt to identify plants that might be indicative of different growing conditions, informal vegetation surveys have been conducted within easy walking distance of the site (up to about 2 km) during the late spring and summer seasons of 1988, 1989, and the late spring and early summers of most years thereafter.

I have carried out much of my collecting activity within the barbed wire enclosure of the approximately 2-ha main excavation area. In the spring of 1996, the authorities erected a fence around Tumulus MM, so beginning in 1997 (and subsequent years), I began a more formal vegetation survey within that protected area (Miller 1999; Miller and Bluemel 1999; Appendix C). The unprotected areas in which I collect tend to be near grazed land or fields. Peering into the grain fields from the edge (to avoid trampling them), it is clear that weed-killing chemicals are in use, so I have been unable to investigate weeds growing in the fields themselves.

Even though the modern vegetation is by no means "natural," there are some generalizations that are probably valid. I was particularly interested in recognizing the following contrasting situations: steppe/disturbed steppe; steppe/agricultural field; unirrigated field/irrigated field or garden (Table 5.8). It would also be of interest to be able to distinguish plants whose seeds ripen in spring or fall, for that might enable one to recognize summer cropping, e.g., of millets or sesame (cf. Nesbitt and Summers 1988). The *Flora of Turkey* has general indications of flowering and fruiting times.

Due to the difficulties inherent in identifying charred seeds (namely, one is delighted to determine genus, let alone species), I have not been able to isolate many types that would be indicators of these situations. Some of the most common archaeological seeds (e.g., *Galium*) have extremely broad ecological tolerance. Most seed types occur in small numbers, making determinations even more uncertain.

Distribution of the Taxa in Time and Space

Among the more than 33,000 seeds counted, the two best-represented families are grasses and legumes (more than 7000 seeds apiece), followed by the sedges (more than 4000). If you add Chenopodiaceae, mustards, mints, and composites (daisy family), these eight

Table 5.8 Ecological grouping for archaeologically common or diagnostic types

Archaeobotanical designation	Irrigated, streamside	Flood-plain	Segetal	Ruderal	Over-grazed	Steppe
Apiaceae: *Eryngium*	·	·	·	√	√	·
Asteraceae: *Artemisia*	·	·	·	·	√	·
Onopordum	·	√	·	√	√	·
Boraginaceae: *Heliotropium*	·	√	·	√	·	·
Cistaceae: *Helianthemum*	·	·	·	·	·	√
Cyperaceae: *Carex*	√	·	·	·	·	·
Carex 3	√	·	·	·	·	·
Cyperaceae (includes Cyperaceae 1–8 and indet.)	√	·	·	·	·	·
Eleocharis	√	·	·	·	·	·
Fimbristylis	√	·	·	·	·	·
Dipsacaceae: *Scabiosa*	·	·	·	·	·	√
Fabaceae: *Alhagi*	·	·	√	√	√	·
Medicago	·	·	·	√	·	√
Onobrychis	·	·	·	·	·	√
Trifolium/Melilotus	√	·	·	·	·	·
Trigonella	·	·	·	·	·	√
Trigonella astroites type	·	·	·	·	·	√
Fumaricaceae: *Fumaria*	√	·	·	·	·	·
Lamiaceae: *Teucrium*	·	·	·	·	·	√
Ziziphora	·	·	·	·	·	√
Papaveraceae: *Glaucium*	·	·	·	√	√	·
Plantaginaceae: *Plantago*	√	·	·	·	·	·
Poaceae: *Aegilops*	·	√	·	√	·	·
Eremopyrum	·	·	√	·	·	√
Hordeum cf. *murinum*	·	√	·	√	√	·
Stipa	·	·	·	·	·	√
Taeniatherum	·	·	·	√	·	·
Polygonaceae: *Polygonum*	√	·	·	·	·	·
Rumex	√	·	·	√	·	·
Polygonaceae/Cyperaceae Polygonum/Cyperaceae	·	√	·	·	·	·
Portulacaceae: *Portulaca*	√	·	·	·	·	·
Primulaceae: *Androsace*	·	·	·	·	·	√
Ranunculaceae: *Adonis*	·	√	·	√	√	·
Thymeleaceae: *Thymelaeaceae*	·	·	·	·	·	√
Zygophyllaceae: *Peganum harmala*	·	·	·	·	√	·

families account for about 75% of the seeds (90% of the seeds identified at least to family). Because of the inherent variability of the samples in quantity of remains and taxa, the strongest conclusions tend to be based on multiple lines of evidence, such as seeds and charcoal. This section attempts to reconcile results based on different types of quantification. Note further that some periods are characterized with very few samples. There are only two Middle Bronze (YHSS 10) samples, so I exclude them from the discussion and illustrations.

Ubiquity (frequency)

The first and simplest measure of taxon importance is percent ubiquity (the percentage of samples containing at least one exemplar of a given taxon) (Hubbard 1975) (see Tables 5.9, 5.10; Figs. 5.11–5.15; Appendix G1). This measure is inappropriate to track rare types over time, and it is inappropriate if there are very few samples in a phase. Of the seven phases, three have relatively few samples and seeds (YHSS 6B, 5, and 1), so low ubiquity values for some taxa in those samples may simply indicate that there were few seeds of any sort, and low values are due to chance preservation. Even though the number of seeds per sample in the other phases is high enough so that ubiquity values do not misrepresent the sample population, ubiquity is not as valuable an indicator as we might like. Represented by many samples containing many

seeds, changes between YHSS 8/9 and 7 and between YHSS 4 and 3 are most likely to be stable even if more samples are analyzed.

The two most important crop plants (in terms of total amount), *Hordeum* and *Triticum aestivum/durum*, have similar frequencies, and for all time periods appear in at least 80% of the samples. *Triticum dicoccum* and *Lens* occur in smaller amounts in many fewer samples. *Triticum monococcum* appears to decline over time, as does, arguably, *Vicia ervilia*.

Ubiquity of even the most numerous seeds of wild plants is not as informative as we would hope; I include graphs so readers may judge for themselves. Possible exceptions are small but noticeable long-term increases in antipastoral vegetation (*Peganum harmala*, with its hallucinogenic alkaloid, harmaline, and *Alhagi*, with its spiny stems) from YHSS 8/9 and 7 to YHSS 4 and 3.

Ratios

Seed:Charcoal (Table 5.11, Fig. 5.16). In arid or relatively treeless regions, dung is a commonly used alternative fuel. This observation leads to the suggestion that the seed to charcoal ratio could provide a way to monitor the use of dung and wood fuel (Miller 1984, 1988; Miller and Smart 1984). This approach presumes that the bulk of the material from occupation debris originated as fuel, rather than, say, as a result of cooking accidents or trash disposal. Samples that are

Table 5.9. Ubiquity (%) of cultigen taxa appearing in 25% or more of the samples. Number in parentheses for cultigens is total in grams (YHSS 10 excluded; only 2 samples)

YHSS phase	8/9	7	6	5	4	3	1
No. of samples	32	66	8	15	53	36	15
Hordeum	100 (7.90)	97 (12.41)	80 (0.13)	87 (1.53)	98 (18.33)	91 (6.27)	80 (0.68)
Triticum aestivum/durum	94 (5.36)	94 (18.86)	80 (0.08)	93 (0.66)	94 (9.07)	94 (4.07)	80 (0.54)
Triticum monococcum	41 (0.17)	65 (1.79)	40 (0.01)	13 (+)	23 (0.13)	6 (0.03)	7 (+)
Triticum dicoccum	9 (0.03)	15 (0.35)	40 (0.01)	7 (+)	8 (0.07)	19 (0.12)	13 (0.02)
Vicia ervilia	72 (1.02)	38 (1.10)	40 (+)	7 (0.03)	45 (0.36)	25 (0.25)	27 (0.20)
Lens	16 (0.06)	11 (0.06)	0 (0)	20 (0.05)	25 (0.14)	16 (0.46)	13 (0.08)

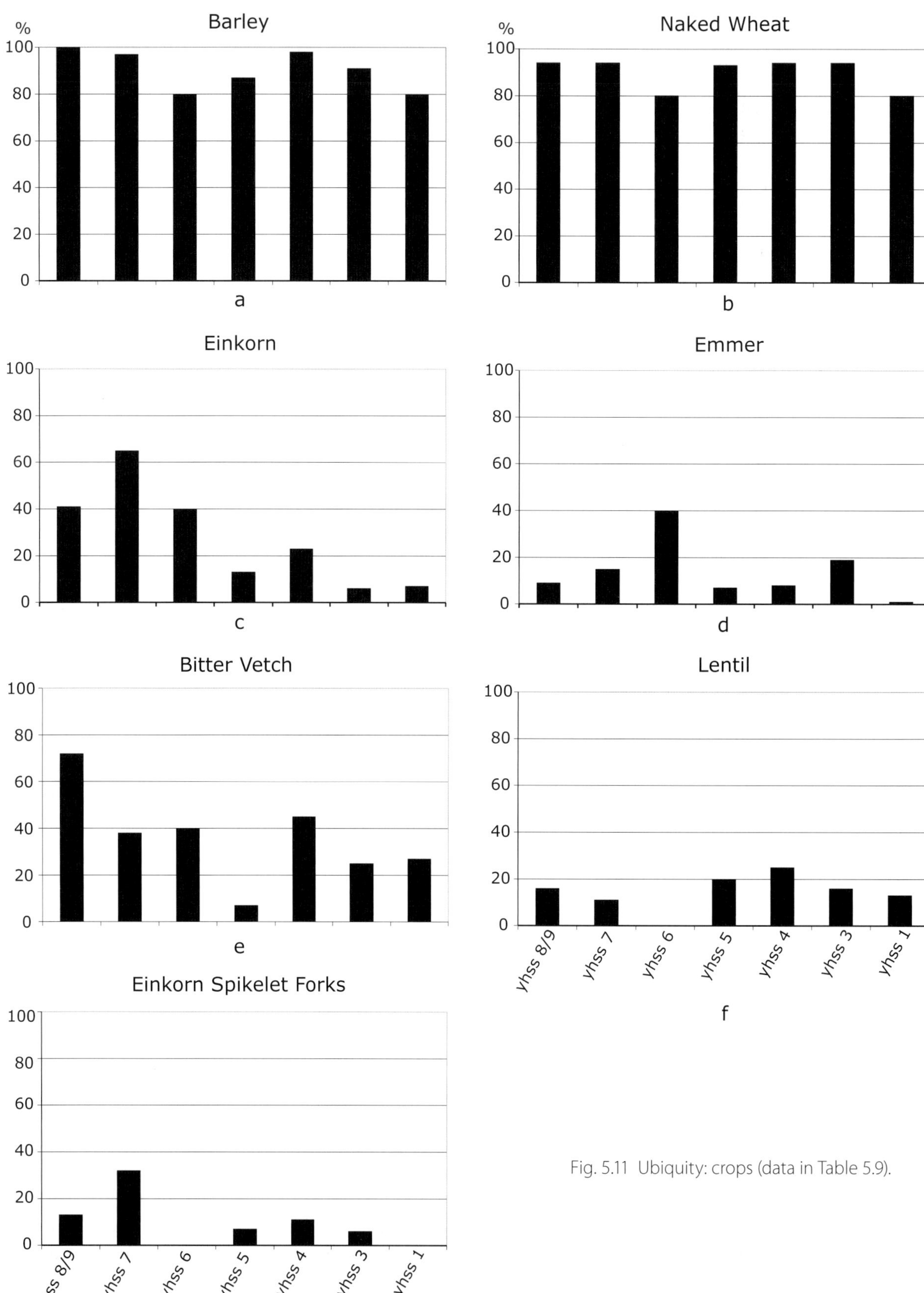

Fig. 5.11 Ubiquity: crops (data in Table 5.9).

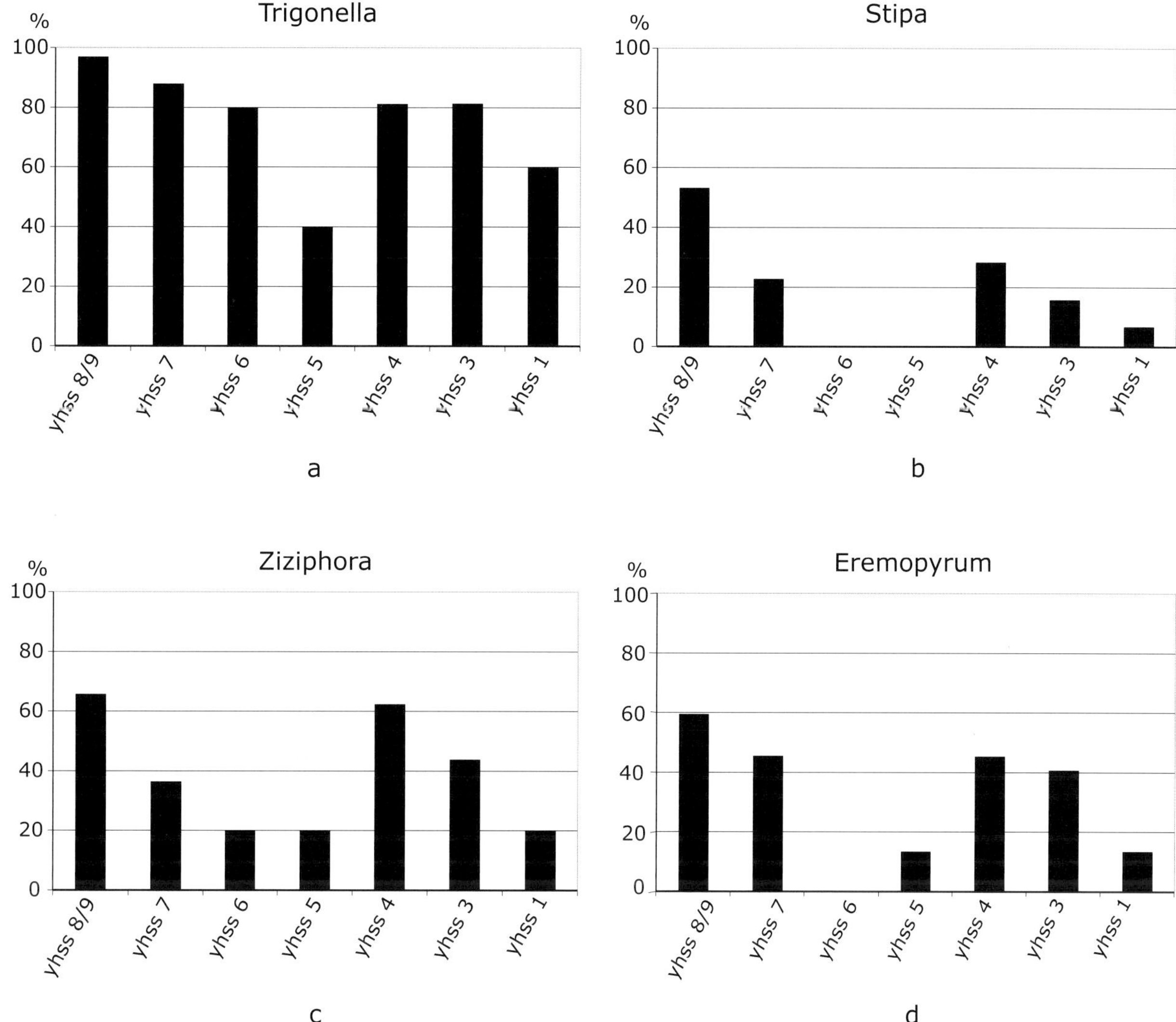

Fig. 5.12 Ubiquity: plants of steppe (data in Table 5.10).

clearly not fuel, such as structural wood from burnt buildings, are excluded. The use of this ratio has helped document long-term vegetation shifts in southeastern Turkey and southern Iran (Miller 1990).

The seed:charcoal ratio reported here is based on the material larger than 2 mm, and is effectively a cereal:charcoal ratio, since only a few legumes and weed seeds were larger than 2 mm. It could not be calculated for the few samples with virtually no charcoal, since the denominator cannot be zero. Here, too, the distribution is not normal, and so the mean value is not relevant. Assuming the results are not due to sample numbers too small to overcome chance between-sam-

ple variability, values for the seed:charcoal ratio do not show a simple trend. Rather, the relatively low average values (between 0.06 and 0.28) are very similar to those for sites thought to be located in steppe-forest and open woodland environments where dung fuel is inferred to have supplemented wood (Table 5.12). This is consistent with the conclusion based on the wood charcoal analysis that despite some loss of arboreal vegetation, wood fuel was available throughout the sequence. The statistical distribution of the seed:charcoal ratio does not follow a normal distribution, however, so the mean was calculated only to provide some rough comparability with other sites. Typical of many archaeobotanical

Table 5.10. Ubiquity (%) of wild and weedy taxa (YHSS 10 excluded; only 2 samples) appearing in 50% or more samples (and *Peganum*); for Cyperaceae, all genera grouped. Number in parentheses for wild plants is total number of that taxon

YHSS phase	8/9	7	6	5	4	3	1
No. of samples	32	66	8	15	53	36	15
Caryophyllaceae *Gypsophila*	53 (95)	47 (67)	20 (1)	33 (6)	40 (63)	28 (23)	47 (15)
Chenopodiaceae *Chenopodium*	59 (139)	26 (47)	0 (0)	20 (7)	17 (15)	41 (99)	33 (46)
Suaeda	66 (82)	47 (95)	20 (1)	47 (17)	36 (41)	41 (28)	20 (8)
Cyperaceae *Carex*	91 (143)	52 (129)	0 (0)	27 (7)	77 (547)	75 (181)	60 (123)
Cyperaceae (including *Carex*)	94 (328)	85 (592)	40 (2)	80 (50)	96 (969)	97 (1029)	73 (930)
Fabaceae *Alhagi*	3 (1)	6 (8)	0 (0)	7 (3)	57 (219)	50 (398)	13 (4)
Trifolium/Melilotus	69 (95)	45 (209)	20 (1)	27 (10)	45 (84)	41 (96)	27 (13)
Trigonella	97 (1049)	86 (827)	80 (6)	40 (22)	75 (854)	78 (1118)	53 (131)
Trigonella cf. *astroites*	78 (216)	44 (110)	20 (1)	0 (0)	32 (112)	22 (87)	33 (20)
Lamiaceae *Ziziphora*	66 (89)	36 (204)	20 (1)	20 (4)	62 (159)	44 (123)	20 (7)
Poaceae *Eremopyrum*	59 (79)	45 (106)	0 (0)	13 (2)	45 (231)	41 (26)	13 (3)
Hordeum cf. *murinum*	50 (36)	48 (114)	40 (2)	0 (0)	36 (63)	44 (26)	4 (104)
Stipa	53 (42)	23 (34)	0 (0)	0 (0)	28 (73)	22 (11)	7 (2)
Rubiaceae *Galium*	81 (165)	86 (420)	60 (4)	67 (47)	74 (696)	53 (76)	53 (19)
Zygophyllaceae *Peganum*	22 (22)	6 (23)	0 (0)	7 (1)	15 (426)	31 (79)	13 (9)

data, the distribution is skewed left (i.e., most samples are characterized by relatively low values). If the median values are plotted by period, a low-point in the seed:charcoal ratio occurs during the Middle Phrygian period (Fig. 5.16a; source: file YH F1).

The wild:charcoal ratio is based on the count of seeds of wild plants eaten by the herds, and includes plants of disturbed ground (field weeds, ruderals), as well as native steppe vegetation. It shows a similar, though not identical trend (Fig. 5.16b). Both ratios calculated as mean or median consistently show lowest values for the Middle Phrygian (YHSS 5) samples.

This suggests that wood fuel was most available at that time, but the shifts are not that large. That is, the changes that did occur, e.g., in species composition, were easily accommodated in the fuel economy.

Wild:Cereal. Insofar as the seeds come from dung fuel, the wild:cereal ratio allows one to assess grazing and foddering practices. Along the Euphrates, where herding is an increasingly important subsistence strategy in the rainfall agriculture zone as one goes from the moister north (precipitation greater than 350 mm/year) to the drier south (precipitation under 300 mm/year). Archaeologically, this is reflected in the wild:cereal ratio and proportion of sheep and goat relative to cattle and pig (Miller 1997b). In all periods at Gordion, sheep and goat are the predominant domestic herbivores (Zeder and Arter 1994), so their feeding habits would provide the predominant impression of fodder and wild plant cover. Similar to the seed:charcoal ratio, the statistical distribution of the wild:cereal does not follow a normal distribution, so it is not appropriate to compare the mean values by period; most samples are characterized by relatively low values. Some patterning appears when the data are organized in two slightly different ways. First, I recognized four "natural" groupings of samples (Fig. 5.17), those with wild/cereal values of 0 to ≤375; 375 to ≤775; 775 to ≤1900, and ≥1900. The proportion of samples with relatively high wild:cereal ratios declines from the beginning of the sequence until Middle Phrygian times, and then increases (Fig. 5.18). A similar pattern appears if one simply plots the median by time period (Fig. 5.19). These results are consistent with an interpretation that herding was least important relative to farming during the Middle Phrygian period (see discussion in Chapter 6).

Percentages

Percentages allow comparisons within categories that are homogeneous with regard to a particular question. For example, they may suggest relative importance of a taxon within an ecological or economic category (e.g., percent *Trigonella* relative to wild seeds, percent *Triticum* relative to cereals or relative to field crops). Some variables, such as volume of soil, weight of charcoal or seeds, or number of wild seeds, always or nearly always have a measurable amount. Therefore, their values can be used to calculate various ratios for each sample, and if the resulting distribution is close to normal, average values per period could have meaning. For a given plant taxon, however, a few samples may have many of that type, but most samples will contain none. In this context, average per sample is meaningless. Therefore, to detect temporal patterning, regardless of statistical significance, archaeobotanical analysis has to use less than perfect approaches. It is in that spirit that I consider the cereals and the seeds of some of the more common wild taxa relative to wild seeds as a group, by time period (*Trigonella,* Cyperaceae). I also consider

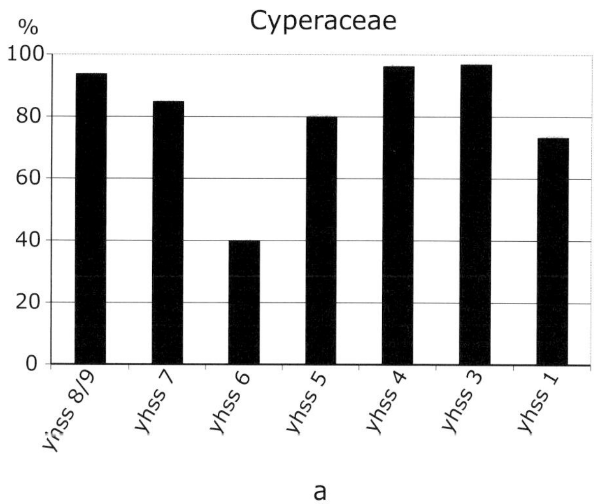

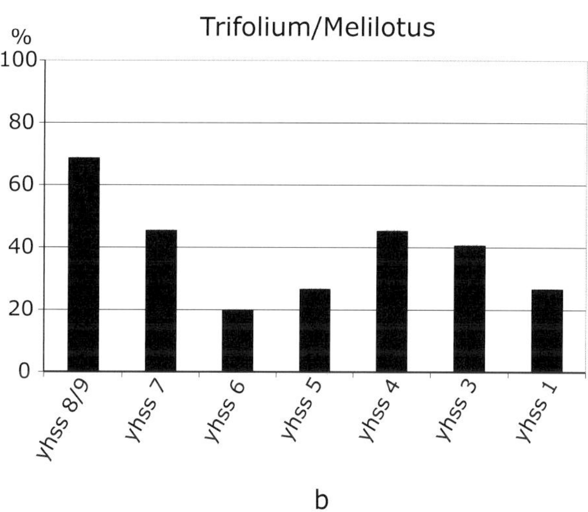

Fig. 5.13 Ubiquity: plants of moist areas (data in Table 5.10).

groups of taxa that individually are not common, but might be indicative of particular environmental conditions (Tables 5.8, 5.13; Hans Helbaek [1969] introduced this approach in his study of Ali Kosh). Although many taxa can grow under a fairly broad range of conditions, for this report I have assigned taxa to ecological group based on personal observation since 1988, as well as information in the *Flora of Turkey* (Davis 1965–1988).

Ecological groups are indicative of steppe, overgrazed steppe, roadsides and disturbed ground (ruderals), floodplain, and irrigated and streamside. Some taxa occur commonly under more than one condition (e.g., ruderal and overgrazed, floodplain and rud-

Table 5.11. Summary chart based on averages by phase

YHSS phase	10	8/9	7	6	5	4	3	1
No. of samples	2	32	66	8	15	53	36	15
Density g/liter	0.45	0.89	1.33	0.11	3.49	1.23	1.70	0.56
Seed:charcoal g/g	0.21	0.17	0.28	0.07	0.03	0.21	0.22	0.06
Wild:charcoal #/g	40	45	51	10	3	67	97	50
Wild:cereal #/g (divide by 100 for approx. #/cereal grain)	582	257	199	168	109	267	451	1213
Median seed:charcoal	0.21	0.14	0.11	0.08	0.02	0.05	0.08	0.04
Median wild:charcoal	40	35	29	7	2	7	16	23
Median wild:cereal*	468	206	152	150	97	132	184	650

*wild:cereal excludes samples with no measureable cereal (0 in denominator): YH 30039 (YHSS 4), YH 32692
    (YHSS 5), YH 25869 (YHSS 6), YH 27277 (YHSS 7)

Table 5.12. Comparison of average seed:charcoal ratios, southeast Turkey and northwest Syria*

Site	No. of samples	Period	Vegetation inference	Ratio
Gritille	18	Medieval—later	depleted oak woodland	2.40
Gritille	14	Medieval—early	open oak woodland	0.12
Hacınebi	26	Chalcolithic	steppe-forest	0.24
Sweyhat	17	Early/Mid Bronze	steppe	1.13

*Source: Gritille (Miller 1998); Hacinebi: Stein et al. (1996); Sweyhat: Miller (1997b)

eral). *Trigonella* constitutes the bulk of the seeds of healthy steppe plants (Fig. 5.20a). The category "overgrazed steppe" includes taxa that are minor natural components of steppe, but significant components of disturbed steppe, especially *Peganum harmala* (Fig. 5.21). Never common, the later part of the sequence arguably has more of these types. Ruderal types show a similar distribution, due in part to overlap in the types represented (Fig. 5.22). Combining taxa, including *Galium,* that are characteristic of ruderal and overgrazed areas suggests a general indicator of disturbance (Fig. 5.23). Here, too, the end of the sequence has more of those types, but they are never very numerous. The present-day floodplain is severely overgrazed, but the vegetation there does include types that are not common elsewhere (Fig. 5.24). Archaeologically, seeds of this zone are few. The sedge family (Cyperaceae) constitutes most of the seeds of irrigated fields and streamsides (Fig. 5.25). Overall, there seems to be an increase in these taxa.

The percentages of crop plants are interesting, too. For example, the percent of barley relative to wheat is highest just when the wild:cereal ratio is the lowest, during the Middle Phrygian period (YHSS 5), suggesting that maximum foddering (high barley percent) is associated with minimum pasturing (low wild:cereal). This association is weakened by the observation that the proportion of barley rachis fragments mostly follows the wild:cereal ratio rather than the percent of barley (compare Figs. 5.7 and 5.21). The lack of correlation between grain and rachis occurs at the level of individual sample, too: deposits with a lot of barley do not necessarily have many barley rachis fragments. From a taphonomic perspective, this suggests that the grain and straw were already separate by the time the material was burned. In turn, this suggests animals were fed grain or straw that was already processed, not a surprising conclusion.

This chapter has introduced several ways of quantifying the plant remains from flotation samples. By

Table 5.13. Summary chart based on amounts of cultigens and summed percents of wild and weedy types (data in Appendix F)

YHSS phase	8/9	7	6	5	4	3	1
No. of samples	32	66	5	15	53	32	15
Cereals (inc. rice), total wt. (g, sum)	21.28	50.00	0.44	4.07	40.63	16.18	1.95
Wheat (g, sum) (inc. einkorn)	6.32	24.07	0.11	0.81	11.44	5.34	0.70
Barley (g, sum)	7.90	12.42	0.13	1.53	18.33	6.31	0.68
Einkorn (g, sum)	0.17	1.79	0.01	0	0.13	0.03	0
Bread or hard wheat	5.36	18.86	0.08	0.66	9.07	4.07	0.54
Italian millet (count)	0	9	0	3	161	77	91
Bitter vetch (g, sum)	1.02	1.10	+	0.03	0.36	0.25	0.20
Pulse (g, total, inc. bitter vetch)	1.43	1.38	1.00	2.18	0.73	0.99	0.38
Einkorn rachis fragments (est. no. spikelet forks)	32	433	0	4	292	11	0
Total wheat rachis	390	1000	7	15	3605	863	22
Total barley rachis	264	486	0	8	815	349	87
% barley (B/(B+W))	56	34	54	65	62	54	49
% barley rachis (Brf/(Brf+Wrf))	40	33	0	35	18	29	80
Wild & weedy (based on total per phase)	5060	7710	58	368	8765	6573	2557
% ruderal	3	3	7	5	6	10	6
% overgrazed	2	2	5	2	10	9	5
% overgrazed+ruderal	3	4	7	6	11	11	6
% steppe, including *Trigonella*	30	18	17	9	17	23	7
% Trigonella	25	12	12	6	11	18	6
% floodplain	2	4	3	4	3	3	6
% Cyperaceae (combined)	4	6	3	11	10	15	32
Other irrigated, steamside	3	2	2	7	2	3	3
% Galium	3	5	7	13	8	1	1

Ruderal: *Adonis, Aegilops, Alhagi, Eryngium, Glaucium, Heliotropium, Hordeum* cf. *murinum, Medicago, Onopordum, Rumex, Taeniatherum*

Overgrazed: *Adonis, Alhagi, Artemisia, Eryngium, Glaucium, Hordeum* cf. *murinum, Onopordum, Peganum*

Steppe: *Androsace, Eremopyrum, Helianthemum, Medicago, Onobrychis, Scabiosa, Stipa, Teucrium, Thymelaea, Trigonella, Trigonella astroites-type, Ziziphora*

Floodplain: *Onopordum, Eleocharis, Adonis, Aegilops, Heliotropium, Hordeum* cf. *murinum, Polygonum*/Cyperaceae

Irrigated, streamside: *Carex,* Cyperaceae, *Fimbristylis, Eleocharis,* other Cyperaceae, *Fumaria, Plantago, Polygonum, Portulaca, Rumex, Trifolium/Melilotus*

Table 5.14. Sample distribution of the wild:cereal ratio. (data in Appendix F)

		Range of Values of the Wild:Cereal Ratio			
YHSS phase	No. of samples	0 to ≤ 375	375 to ≤775	775 to ≤1900	≥1900
1	15	6	2	3	4
3	32	22	4	5	1
4	52	43	6	2	1
5	14	14	0	0	0
6	5	5	0	0	0
7	65	59	4	2	0
8/9	31	25	6	0	0

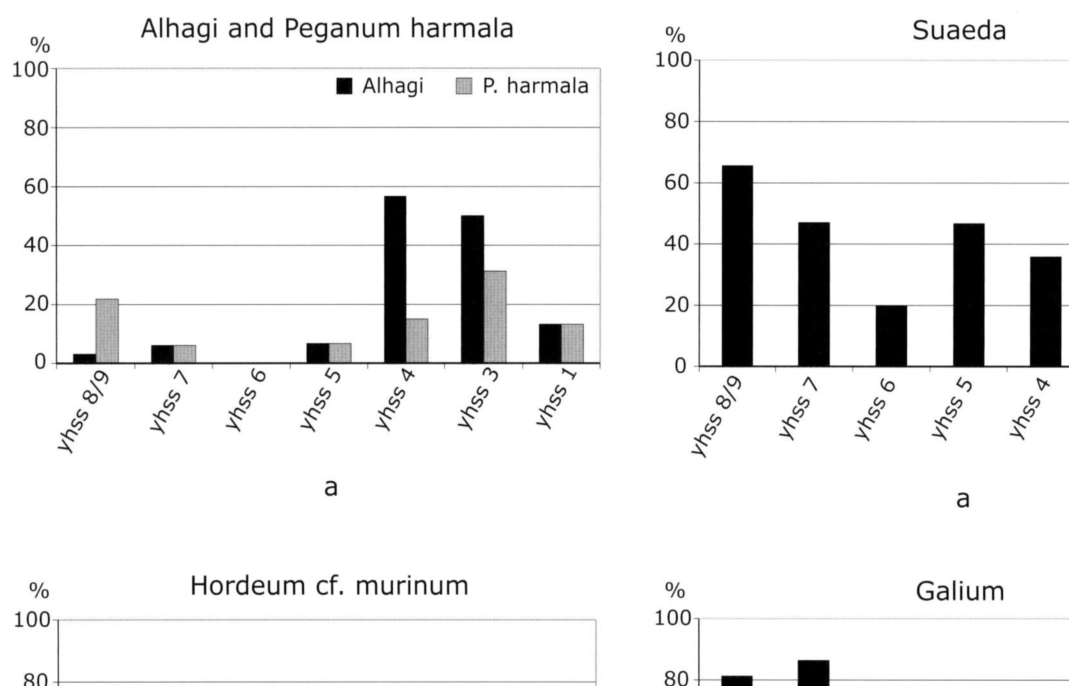

Fig. 5.14 Ubiquity: plants of disturbance (data in Table 5.10).   Fig. 5.15 Ubiquity of other common taxa (data in Table 5.10).

themselves, the plant remains show few clear chronological trends. Measures of ubiquity are disappointingly uninformative, and for the most part are not noticeably consistent with other kinds of percentage data. As will be seen in the concluding chapter, when the plant remains are viewed in the context of the broader agropastoral system, interpretable patterns emerge.

Flotation Samples from Burned Buildings

Samples from the burned buildings excavated in 1988 and 1989 are considered separately (Fig. 1.1). Floor deposits from the Burnt Reed House (BRH; YHSS 725) and Terrace Building 2 of the Destruction Level (YHSS 620) both had *in situ* concentrations of crop plants along with construction debris. The floor deposits from the Abandoned Village (YHSS 350) had roofing debris.

The Burnt Reed House (YHSS 7)

The Burnt Reed House was a wattle-and-daub structure. Its construction material is discussed in Chapter 4. Excavators found traces of basketry and associated crop remains. Concentrations of bitter vetch (YH 33335), barley (YH 33368), and bread or hard wheat (YH 33382, YH 33402) were found on the floor and some bread or hard wheat (YH 33394) was found in a pit. Parts of YH 30416 (mostly barley) and YH 33379 (mostly wheat) were sent for radiocarbon dating. The remainder did not have noticeable amounts of crop seeds.

The Destruction Level, Terrace Building 2A (TB2A; YHSS 6A)

The end of the Early Phrygian (YHSS 6) period is marked by the catastrophic fire that destroyed much of the central part of the Citadel Mound, including a row of eight attached buildings that backed onto the "palace" precinct. Most of these Terrace Buildings were excavated by Rodney Young's team. The contents varied, but the basic structure was repeated: each building had a front room and a back room; the back rooms had grinding stones along the back wall. Seed

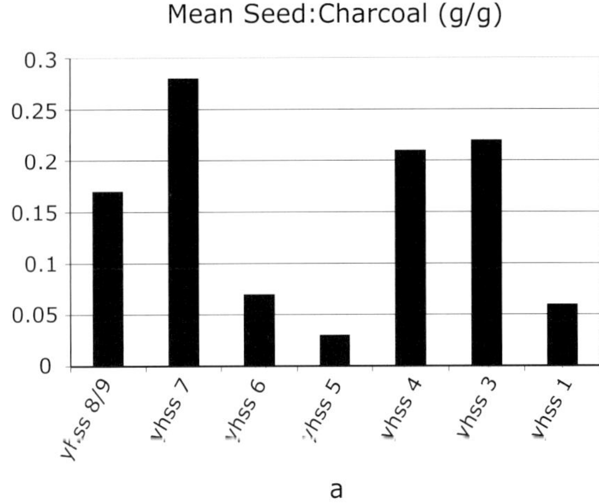

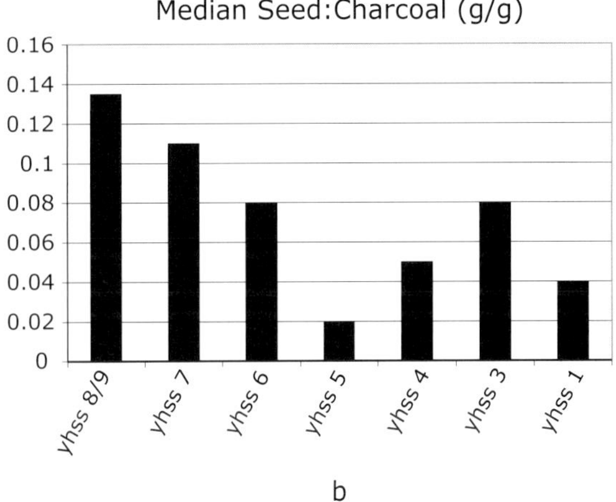

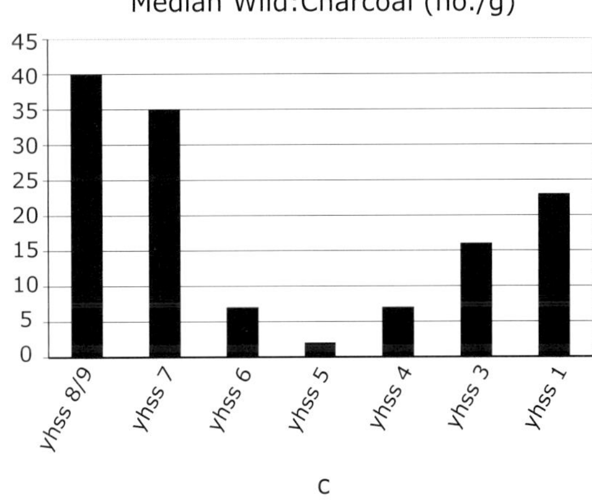

Fig. 5.16 Comparison of measures of seeds to wood charcoal. a. Mean seed:charcoal (g/g); b. Median seed:charcoal (g/g); c. Median wild:charcoal (no./g) (data in Table 5.11).

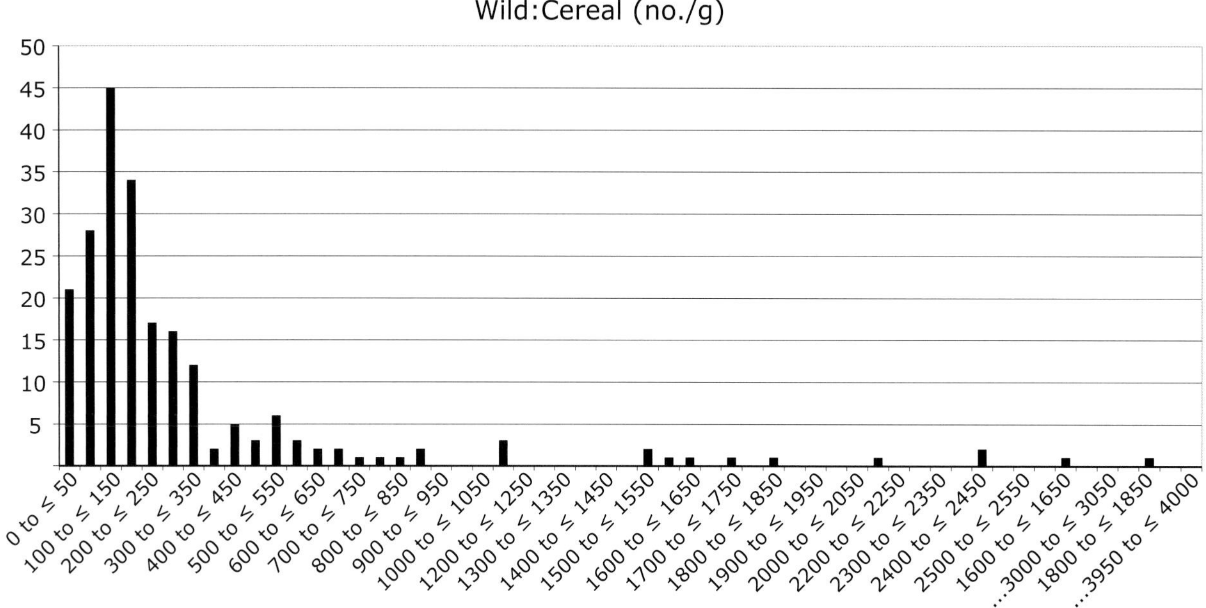

Fig. 5.17 Distribution of values of wild:cereal (no./g) for all samples with that ratio calculated (data in Appendix F; includes 216 samples).

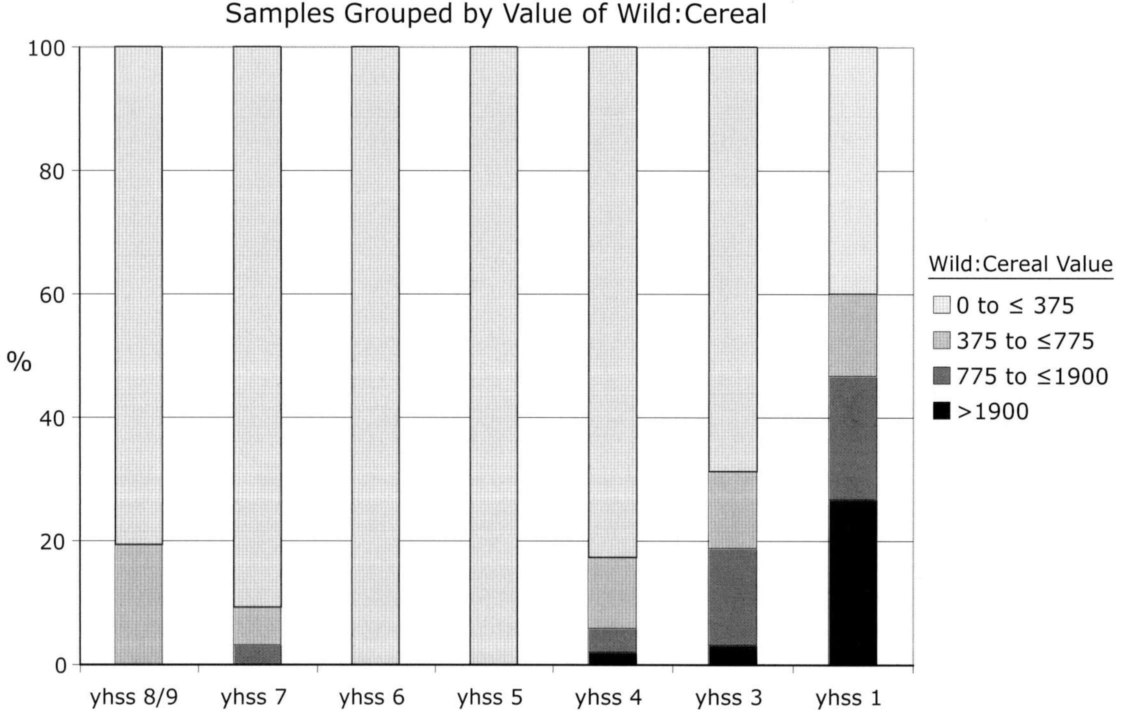

Fig. 5.18 Wild:cereal (percent of samples for each period by value) (data in Appendix F).

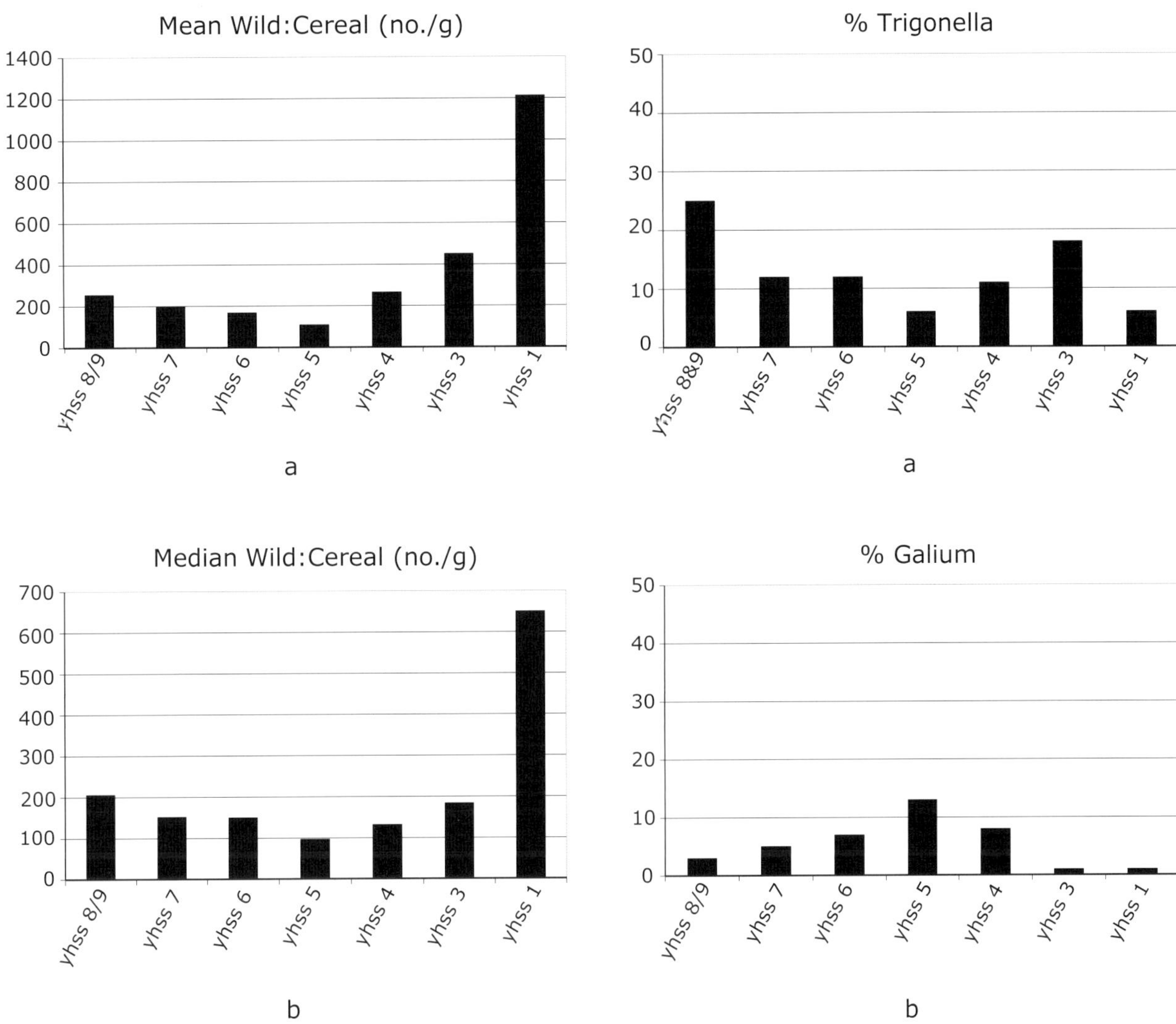

Fig. 5.19 Mean and median wild:cereal (data in Table 5.14).

Fig. 5.20 Common types (percent of total number of seeds per period): a. *Trigonella* and *Trigonella astroites*-type; b. *Galium* (data in Table 5.13).

remains from the Terrace Buildings included free-threshing wheat, hulled six-row barley, lentils, and bitter vetch (Mark Nesbitt, letter 22 January 1989, Gordion Archive). The samples from the antechamber of Terrace Building 2 analyzed for this report had concentrations of barley on the floor (YH 33230, YH 33573, YH 33587, YH 33613) and from pottery jars (YH 33554, YH 33590), free-threshing wheat on the floor (YH 33246) and from a jar (YH 33580), and lentil on the floor (YH 33575) and in a pottery jar (YH 33243). Other seed concentrations in TB2 in-cluded barley (YH 33574, YH 33600, YH 33602), lentils from a jar (YH 33243, YH 33521), and flax from a jar (YH 33595). Portions of the last three were sent for radiocarbon dating (DeVries et al. 2003).

The Abandoned Village Structure (YHSS 3A)

Samples from the floor of a burned later Hellenistic domestic structure had quite a bit of wood charcoal and straw, presumably from roofing debris. There were no *in situ* seed concentrations.

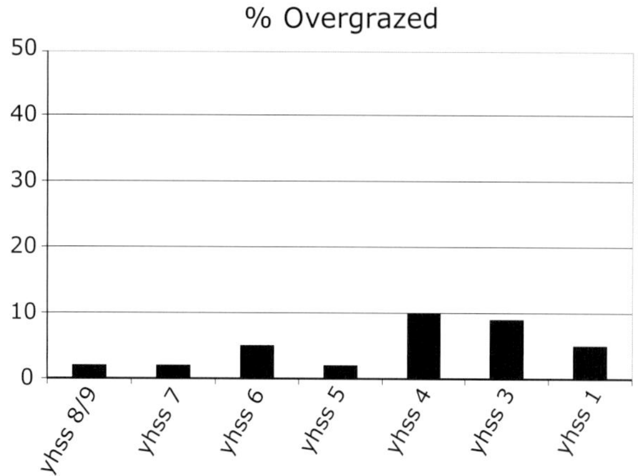

Fig. 5.21 Plants of overgrazed steppe (percent of total number of seeds per period) (data in Table 5.13).

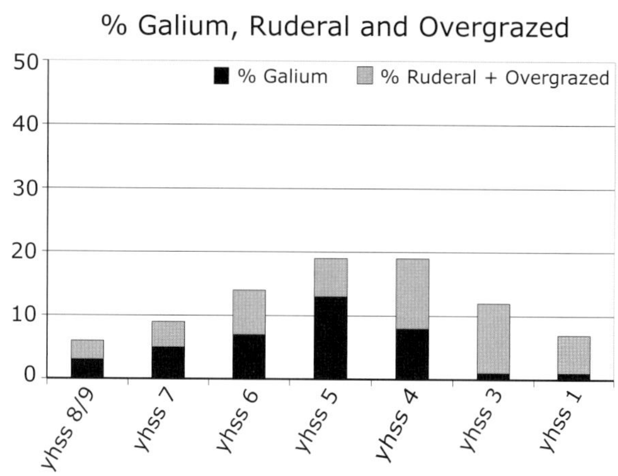

Fig. 5.23 *Galium,* ruderal, overgrazed, combined (percent of total number of seeds per period) (data in Table 5.13).

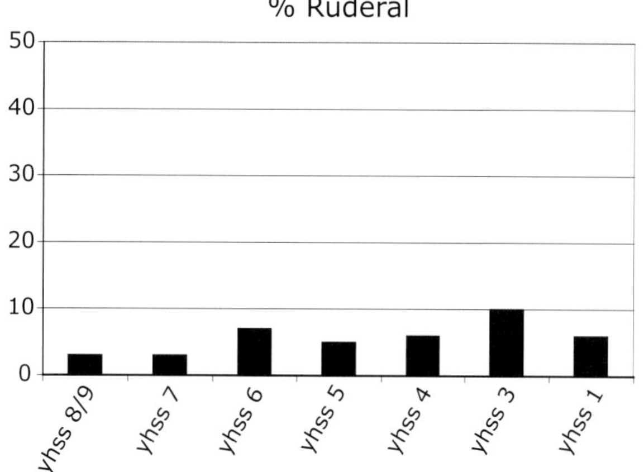

Fig. 5.22 Ruderal plants (percent of total number of seeds per period) (data in Table 5.13).

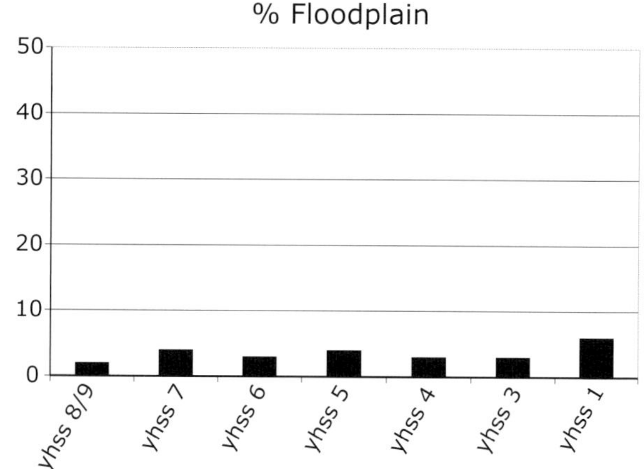

Fig. 5.24 Floodplain types (percent of total number of seeds per period) (data in Table 5.13).

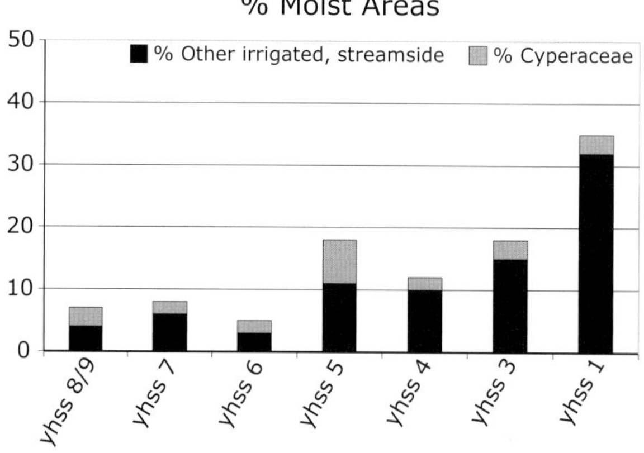

Fig. 5.25 Indicators of irrigation and streamsides (percent of total number of seeds per period) (data in Table 5.13)..

6

Interpretation—Summary and Conclusions

There are a number of questions one can ask of macroremains from an archaeological site. At the most basic level, one can record the plants growing nearby that were used for food, fuel, fodder, and construction in different time periods. Archaeobotanical data also speak to land-use practices and consequent long-term human impact on the vegetation and landscape. Given the long, well-dated sequence at Gordion, there are several questions specific to the agropastoral economy, historical events, movement of peoples, and other cultural trends that are worth addressing with archaeobotanical data. Some of the broad conclusions drawn from these data support interpretations based on archaeology, history, and other natural sciences.

Vegetation Cover and Land Use

Broadly, the modern zones of vegetation are similar to those of the past. The area immediately surrounding Gordion probably supported grassy steppe with isolated trees, except along the river, which would have been home to riparian types such as willow. Where oak, juniper, and pine grow today, it is reasonable to assume that similar climatic and edaphic conditions allowed them to grow in the past, as well. All three, especially juniper and pine, would have been more prominent in the landscape than they are today.

Even under the most benign conditions of the past 4000 years, changes in vegetation cover have been determined by human activity more than by climate within a 50-km radius of the site. Some change has been irrevocable, however. In central Anatolia, oak replaces pine with cutting (Bottema and Woldring 1984:139), and oak also recovers from cutting faster

than juniper. A more severe result of tree-cutting and land clearance on once-wooded slopes above Çekerdeksiz and along the Porsuk and Sakarya valleys is soil erosion, which has left bedrock or at best a thin soil layer in many areas. That means that areas at present bare or treeless once supported some woody vegetation. This loss of vegetation and soil adversely affected surface runoff and the water table, with a corresponding longlasting impact on the ability of vegetation to regenerate. Even local climate conditions may have changed as a result of vegetation, soil, and groundwater loss.

Fuel-cutting

The dominant forest tree genera in the region are juniper (*Juniperus oxycedrus* and *J. excelsa*), oak (primarily *Quercus pubescens,* but also *Q. cerris*), and pine (*Pinus nigra*). One or another of these taxa predominate throughout the sequence, representing over 80% of the charcoal. Along with the increase in taxa of secondary forest and streamside, the charcoal data support the view that tree cover overall did not suffer greatly with long-term exploitation of woodland. Consistent with this view, geomorphological studies indicate that severe soil erosion on the slopes is a relatively late phenomenon, occurring after AD 600 (Marsh 2005). The early importance and subsequent decline in juniper as a fuel wood suggests local changes in availability. In particular, it seems likely that scrubby juniper initially grew on the ridges at the edge of the valley within 0.5 km of the site. The maximum use of oak as fuel appears to be during the Middle Phrygian period. This is also the time of maximum wood fuel use relative to dung (i.e., low seed to charcoal ratios). This suggests that oak, unlike juniper, was sustainably harvested at that time. Scrubby oak and juniper today co-occur on the basaltic soils of Çile Dağı above Şabanozu; one imagines that cover extended along the hills to the south (Dua Tepe), and that the wooded area in general was less impacted by fuel-cutting.

Farming and Herding

The agropastoral nature of the subsistence economy at Gordion is easily inferred from the Gordion seed and animal bone assemblage. That evidence shows that the human diet was based on domesticated species. Barley, bread wheat, emmer, einkorn, lentil, and bitter vetch constitute the small group of cultigens grown in all periods (though bitter vetch and barley may have served as animal fodder rather than food, at least sometimes). In addition, occasional finds show that chickpea (YHSS 3, 6), millets (YHSS 7; 5–1), and rice (YHSS 1) were grown. Fruit and nuts included sporadic cherry, fig, wild almond and pistachio, a single coriander seed (YHSS 4), and a fenugreek seed (YHSS 1). The single pot of flax seed (YHSS 6A Destruction Level) and the six cotton seeds (YHSS 1) probably represent fiber plants, but might have been grown for oil. Hazelnut, found in the Destruction Level, was almost undoubtedly imported from the Black Sea region.

Throughout the Gordion sequence, both farming and herding were practiced within an integrated subsistence system, but they would have had somewhat different effects on the pre-existing vegetation cover. Clearance of woodland and steppe for agricultural fields changes species composition, with plants that tolerate soil disturbance (disproportionately weedy annuals, but also deep-rooted perennials like camelthorn) replacing shallow-rooted perennials (like native grasses) and trees. As it takes time for crops to grow from seed, there is more opportunity for winds to carry off topsoil, even on flat ground. Herding, too, promotes changes in species composition; as the animals preferentially eat the palatable types, spiny and/or unpalatable types tend to proliferate, as do species that can withstand trampling. Even in the absence of human occupation, wild herbivores graze and affect vegetation. The difference between the environmental impacts of low-intensity exploitation of wild game and heavy investment in herding is one of degree. Overgrazing can create bare ground that is

then subject to wind erosion. Marsh (2005) identifies erosional processes as they affected the sediment load and course of the Sakarya River. He posits plowing as an important factor (p. 165), with grazing and fuel-cutting also having an impact. He infers erosion that is a probable result of agricultural activity as early as the Bronze Age, but precise dating eludes us. Charred seed remains that show shifting emphases on farming and herding can enrich our understanding of these processes.

Integrating agricultural and pastoral production produces benefits for humans, crops, and even domesticated animals. Dung left by stubble-grazing animals fertilizes fields, the animals are protected from predation (though, of course, the males pay a disproportionate price through slaughter) and are foddered even if there is snow cover. The balance between farming and herding is not constant, however. Broadly, it may be determined by climate or other environmental factors, but even over decades and centuries, a wide variety of social factors can play a role in the emphasis people place on one subsistence activity or the other.

Based on analogies with sites along the Euphrates, I propose that the wild:cereal ratio may serve as an indicator of relative dependence on herding (high values) and farming (low values). The lowest value at Gordion dates to the Middle Phrygian period, a time of maximum settlement size when the city of King Midas was at its wealthiest (Fig. 5.19) (Voigt 2007; Voigt and Henrickson 2000).

Trigonella is the single most numerous genus of wild plant seed at Gordion, so its distribution can be considered apart from the other wild seeds. The plant grows most commonly in protected areas. Like clover, *Trigonella* is a prime fodder plant. It appears to be sensitive to grazing, so its seeds serve as a proxy for healthy steppe. The percent of *Trigonella* relative to other wild seeds mostly follows that of the wild:cereal ratio. An exception is at the end of the sequence, when the wild:cereal ratio is highest, but the proportions of *Trigonella* shrink (Fig. 5.20a). The best explanation is long-term decline in pasture quality that was manifest by the Medieval period. Over time, the indicators of disturbance (primarily *Galium*) are inversely distributed relative to the wild:cereal ratio (Figs. 5.22, 5.23), with maximum value for the time thought to have maximum dependence on farming. These gratifying results would be strengthened if the faunal data showed corresponding trends (see below).

Climate

Given the fairly consistent subsistence strategies evidenced by the plant (and animal) remains, climate fluctuations over the past 3000 years do not appear to have adversely affected the Gordion agropastoral system: the weather was always erratic.

Where the landscape over time has been so influenced by a human presence, distinguishing "natural" from human-induced changes in the vegetation is not straightforward. Even if dryness has prevented non-riverine trees from growing in central Anatolia below 700 m, the precise boundaries between the treeless steppe, steppe-forest, and forest zones will probably never be known with certainty, much less the shifts that have occurred over the centuries. For the periods under discussion, major global climate change does not appear to be the primary factor in vegetation changes. Rather, as Bottema and Woldring's (1990) review of Holocene pollen data suggests, many vegetation changes are most readily explained as the result of human interference with the vegetation.

Even in the absence of major world-wide or regional climate shifts, highly localized habitat shifts might occur in a landscape that would already be sensitive to small changes in moisture available to plants, since the vegetation cover itself influences available moisture. Depending on the scale of the disturbance, cutting down trees in central Anatolia could have some potentially far-reaching effects on the water balance. As Kuniholm (1977) points out, deforested land in the region suffers soon and severely from erosion (dust storms from the winds across the plateau can be quite dramatic), and forest soils would not last long. In turn, severe erosion in the hills would leave bare ground, increased run-off would follow, and the water table could be lowered as well. Loss of vegetation could allow temperatures to rise, which would intensify the effects of summer drought.

Cause and effect are difficult to assign, but land use (and vegetation) inferred from the Gordion archaeobotanical data may have responded to short-term climate fluctuations. Although pollen, speleothem, and other proxy records for climate are not available for the Sakarya basin specifically, commonly observed decreases in the arboreal/non-arboreal (AP/NAP) ratios in the Holocene pollen record of the eastern Mediterranean are probably the result of human exploitation of the woodland. Even so, at about 3200 BP (ca. 1500 calib. BC), a decrease in some Compositae pollen (*Centaurea solstitialis*-type) suggested "moisture conditions more favourable for tree growth" on the Anatolian plateau (Bottema and Woldring 1990:262); as mentioned earlier, in a moister environment, perhaps pine grew closer to the settlement in denser stands than it does today, which would account for its relatively important contribution to the otherwise local Late Bronze Age charcoal assemblage.

Finally, there is increasing evidence of a moist phase dated to the mid-9th century BC across Eurasia, from Greece (Morris 2004:730) to the steppes of Central Asia (van Geel et al. 2004). The plant evidence for the Middle Phrygian period supports the view that somewhat more favorable conditions prevailed at that time. This interval of climate amelioration may partially account for the successful increased reliance on cultivation at that time. Nevertheless, at all times, the people of Gordion had to contend with a high degree of uncertainty due to very high interannual variability in precipitation.

Irrigation

Although we have not excavated ancient field surfaces or found traces of canals, irrigation leads to changes in species composition by changing the water balance in and around the fields. Seeds of water-loving plants may tentatively be put forward as indicators of irrigation. Certainty eludes us, because at least some of the same taxa might also grow along the river or in uncultivated marshy areas. The benefits of supplemental irrigation for rain-fed crops under conditions of erratic precipitation are obvious; a secure water source reduces one of the most unpredictable and uncontrollable variables in farming. The cost in terms of scheduling and person-hours of labor can be considerable. We might therefore expect evidence for irrigation to be high when farming is important and/or when human population densities are relatively high. In a region with a predictably dry summer, the cultivation of crops that require irrigation at that time of year has the added benefit of using available labor to the fullest in an otherwise slow agricultural season.

The available archaeobotanical, geomorphological, and settlement survey data support the view that irrigation practices instituted by the Phrygians continued and expanded into the Medieval period. For purposes of this analysis, the taxa that are considered to be indicators of wet areas are: Cyperaceae, including *Carex, Fimbristylis,* and *Eleocharis, Fumaria, Plantago, Polygonum, Portulaca, Rumex,* and *Trifolium/Melilotus* (Apps. C, D). Sedges show an interesting distribution through time. In the earlier part of the sequence, sedges constitute less than 10% of the wild and weedy assemblage. They are somewhat more prominent in Middle Phrygian and later deposits, with an apparent steep increase in the latest deposits (33% in the YHSS 1). Since sedges grow in moist, low-lying areas—along the river, on the old floodplain, and along irrigation ditches—it is conceivable that higher proportions of sedges relative to other wild and weedy plants are an indication of expansion of these moist habitats during the Medieval occupation.

Geomorphological studies and archaeological surveys show that regional occupation was oriented towards springs and surface streams, but over time, the water table dropped (Kealhofer 2005:144–45; Marsh 2005). Maximum regional population as well as the maximum size of Gordion itself occurs in the Middle Phrygian/YHSS 5 period, and indeed, there is a small peak in indicators of irrigation at that time (Fig. 5.25). There do not appear to be many summer-irrigated crops, but irrigating staple crops would have reduced the risk of crop failure at Middle Phrygian Gordion, even if the climate was moister than it was before. By Late Phrygian/YHSS 4 times, regional and local population densities had declined, reducing the labor supply available for maintaining the irrigation works. Subsequently, the summer-irrigated crop, millet, shows a fairly steady increase relative to other cereals, reaching its maximum at the end of the sequence (Fig. 5.8). Rice and cotton first appear in the Medieval/YHSS 1 period, and may have been grown as cash crops (see Samuel 2001:428).

The first peak in the proportion of wet-habitat plants occurs during the periods when crop acreage is presumed to be the greatest and the population could support the labor requirements of irrigation (YHSS 5). The second peak occurs after the introduction of several summer crops (YHSS 1), whose cultivation would make the available labor force more productive by lengthening the agricultural year and eliminating a 'slow' season.

Catchment

For studies of complex societies, plant remains from a single archaeological site, even one of central importance like Gordion, can merely begin to provide a general picture of off-site activities involving plants. Inferences start with current plant distribution, soils, and climate, which allow the archaeobotanist to assess the likelihood of the source of a particular taxon. Based on the Gordion plant remains, the people probably had a fairly local diet, as wheat, barley, lentils, chickpea, and the other crops could all have been farmed nearby. One exception is hazelnut, a product of the Black Sea region, found only in Middle Phrygian deposits.

More difficult is identifying the type of botanical evidence that could attest to the reach of the Gordion rulers, especially in the Middle Phrygian period. At that time of maximum power, they might have been able to extract staple foods as tribute or tax from beyond the valley, thereby substantially enlarging the catchment of Gordion for food procurement. On present evidence, however, the most reasonable conclusion is that Gordion fed its people by intensifying agriculture locally rather than by importing grain. The percent of seeds associated is higher in the Middle Phrygian period deposits than at any point until the introduction of irrigated summer cropping in Medieval times. The only sources of irrigation water are nearby (the Sakarya and Porsuk rivers), so if the land under irrigation increased, that land was near the settlement. A second, less persuasive argument posits the following conditions: population density and consequent fuel needs were high; renewable oak is the most important wood fuel; the seed:charcoal ratio is low (so dung fuel is not used as much). Therefore, the comparatively strong emphasis on farming (high numbers for cattle and pig relative to sheep and goat, low wild:cereal ratio) resulted in less dung for fuel because there were fewer animals or the dung was needed to fertilize the intensively farmed fields. Were animals on the hoof brought in from elsewhere, the bones of sheep and goat would not show such a sharp decline.

Surprisingly, the medieval deposits may show stronger archaeobotanical evidence for ties to the wider world. While caprine herding remained a mainstay of the subsistence system, the intensive summer irriga-

Table 6.1. Bone counts (Zeder and Arter 1994; Zeder, pers. comm.)

YHSS phase	8/9	7	6	5	4	3	1
Caprid	1816	2639	1300	1002	1421	1057	685
Cattle	277	339	195	352	341	105	88
Pig	146	118	96	414	371	145	99
Deer	9	198	16	6	7	6	4
Hare	21	9	21	53	29	9	5
Equid	81	30	12	8	48	27	28
Fish	17	17	6	13	13	22	3
Bird	9	13	22	37	70	37	45
Canid	14	22	9	5	5	10	10
Rodent	1	0	3	2	1	1	6
Reptile	6	43	7	0	6	1	6

tion of cash crops suggests that new considerations of tax obligation or commerce may have influenced decisions about agricultural production.

Integrated Economies and Archaeobiological Data

One of the reasons it has been very difficult to reconcile interpretations of plant and animal data is taphonomic. Insofar as charred macrobotanical remains are the remains of fuel, and uncharred animal bones the remains of food refuse, there is no particular reason to suppose them to be correlated within deposits. That correlations are discoverable at the level of site and time period is surely a result of the integrated functioning of the agropastoral economy. At Kurban Höyük along the Euphrates, for example, high wild:cereal ratios were associated with a more pastoral orientation expressed in higher percentages of sheep and goat bone relative to cattle and pig (see Miller 1997b). This observation seems to apply to Gordion, too.

If one assumes that the Gordion wild seed:cereal ratios reflect animal diet, higher values would reflect animals sent out to pasture and lower ratios reflect crop-foddering. And indeed, our initial results bear this out (Tables 6.1, 6.2; Miller et al. 2009). Patterning of the plant and animal remains could be interpreted along a continuum that can be thought of as an economic orientation away from or toward the settlement, which roughly reflects emphasis on pastoralism vs. emphasis on farming (Table 6.3). The numerical values of the various economic indicators is not stable, in the sense that additional samples could easily change the details. It will be seen, however, that for most of the measures calculated here, the assemblage of the Middle Phrygian period (YHSS 5) stands out for its emphasis on farming.

Table 6.2. Percent of food animals (from bone count)

YHSS phase	8/9	7	6	5	4	3	1
Sheep/goat	80.0	79.9	79.9	54.8	65.5	80.0	77.8
Cattle	12.2	10.3	12.0	19.3	15.7	7.9	10.0
Pig	6.4	3.6	5.9	22.7	17.1	11.0	11.2
Deer	0.4	6.0	1.0	0.3	0.3	0.5	0.5
Hare	0.9	0.3	1.3	2.9	1.3	0.7	0.6

Table 6.3. Economic indicators associated with pastoral and agricultural pursuits

Spatial focus relative to settlement	Distant (pastoral)	Close (agricultural)
Fodder source (wild:cereal ratio)	Grazed pasture (wild:cereal ratio higher)	Cultivated plants (wild:cereal ratio lower)
Domestic food mammals (bone count)	Sheep and goat	Cattle and pig
Wild food mammals (bone count)	Deer	Hare
Work animals (bone count)	Dog, equid higher	Dog, equid lower

Zooarchaeologists quantify animal remains in several ways to develop interpretations of ancient diet: counts of identifiable bone, minimum number of individuals, available meat equivalent. I use percent bone counts by time period as a relative indicator of animal taxa consumed, because this measure most directly quantifies the archaeological materials with the fewest additional assumptions. For purposes of this analysis, I compare the percentages of the food animals: caprid (sheep/goat), cattle, pig, deer, and hare (Fig. 6.1). (Miller et al. 2009 consider deer and hare separately.) I exclude mammals unlikely to have been eaten (equid, many of which are donkeys or horses; canid, which includes domestic dog; commensal rodents). I also exclude birds, fish, and amphibians, be-cause the way they are used and the number of bones for the various taxa are not comparable to those of mammals. Typical of most sites in west Asia, caprids constitute the bulk of the food bone assemblage, never less than 55%. The percentage bone counts for cattle and pig roughly follow each other (Miller et al. 2009). Deer and hare combined do not exceed 6% of the assemblage, yet a pattern emerges: deer bone count percentages tend to rise and fall with sheep-goat, and rabbit with cattle and with pig. Although we cannot assume that all dogs and equids worked with shepherds, there also appears to be some association of caprids with canids and equids (Fig. 6.2), as well as with the primary plant indicator of herding, the wild:cereal ratio.

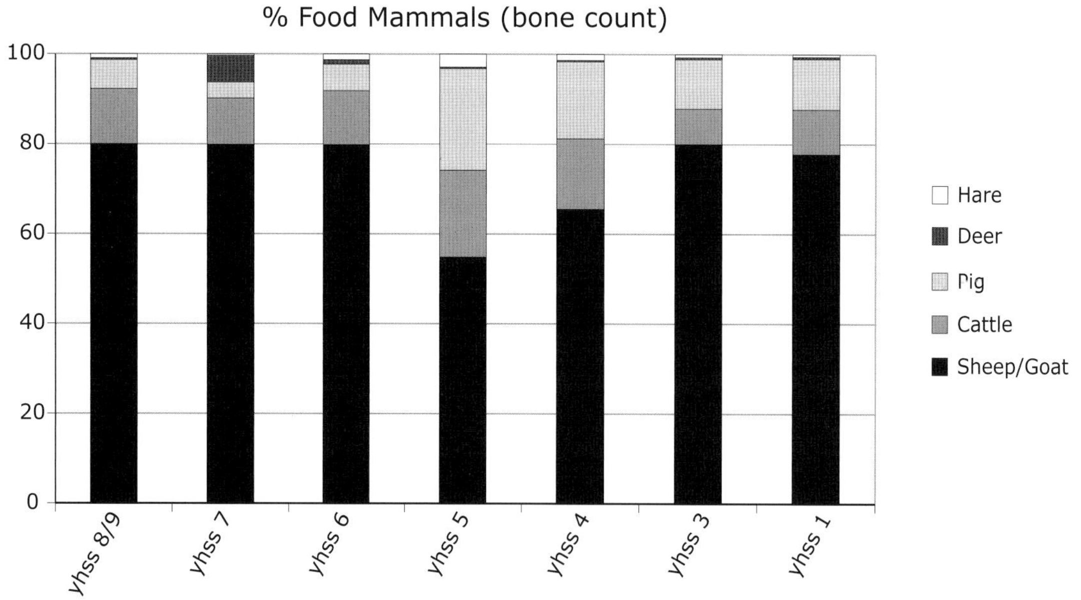

Fig. 6.1 Major food mammals (bone count) (data in Table 6.1).

Table 6.4 Other indicators of plant use. Summary chart based on grain weight (g) and rachis count for each phase

YHSS phase	8/9	7	6	5	4	3	1
No. of samples	32	66	8	15	53	32	15
Wheat grain sum (includes einkorn)	6.32	24.07	0.11	0.81	11.44	5.34	0.70
Einkorn grain sum	0.17	1.79	0.01	0	0.13	0.03	0
Barley grain sum	7.90	12.42	0.13	1.53	18.33	6.31	0.68
Cereal grain sum	21.28	50.00	0.44	4.07	40.63	23.18	1.95
Einkorn rachis	8	103	0	1	35	5	0
Wheat rachis (includes einkorn)	390	1000	7	15	3605	863	22
Bitter vetch sum	1.02	1.10	0	0.03	0.36	0.25	0.20
Pulse sum (includes bitter vetch)	1.43	1.38	1.00	2.18	0.73	0.99	0.38
Wheat/(Wheat + Barley) (% grain)	44	66	44	35	38	29	51
Wheat/Cereal (% grain)	30	48	25	20	28	23	36
Barley/Cereal (% grain)	37	25	32	38	45	57	35
Einkorn/Wheat (% grain)	2.7	7.4	9.1	0	1.1	0.6	0
Einkorn rachis/Wheat rachis (%)	8	43	43	27	8	1	0
Bitter vetch/Pulse (%)	71	80	0	1	49	13	53

It is in this context that the fuel economy can be best understood. First, throughout the sequence the seed:charcoal ratios are similar to those of sites in steppe-forest or open woodland (Table 5.12), so although people altered the forest composition, mainly by removing juniper, fuel wood was always available within 50 km of Gordion. Even at Gordion's maximum population, not only was wood fuel available (renewable oak being most prominent), the use of dung appears to be at a nadir. During the other periods, when pastoral production prevailed, dung fuel appears to have been more convenient to use. In the later part of the sequence, pine appears to be associated with the indicators of pastoralism and orientation away from the settlement; the proximity of juniper in YHSS 8/9 and 7 may explain the comparatively low, but not insignificant, pine percentages at the beginning of the sequence (Fig. 4.1).

Cultural Affiliation

Crop choice could speak to questions raised by ancient histories about population movements of Phrygians and, later, Galatians (European Celts) to Anatolia.

In particular, einkorn wheat was probably first domesticated in Anatolia and spread to adjacent regions (see Heun et al. 1997). Over time, its popularity declined, and by the end of the Bronze Age, einkorn was relatively more important in southeastern Europe than in Anatolia (Hubbard 1976). The distribution of einkorn in Late Bronze Age and Iron Age sites in Turkey suggests that by the 1st millennium BC, its cultivation was restricted to the uplands of Anatolia west of the Euphrates. (See einkorn distribution maps in Simone Riehl's database, http://cuminum.de/archaeobotany/.) It would be satisfying to think that Phrygian immigrants brought this taste of home from southeastern Europe. Einkorn grains and rachis fragments are never very numerous, so the data are weak. They do, however, show a small increase in ubiquity in YHSS 7 (Fig. 5.11c, g), and so the einkorn distribution is at least consistent with the other lines of evidence for this population movement (Voigt and Henrickson 2000a:42).

From the Early Iron Age appearance of the Phrygians to the Middle Phrygian period (YHSS 7–5), there is strong continuity in the cultural affiliation of the inhabitants (Voigt and Henrickson 2000b). Therefore, any changes in food choice reflect internally generated developments, not the immigration of newcomers.

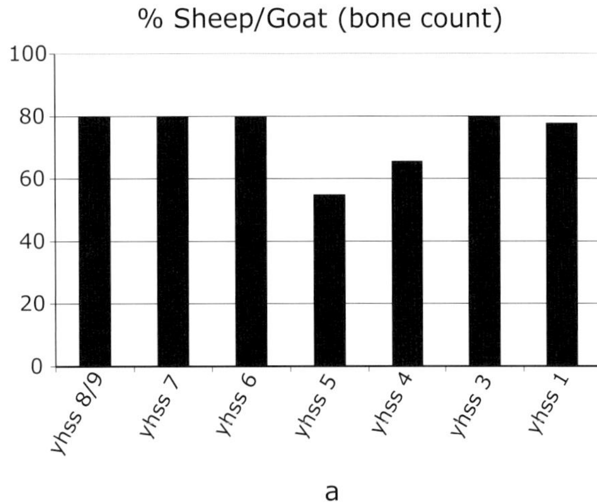

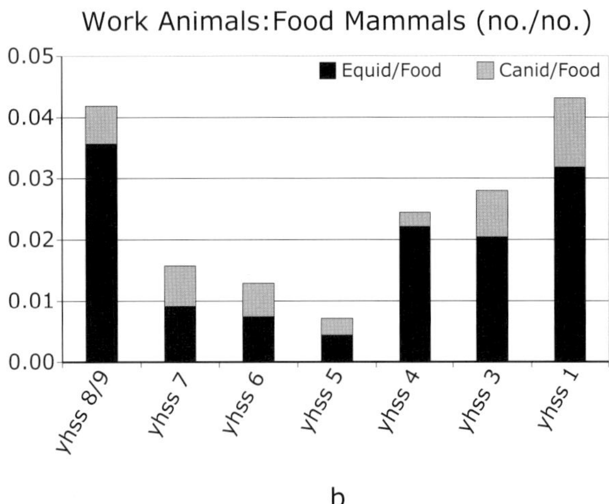

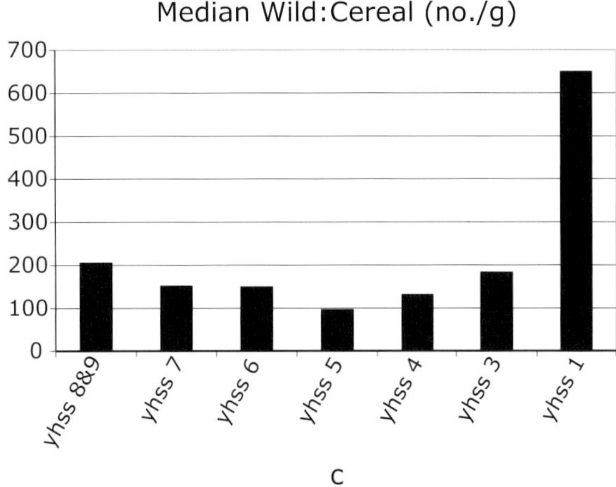

Fig. 6.2 Caprids, herder animals, and wild:cereal.

In the absence of a positive culinary hypothesis to test, the other key ethnic change, i.e., the arrival of the Galatians at Gordion, does not appear to have direct archaeobotanical correlates. On the other hand, if the Galatian element was primarily military, non-farming mercenaries granted land might well have deferred to the agricultural choices of their local wives and farmers. For the combined Hellenistic deposits, the differences in the various indicators of economic activity generally show a continuation of trends that began during the Late Phrygian period (i.e., a shift in the agropastoral balance toward the pastoral).

Summary of Results

Analysis of the Gordion archaeobotanical assemblage remains provisional. The flotation samples are unevenly distributed over the periods represented in the 1988 and 1989 deep soundings, and the diversity of the seed assemblage makes generalizations difficult. In earlier chapters I have provided alternative ways to calculate the data, partly to provide comparability with other reports, and partly to demonstrate that some variables are more reliable or stable than others. For many variables, the numerical values are less interpretable than the direction of change between periods. For most of the variables considered, values for the prosperous Middle Phrygian period (YHSS 5) stand out as being at the extreme end of the range for the sequence (lowest mean and median seed:charcoal, wild seed:charcoal, and wild:cereal; lowest percent wheat; highest indicators of disturbance). Some variables show overall temporal trends over the sequence (decline in juniper, increase in successional trees, increase in millet). And some variables have a more complex distribution that might be due to chance or might have some significance (for example, irrigation in the Middle Phrygian period vs. the Medieval period, the distribution of pine and oak). A key finding has been that the faunal data share the Middle Phrygian anomaly, and thereby inform and strengthen the archaeobotanical analysis.

A narrative summary would begin during the Middle Bronze Age (YHSS 10), when Hittite influence was highest, settlement in the region was relatively high, with subsistence dependent on farming (Kealhofer 2005), but there are too few botanical remains for useful interpretation. By the Late Bronze Age (Hittite empire, YHSS 8/9), settled population declined. The pithouse dwellers of the Early Iron Age (YHSS 7) are thought to represent a new immigrant element (i.e., the earliest Phrygians; Voigt and Henrickson 2000:42), and the uptick in einkorn could reflect that. The Early Phrygian (YHSS 6) descendents of the pithouse dwellers established Gordion as the Phrygian capital. Economic indicators for YHSS 8/9 to YHSS 6 are similar to one another: they share high proportions of caprids and somewhat high wild:cereal ratios, low indicators of irrigation, decreasing indicators of pasture quality, and increasing indicators of disturbance. The Middle Phrygian (YHSS 5) period stands out as a time when farming was more important than in previous or subsequent times; archaeological settlement survey reached the same conclusion with that independent data set (Kealhofer 2005:148). This reorientation does not reflect population replacement, as the periods are characterized by cultural continuity. A factor contributing to Middle Phrygian prosperity could be a short-lived period of ameliorating climate. Cereals, especially barley, were important for fodder; cattle and pig, both dependent on the river for surface water, were husbanded; and hare was trapped or tended close to home. With the shrinking of the Middle Phrygian city, pastoral production again became increasingly important (Late Phrygian, YHSS 4), a trend that continued into the Hellenistic period (YHSS 3), when subsistence acquisition was again more oriented toward the steppe and woodland; herding, with sheep and goat grazing in uncultivated tracts, and deer-hunting characterize the assemblage. Unlike the Middle Phrygian period, the Medieval period (YHSS 1) shows continuing reliance on pastoral production, but the introduction of new crops requiring irrigation in the summer enabled people to increase their productivity by farming year-round, even as they then had to work harder and longer.

Appendix B

Wood Charcoal Identification Criteria

The Taxa

Identifications are based on comparison with the author's incomplete wood and charcoal collection, and illustrations and descriptions of woods in Panshin and de Zeeuw (1970), Fahn et al. (1986), and Schweingruber (1990). It is difficult, and frequently impossible, to determine a wood sample to the species level. If there are distinguishing characteristics, they may not be preserved in charred specimens, due to size or color changes, or the destruction of delicate features. Even between genera, some types are easily confused. Some of the previously published specific determinations of wood from Gordion seem to be based on features apparent in uncharred wood but not in charcoal; others are inferred on geographical grounds. I have found no anatomical grounds for giving determinations to species, although occasionally one might use a phytogeographical argument to go beyond the genus level. To enable interested wood anatomists to assess my reasons for assigning items to a given taxon, the features I have used to distinguish the taxa are listed. Features indicated with an asterisk are ones that were looked for on all pieces. Features without an asterisk were used to check, confirm, or delimit an identification.

Conifers

Two types of conifers were distinguished, pine (*Pinus*) and juniper (*Juniperus*). A few small or distorted pieces remain indeterminate and are referred to as "conifer," though they are more likely to be juniper than pine or any other type. Kayacık and Aytuğ (1968) report several conifers from the Tumulus MM and its furnishings: *Pinus sylvestris, Juniperus foetidissima, Cedrus libani,* and *Taxus baccata*; the pieces originally thought to be yew (*T. baccata*) are now understood to be pine and Lebanon cedar (Blanchette and Simpson 1992). Criteria for the specific identification of pine and juniper could not be developed with the charred specimens available from the City Mound, and cedar and yew were not seen.

Pinus (Pine)

Low magnification:
    x-section: *resin ducts, usually in later half of growth ring
High magnification:
    x-section: intercellular spaces not seen
    r-section: pinoid cross-field pits, ray tracheids on margins of rays

Juniperus (Juniper)

Low magnification:
    x-section: *resin ducts absent
High magnification:
    x-section: *intercellular spaces frequent
    r-section: pits cupressoid/taxodioid, no ray tracheids on margins of rays (and rays relative low height), tangential walls of ray cells thin and finely nodular
    t-section: ray height less than 12 cells, and usually less than 6
Possible confusions:
Abies (fir): like juniper, fir does not have resin ducts. Anatomy manuals and a comparative piece (*Abies alba*) do not have intercellular space, ray height seems to be over 10 cells, and tangential walls of ray cells are dentate.
Taxus (yew): like juniper, yew does not have resin ducts, but it does have spiral thickenings, absent from

all coniferous wood examined for this report.
Cedrus libani (Lebanon cedar): like juniper, Lebanon cedar does not have resin ducts, but it does have marginal ray tracheids, absent from all non-pine coniferous charcoal examined for this report.

Dicots

Alnus cf. *viridis* (formerly Unknown 5; YH 30419, "planks") (Alder)

Low magnification:

    x-section: *diffuse porous, *vessels solitary to radial multiples of 4 or more, as many as 9 seen, and some pore groups, distributed throughout growth ring. Rays thin, but visible at low power, pores not particularly small. Note that rings are fairly wide.

High magnification:

    x-section: same as above, *scalariform perforation plates visible

    r-section: *scalariform perforation plates, ca. 8–9 bars

    t-section: rays 1-seriate, vessels with densely alternate pits

Possible confusions:

Maple and boxwood are two of the woods, identified from Tumulus MM furnishings, that were not seen in these samples. They do not correspond to any of the unknowns; they seem closest to *Alnus,* but are not. (The third wood known from the tomb furnishings but not from the 1988/89 excavations is walnut, *Juglans regia.*) Robert Blanchette (University of Minnesota Forest Pathology and Wood Microbiology Research Laboratory, pers. comm. November 8, 1993) confirmed the identification of this sample as *Alnus.*
Acer: Alder is reminiscent of maple, except that maple does not have scalariform perforation plates.
Note: few woods have scalariform perforation plates. These specimens do not resemble other wood types with scalariform perforation plates, namely birch (*Betula*), *Viburnum,* holly (*Ilex*), plane (*Platanus*), cornelian cherry (*Cornus*), beech (*Fagus*). Boxwood, hazel, and alder still need to be considered.
Buxus: boxwood pores are small and solitary
Corylus: has aggregate rays
Alnus orientalis, A. glutinosa, and *A. viridis:* This wood looks like alder at low magnification, and alder has the

densely alternate pits characteristic of this type. Unfortunately, only the first two species are reported for Turkey, but both of them have aggregate rays. *A. viridis* does not have aggregate rays, but is not mentioned in the *Flora of Turkey* (Davis 1982).

Fraxinus (Ash)

Low magnification:

    x-section: *ring porous, *growth rings distinct, *large early wood vessels, solitary or radial multiples, tyloses common, *small, sparse latewood vessels, usually in radial pairs

High magnification:

    x-section: same as above, vessels smaller than "*Morus*"

    r-section: *homocelllar, no spiral thickenings

    t-section: *rays biseriate, no spiral thickenings

Populus (Poplar)

Low magnification:

    x-section: *diffuse porous, growth ring usually distinct, *pores rounded in cross-section, solitary or radial multiples or small groups, *frequently with occluded vessels, *fine rays

High magnification:

    x-section: as above, fibers thin-walled

    r-section: *homocellular (*Populus*), *heterocellular (*Salix*), large pits in vessels congregate just outside ray margins

    t-section: *1-seriate

Possible confusions:

Salix (willow): except for one piece from YH phase 0 (YH 20801), all these were poplar.
Pyrus/Crataegus: At first, I erred in the identification, but poplar pores are more likely to be rounded than angular, and the large pits in vessels congregating on the ray margins are also distinctive.

Prunus (*persica/armenaica/communis*-type) (Peach/Apricot/Almond)

Low magnification:

    x-section: *ring porous, growth rings distinct. *Early wood vessels one deep, occluded/tyloses. *Late wood pores fairly evenly distributed in growth ring, not sparse, seem to have crystals, rays wide

High magnification:
    x-section: same as above
    r-section: rays average 6–7-seriate, up to 8-seriate. Spiral thickenings in vessels, probably homocellular (hard to see in specimen)
    t-section: 1-seriate and multiseriate rays

Pyrus/Crataegus (formerly Unknown 2) (Pear/Hawthorn)

Low magnification:
    x-section: *diffuse porous to semi-diffuse porous, growth ring ±distinct, *vessels mostly solitary and evenly distributed across growth ring, *frequently angular cross-section
High magnification:
    x-section: same as above, vessels 2+ -seriate, fibers may be relatively thick-walled compared to *Populus/Salix*
    r-section: heterocellular and homocellular, vessels usually with thin spiral thickenings
    t-section: rays upt to 6 or so-seriate, though usually 2–3-seriate
Possible confusion:
Populus: thinner-walled fibers, pores more rounded, more likely (but not necessarily) to be in small groups or radial multiples, *uniseriate, no spiral thickenings
Note: According to Schweingruber (1990), it is not possible to distinguish the woods of wild pear and hawthorn. Either are possible at Gordion.

Quercus (Oak)

Low magnification:
    x-section: *ring porous, *growth rings distinct, *wide rays and narrow ones, tyloses in most samples, tangential parenchyma in late wood, flame-like arrangement of vessels in late wood (usually rings were so narrow that only large, early wood pores were present)
High magnification:
    x-section: not checked
    r-section: not checked
    t-section: not checked

Rhamnus (Buckthorn)

Low magnification:

x-section: *diffuse porous, *growth rings distinct, *vessels in oblique, flame-like groups, *no vessels interspersed among fibers. Looks identical to *Rhamnus cathartica* depicted by Schweingruber (1990)
High magnification:
    x-section: not checked
    r-section: not checked
    t-section: not checked

Ulmus (Elm)

Low magnification:
    x-section: *early vessels mostly large, but sometimes associated with narrow ones, *late wood vessels clustered in oblique bands, frequently continuing across rays, ±tyloses
High magnification:
    x-section: same as above
    r-section: *rays homocellular or heterogeneous type I (i.e., cells mostly procumbent, with marginal row of square cells), *spiral thickenings, especially in narrow vessels
    t-section: rays 1- and 2-seriate, frequently 3–5-seriate (one piece 7–8-seriate), up to >20 cells high
Possible confusion:
Celtis: At first, *Ulmus* was identified as *Celtis,* but even though many of the rays were heterogeneous, they had square rather than upright marginal cells, and the type seemed to have mostly homocellular rays with procumbent cells.

Ulmus/Morus (Elm/Mulberry)

Low magnification:
    x-section: *ring porous, *growth rings distinct, *early wood vessels large, solitary or radial multiples, tyloses common, *late wood vessels in small isolated clusters, oblique arrangement, but not continuous bands
High magnification:
    x-section: same as above
    r-section: *heterocellular (heterogeneous type I?), *distinct, fine spiral thickenings, widely spaced in examples seen, especially in smaller vessels
    t-section: *rays wide (4–5-seriate), *spiral thickenings
Possible confusion:
The difference between the descriptions of elm and

mulberry is small. It is possible that I have incorrectly identified the rays as heterocellular. Even so, these pieces do not seem to be *Celtis* or *Ulmus,* because the late wood pore groups are isolated, not arranged in continuous or near continuous bands.

Unknown 1 (YH 22895 #6)

Low magnification:
    x-section: *ring porous, *growth rings distinct, early wood pores solitary but closely packed, late wood pores mostly solitary and round, very occasionally pairs

High magnification:
    x-section: same as above
    r-section: heterocellular (otherwise, would be reminiscent of ash)
    t-section: very fine rays

Unknown 3 (YH 25660 #1)

Low magnification:
    x-section: ring porous, growth rings distinct, early wood pores single, tyloses, late wood pores radial twos and clusters, but sparse, seems like ash but not

High magnification:
    x-section: same as above
    r-section: rays heterocellular: 1–3 rows of uprights on margins of procumbents, no spiral thickenings
    t-section: rays several seriate

Unknown 4 (similar to *Tamarix*; YH 25748 #3, 5; YH 31330 #9)

Low magnification:
    x-section: *ring porous, *growth ring distinct, *early wood vessels solitary or in group, *rays wide, clear at low magnification, *late wood pores sparse, solitary, radial files of 2, and small groups, *±tyloses

High magnification:
    x-section: same as above
    r-section: no spiral thickenings, seems to be heterocellular (rows of procumbent with square cells interspersed?)
    t-section: rays wide, 5–6-seriate, vessels with many minute pits

Appendix C

Vegetation Survey

During 16 field seasons between 1988 and 2007, I informally surveyed the vegetation growing within about 2 km of the site of Gordion and village of Yassıhöyük (Table C1).

Botanical fieldwork took place concurrently with the excavation and archaeological conservation seasons, between the beginning of June at the earliest and the middle of August at the latest; most botanizing was done in June. Therefore, early spring and fall-flowering plants are largely absent from this discussion. A variety of habitats occurs within the surveyed area: degraded pasture, irrigated gardens and fields, the former bed of the Sakarya River, the banks of the river, and roadsides, many of which are field edges. The most intensive and systematic botanical survey was on Tumulus MM, 1997–2007. Whenever I had occasion to leave the valley bottom, there were always at least a few plants I had not seen before, but I did not have time for collecting more than a few specimens.

In 1993, I was able to work on voucher specimens collected up to that date at the Royal Botanic Garden in Edinburgh, where several staff members helped make the determinations; other plants were less securely determined at the botanical laboratory of the British Institute of Archaeology in Ankara (some-times with Mac Marston) and with the help of P. H. Davis's *Flora of Turkey* and various field guides. Mecit Vural and other botanists visiting Gordion have also suggested identifications for present-day plants (Table C2 lists identified taxa). Even so, it is not a simple matter to identify plants in the field, since their distinguishing characteristics are not always present or obvious (i.e., the list of plant taxa should not be taken too literally). In addition to voucher specimens, I have collected seeds and wood from the area when available. Although it is easy to recognize some vegetation patterns, it is hard to identify taxa that are uniquely associated with a particular environmental niche (Davis's *Flora of Turkey;* pers. obs.); not only are most of the plants tolerant of a range of conditions, most of the terrain I examined was adjacent to irrigated fields or overgrazed pasture. There are, however, a few plant taxa that stand out as potentially useful for tracing changes in land-use patterns because they generally grow on the grassy steppe, the degraded steppe, fields, or in the case of the tree taxa, in the forest. Unfortunately, their charred seeds are generally not distinctive beyond the level of genus or family.

See file: YH C2 plant list.

Table C1. Locality descriptions (see Fig. 2.1)

Locality	Description	N	E	Approximate elevation (m)
Tumulus MM	Within fenced area; possibly seen only in a particular sector: north, northwest, etc.	39°39'14"	31°59'52"	700
City Mound protected	Within fenced area of Gordion City Mound	39°39'00"	31°58'53"	680
Irrigated field edge	Various, within about 5 km of Yassıhöyük	39°39'23"	31°59'40"	ca. 680–700
Ditch	Both roadside ditches (typically adjacent to irrigated fields) and a drainage ditch from the uplands to the floodplain			ca. 680–700
Grazed	Unprotected areas within about 5 km of Yassıhöyük	39°39'23"	31°59'40"	ca. 680–710
Roadside, tracks	Along roadsides or the railroad track, within 2 km of the village	39°39'23"	31°59'40"	ca. 680–700
Gardens	Irrigated gardens (at dig house or beyond the village)	39°39'23"	31°59'40"	700
Sakarya bank	Along the Sakarya River	39°39'19"	31°58'45"	680
S ridge	Gypsum ridge south of the village; on top are several tumuli, and area is bordered by irrigated fields. Overgrazed pasture	39°38'35"	31°59'28"	710
N ridge, Kızlarkayası	Gypsum ridge north of village en route to Kızlarkayası and the grassy steppe	39°40'40"	32°00'00"	ca. 700–750
Sakarya plain	Poorly drained area on east side of Sakarya west of south gypsum ridge and the village	39°39'17"	32°58'50"	690
Juniper/Pine	Juniper and pine woodlands above about 1000 m, 20–50 km WNW of Gordion, near Yağ Arslan (Hamidiye)	39°50'	31°41'	ca. 1000–1400
Grassy steppe	Formerly, an area of relatively undisturbed vegetation near Kızlarkayası	39°40'51"	32°01'00"	750
Conglomerate outcrop	Across Sakarya from Gordion (west of railroad tracks)	39°36'00"	31°56'28"	705
Roman road	Ridge between Gordion and Hacı Toğrul (basalt substrate)	39°44'	32°12'	ca. 1000
Ankara Çay	Near confluence of Sakarya and Ankara Çay	ca. 40°04'	ca. 31°50'	ca. 600
Çile Dağı	Mountain above Avşar (basalt substrate)	39°46'25"	32°09'12"	ca. 1100–1200
Dig House	Ungrazed but trafficked courtyard	39°39'15"	31°59'21"	693
Eski	Old bed of Sakarya River (Eski Sakarya)	39°39'19"	31°58'45"	680
Pine	Pine forest near Yağ Arslan/Hamidiye	39°50'46"	31°42'01"	ca. 1400
Juniper	Juniper woodland below pine forest	39°18'	31°13'15"	900–1100
Dölleme ağıl	Abandoned sheepfold near Yeni Köseler	ca. 39°43'	ca. 31°54'	ca. 710
Orta ağıl	Abandoned sheepfold near Çekerdeksiz (basalt substrate)	ca. 39°37'	ca. 32°03'	ca. 850
Magara ağıl	Abandoned sheepfold near Çekerdeksiz (basalt substrate)	ca. 39°37'	ca. 32°03'	ca. 850
Çekerdeksiz	Grazed area above Çekerdeksiz (basalt substrate)	39°37'11"	32°03'21"	ca. 850
Satılmış gün	Abandoned sheepfold near Ankara Çay	ca. 40°04'	ca. 31°50'	ca. 600

Table C2A. Plants collected or seen near Gordion. Additional localities listed on Table C2B

Genus/nm #	Tumulus MM	City Mound	Roadside, tracks	Irrigated field edge	Ditch	Gardens	Sakarya bank	Sakarya plain	Grazed	South ridge
Acanthaceae: *Acanthus* sp.	.	.	.	.	.	.	.	.	.	.
Amaranthaceae: *Amaranthus albus/graecizans*	.	.	x	x	.	x	.	.	.	.
Amaranthus, misc.	.	.	x	x	x	.	.	.	.	.
Anacardiaceae: *Rhus coriaria* L.	.	.	.	.	.	.	.	.	.	.
Pistacia terebinthus L.	.	.	.	.	.	.	.	.	.	.
Apiaceae: *Bifora radians* Bieb.	.	.	x	x	.	.	.	.	.	.
Bupleurum cf. *flavum* Forssk	.	.	.	.	.	.	Eski	.	.	x
Bupleurum cf. *turcicum* Snogerup	x	.	.	.	.	.	.	.	.	x
Caucalis platycarpos L.	.	.	.	x	.	.	.	.	.	.
Conium maculatum L.	x	.	.	x	x	.	.	.	.	.
Daucus carota L.	.	x	x	x	x	.	x	.	.	.
cf. *Echinophora* sp.	.	.	x	x	.	.	.	.	x	x
Eryngium campestre L.	x	x	.	.	.	.	.	.	x	x
Eryngium creticum Lam.	x	.	.	.	.	.	.	.	x	x
Falcaria vulgaris Bernh.	x	.	x	x	.	.	.	.	.	.
Malabaila carvifolia Boiss. & Mal./*Peucedanum palimboides* Boiss.	.	.	.	.	.	.	.	.	.	x
Malabaila cf. *secacul* Banks & Sol.	x	x	.	.	.	.	.	.	.	.
Malabaila sp.	x	.	.	.	.	.	.	.	.	.
Opoponax sp.	.	x	x	.	.	.	.	.	.	.
Scandix sp.	x	.	.	.	.	.	.	.	.	.
Torilis leptophylla (L.) Reichb.	x	.	.	.	.	.	.	.	.	x
Turgenia latifolia (L.) Hoffm.	x	.	.	x	.	.	.	.	.	.
Asclepiadaceae: *Cynanchum acutum* L. subsp. *acutum*	.	.	.	x	.	.	Eski	.	.	.

Table C2A cont'd.

Genus/nm #	Tumulus MM	City Mound	Roadside, tracks	Irrigated field edge	Ditch	Gardens	Sakarya bank	Sakarya plain	Grazed	South ridge
Asteraceae: *Achillea* cf. *biebersteinii* Afan.				×			Eski			
Achillea tenuifolia Lam.										
Achillea wilhelmsii C. Koch	×		×		×			×	×	
Achillea perennial (like *A. teretifolia* Willd.)	×								×	
Achillea, various										
Acroptilon repens (L.) D.C.	×		×	×	×					
Anthemis (nm 1232)	×	×						×		×
Artemisia sp.		×							×	×
Calendula/Pulicaria (nm 1977)	×									
Carduus nutans L.	×	×	×					×	×	
Carduus pycnocephalus L.	×	×	×		×		×	×		
Carduus (nm 2495)	×	×	×					×	×	
Centaurea calcitrapa L.		×	×		×			×		
Centaurea patula DC.	×									×
Centaurea pseudoreflexa Hayek	×									×
Centaurea pulchella Ledeb.	×	×								×
Centaurea cf. *rigida* Banks & Sol.	×							×		
Centaurea solstitialis L. subsp. *solstitialis*	×	×	×		×			×	×	
Centaurea virgata Lam.	×								×	×
Centaurea (nm 1227; blue)	×	×	×	×						
Centaurea, misc.										
Chondrilla juncea L.		×								
Cichorium intybus L.	×	×	×	×	×					
Cirsium sp.				×	×		×	×		

Table C2A cont'd.

Genus/nm #	Tumulus MM	City Mound	Roadside, tracks	Irrigated field edge	Ditch	Gardens	Sakarya bank	Sakarya plain	Grazed	South ridge
Cnicus benedictus L.	X	X	.	.	.	.	.	.	.	.
Cousinia halysensis Hub.-Mor.	X	.	.	.	.	.	.	.	X	X
Cymbolaena griffithii (A. Gray) Wagenitz	X	.	X	.	.	.	.	.	X	.
Echinops sp.	X	.	X	.	.	.	.	.	.	.
Gundelia tournefortii L.	X	.	.	.	.	.	.	.	X	X
Helichrisus sp.	.	.	.	.	.	.	.	.	.	.
Jurinea pontica Hausskn. & Freyn ex Hausskn	.	.	.	.	.	.	.	.	.	X
Koelpinea linearis Pallas	X	X	.	.	.	.	.	.	.	.
Lactuca serriola L.	X	X	.	X	X	.	.	.	.	.
cf. *Leucocyclus* sp.	.	.	.	.	.	.	.	.	.	.
Matricaria aurea (L.) Schultz Bip.	.	.	.	X	X	.	.	X	.	.
Matricaria cf. *macrotis* Rech. fil.	X	X	.	.	.	.	.	.	X	X
Onopordum anatolicum (Boiss.) Eig	X	.	X	.	.	.	.	X	.	.
Pulicaria vulgaris Gaertner	.	.	X	.	.	.	X	.	.	.
Scorzonera laciniata L.	X	X	.	.	.	.	.	.	.	.
Scorzonera/Crepis	X	.	.	.	.	.	.	.	.	.
Senecio cf. *vernalis* Waldst. & Kit.	X	.	.	.	.	.	.	.	.	.
Taraxacum sp.	X	X	.	.	.	.	.	.	.	.
Tragopogon dubius Scop.	X	X	.	X	.	.	.	.	.	.
Xanthium spinosum L.	.	.	X	X	X	.	X	.	.	.
Xanthium strumarium subsp. *cavanillesii* (Schouw) D. Löve & P. Dansereau	.	.	X	X	X	.	X	.	.	.
Xeranthemum inapertum (L.) Miller	X	X	.	.	.	.	.	.	.	.
Asteraceae, various	.	.	.	.	.	.	.	.	.	.

Table C2A cont'd.

Genus/nm #	Tumulus MM	City Mound	Roadside, tracks	Irrigated field edge	Ditch	Gardens	Sakarya bank	Sakarya plain	Grazed	South ridge
Berberidaceae: *Berberis* cf. *vulgaris* L.	.	.	.	.	.	.	.	.	.	.
Boraginaceae: *Anchusa arvensis* (L.) Bieb.	.	.	X	X	.	.	.	.	.	.
Asperugo procumbens L.	.	X	.	.	.	.	.	.	.	.
Buglossoides arvensis (L.) Johnston	.	X	.	.	.	.	.	.	.	.
Echium italicum L.	X	.	X	.	X	.	.	.	.	.
Heliotropium spp.	X	.	X	X	.	.	.	.	.	.
Lappula patula (Lehm.) Aschers. ex Gürke	X	X	.	.	.	.	.	.	.	.
Moltkia/Alkanna	X	.	.	.	.	.	.	.	X	X
Nonea caspica (Willd.) G. Don	X	X	.	.	.	.	.	.	X	.
Rochelia disperma (L.fil.) C.Koch	X	X	.	.	.	.	.	.	X	.
Boraginaceae, various	.	.	.	.	.	.	.	.	.	.
Brassicaceae: *Alyssum* cf. *linifolium* Steph. ex Willd.	X	X	.	X	.	.	.	.	.	.
Alyssum sibiricum Willd.	.	.	.	.	.	.	.	.	X	X
Alyssum (nm 1648)	X	.	.	.	.	.	.	X	X	.
Alyssum (nm 2393)	X	.	.	.	.	.	.	.	.	.
Alyssum, various	X	.	.	.	.	.	.	X	.	X
Boreava orientalis Jaub. & Spach	.	.	.	X	.	.	.	.	.	.
Camelina rumelica Vel.	.	X	X	.	X	.	.	.	X	.
Capsella bursa-pastoris (L.) Medik.	.	.	X	.	X	.	.	.	.	.
Cardaria draba (L.) Desv.	X	.	X	X	.	.	.	.	.	.
Conringia orientalis (L.) Andrz.	.	.	.	X	.	.	.	.	.	.
Descurainia sophia (L.) Webb & Berth. ex Prantl	X	X	X	X	.	.	.	.	X	.
Erysimum repandum L.	.	.	.	.	.	.	.	X	.	.
Isatis cf. *cappadocica* Desv.	X	.	X	X	.	.	.	.	.	.
Lepidium latifolium L.	.	.	.	.	.	X	.	.	.	.

Table C2A cont'd.

Genus/nm #	Tumulus MM	City Mound	Roadside, tracks	Irrigated field edge	Ditch	Gardens	Sakarya bank	Sakarya plain	Grazed	South ridge
Lepidium perfoliatum L.	.	X	X	.	.	.	.	.	.	.
cf. *Lepidium* (nm 2682)	.	.	X	X	.	.	.	.	.	.
Matthiola longipetala (Vent.) DC. subsp. *bicornis*	X	.	.	.	.	.	.	.	.	.
Myagrum perfoliatum L.	.	.	.	.	X	.	.	.	.	.
Rapistrum rugosum (L.) All.	.	.	X	X	X	.	.	.	.	.
Sinapis arvensis L.	.	.	X	X	X	.	.	.	.	.
Sisymbrium altissimum L.	X	X	.	X	.	.	.	.	.	.
Sisymbrium sp. (nm 2482)	X	.	X	.	.	.	.	.	.	.
Brassicaceae (nm 2526)	X	.	.	.	.	.	.	.	.	.
Capparidaceae: *Capparis spinosa* L. var. *spinosa*	.	.	.	.	.	.	.	.	X	X
Caryophyllaceae: *Bufonia* cf. *virgata* Boiss.	X	X	X	.	.	.	.	.	.	.
Cerastium dichotomum L.	.	.	.	X	.	.	.	.	.	.
Dianthus zonatus Fenzl	.	.	.	.	.	.	.	.	.	X
Dianthus zonatus Fenzl/*floribundus* Boiss.	X	.	.	.	.	.	.	.	X	X
Gypsophila eriocalyx Boiss.	X	.	.	.	.	.	.	.	.	X
Gypsophila cf. *lepidioides* Boiss.	.	.	.	.	.	.	.	.	.	.
Gypsophila perfoliata L.	.	.	X	.	X	.	.	.	.	.
Gypsophila pilosa Hudson	.	.	.	X	.	.	.	.	.	.
Gypsophila viscosa Murray	X	.	.	.	.	.	.	.	.	.
Gypsophila (nm 2193)	.	.	.	.	X	.	X	.	.	.
Lychnis sp.	.	.	.	X	.	.	.	.	.	.
Minuartia anatolica (Boiss.) Woron	X	.	.	.	.	.	.	.	X	X
Minuartia hamata (Hausskn.) Mattf.	X	.	.	.	.	.	.	.	.	X
Minuartia sp.	.	.	.	.	.	.	.	.	.	.
Silene conica L.	.	X	.	.	.	.	.	.	.	.

Table C2A cont'd.

Genus/nm #	Tumulus MM	City Mound	Roadside, tracks	Irrigated field edge	Ditch	Gardens	Sakarya bank	Sakarya plain	Grazed	South ridge
Silene conoidea L.	.	.	.	X	.	.	.	.	.	.
Silene cf. *otites* (L.) Wibel	.	.	.	.	.	.	.	.	.	.
Silene cf. *subconica* Friv.	X	.	.	.	.	.	.	.	X	.
Silene supina Bieb.	X	.	.	.	.	.	.	.	.	.
Spergularia media (L.) C. Presl	.	.	.	.	.	.	X	.	.	.
Stellaria media (L.) Vill.	X	X	.	.	.	.	.	.	.	.
Velezia sp.	.	.	.	.	.	.	.	.	.	.
Caryophyllaceae (nm 2530)	X	.	.	.	.	.	.	.	.	.
Chenopodiaceae: *Atriplex laevis* C.A. Meyer	.	.	.	X	.	.	.	.	.	.
Atriplex cf. *leucoclada* Boiss.	.	.	X	.	.	.	.	.	X	.
Atriplex, various	X	X	X	X	X	.	X	X	.	.
Beta vulgaris L.	.	.	.	X	X	.	.	.	.	.
Camphorosma sp.	.	.	.	.	.	.	Eski	X	X	.
Chenopodium album L.	.	.	X	X	X	X	.	.	.	.
Chenopodium cf. *vulvaria* L.	.	.	X	.	.	X	.	X	.	.
Chenopodium, various	.	.	X	X	.	X	.	.	.	.
Kochia prostrata (L.) Schrad.	X	.	.	.	.	.	.	.	.	.
Krascheninnikovia ceratoides (...) Güldenst.	X	.	.	.	.	.	.	.	.	.
Noaea mucronata (Forssk.) Aschers. & Schewinf.	X	.	X	.	.	.	.	.	.	X
Petrosimonia sp.	.	.	.	.	.	.	.	X	.	.
Salsola laricina Pallas	X	.	.	.	.	.	.	.	X	.
Suaeda altissima (L.) Pall.	.	.	.	X	.	X	.	X	.	.
cf. *Suaeda confusa* Iljin	.	.	.	.	.	.	.	.	.	.
Chenopodiaceae, various	.	X	X	X	X	X	.	X	.	.

Table C2A cont'd.

Genus/nm #	Tumulus MM	City Mound	Roadside, tracks	Irrigated field edge	Ditch	Gardens	Sakarya bank	Sakarya plain	Grazed	South ridge
Cistaceae: *Cistus laurifolius* L.	.	.	.	.	.	.	.	.	.	.
Fumana paphlagonica Bornm. & Janchen	.	.	.	.	.	.	.	.	.	x
Helianthemum salicifolium (L.) P. Mill.	x	.	.	.	.	.	.	.	x	x
Convolvulaceae: *Convolvulus arvensis* L.	.	.	x	x	.	.	x	.	x	.
Convolvulus galaticus Rostan ex Choisy	.	.	x	.	x	x	x	.	.	.
Convolvulus scammonia L.	.	.	.	x	.	.	x	.	.	.
Convolvulus (nm 2395)	x	.	.	.	.	.	.	.	x	x
Convolvulus, various perennial	.	.	.	.	.	.	.	.	.	x
Cupressaceae: *Cupressus sempervirens* L.	.	.	.	.	.	x	.	.	.	.
Juniperus excelsa Bieb.	.	.	.	.	.	.	.	.	.	.
Juniperus oxycedrus L.	.	.	.	.	.	.	.	.	.	.
Cuscutaceae: *Cuscuta* sp.	x	.	.	.	.	.	.	x	.	.
Cyperaceae: *Cyperus* (nm 1408)	.	.	.	.	.	.	x	.	.	.
Cyperus/Scirpus nm 2130	.	.	.	.	x	.	.	.	.	.
Eleocharis mitrocarpa Steudel/*palustris* (L.) Roemer & Schultes	.	.	.	.	x	.	.	x	.	.
Scirpoides holoschoenus (L.) Sojak	.	.	.	.	.	.	x	.	.	.
Dipsaceae: cf. *Dipsacus cephalarioides* Matthews & Kupicha	.	.	.	.	.	.	.	.	.	.
Dipsacus cf. *laciniatus* L.	.	.	.	.	x	.	x	.	.	.
Scabiosa cf. *rotata* Bieb.	x	.	.	.	.	.	.	.	.	x
Scabiosa cf. *argentea* L.	x	x	.	.	.	.	.	.	.	x
Scabiosa, various	x	.	.	.	.	.	.	.	.	.
Elaeagnaceae: *Elaeagnus angustifolia* L.	.	.	.	x	.	cultiv.	.	.	.	.
Equisetaceae: *Equisetum fluviatile* L.	.	.	.	.	.	.	x	.	.	.

Table C2A cont'd.

Genus/nm #	Tumulus MM	City Mound	Roadside, tracks	Irrigated field edge	Ditch	Gardens	Sakarya bank	Sakarya plain	Grazed	South ridge
Euphorbiaceae: *Euphorbia macroclada* Boiss.	X	.	.	.	.	.	.	.	.	X
Euphorbia (nm 1681)	X	X	.	.	.	.	.	.	.	.
Euphorbia (nm 1772)	.	.	.	.	.	.	.	.	.	.
Fabaceae: *Alhagi pseudalhagi* (Bieb.) Desv.	X	.	X	X	X	.	.	X	.	X
Astracantha sp.	X	.	.	.	.	.	.	.	.	X
Astragalus asterias Stev. ex Ledeb.	X	.	.	.	.	.	.	.	.	X
Astragalus hamosus L.	.	X	X	.	.	.	.	.	X	.
Astragalus lycius Boiss.	X	.	.	.	.	.	.	.	X	X
Astragalus macrocephalus Willd.	X	.	.	.	.	.	.	.	.	.
Astragalus microcephalus Willd.	.	.	.	.	.	.	.	.	.	.
Astragalus odoratus Lam.	.	.	X	X	.	.	.	.	.	.
Astragalus triradiatus Bunge	X	.	X	.	.	.	.	.	.	.
Astragalus (nm 2347)	X	X	.	.	.	.	.	.	.	.
Cicer arietinum L.	.	.	.	.	.	.	.	.	.	.
Galega officinalis L.	.	.	X	X	X	.	Eski	.	.	.
Genista sessilifolia DC.	.	.	.	.	.	.	.	.	.	.
Glycyrrhiza echinata L.	.	.	X	.	X	X	X	.	.	.
Glycyrrhiza glabra L.	.	.	.	.	.	.	.	.	.	.
Hedysarum cappadocicum Boiss.	X	.	.	.	.	.	.	.	.	X
Hedysarum varium Willd.	X	.	.	.	.	.	.	.	X	X
Lotus corniculatus L.	.	.	X	.	X	.	Eski	.	.	.
cf. *Lotus* (nm 1845)	.	.	.	.	.	.	.	.	.	.
Medicago constricta Dur.	X	X	X	.	.	.	.	.	X	.
Medicago minima (L.) Bart.	.	.	.	.	.	.	.	.	.	.
Medicago sativa L.	.	.	X	X	X	.	.	.	.	.

Table C2A cont'd.

Genus/nm #	Tumulus MM	City Mound	Roadside, tracks	Irrigated field edge	Ditch	Gardens	Sakarya bank	Sakarya plain	Grazed	South ridge
Melilotus alba Desr.	.	.	×	.	×	.	×	.	.	.
Melilotus officinalis (L.) Desr.	.	.	×	.	×	×	.	.	.	.
Onobrychis armena Boiss. & Huet	×	.	.	.	.	.	.	.	×	×
Onobrychis tournefortii (Willd.) Desv.	×	.	.	.	.	.	.	.	.	×
Ononis spinosa L.	.	.	×	.	.	.	×	.	.	.
Trifolium resupinatum L.	.	.	.	.	.	.	×	.	.	.
Trifolium (nm 2452)	.	.	.	.	.	.	.	.	.	.
Trifolium sp.	.	.	.	.	.	.	×	.	.	.
Trigonella astroites Fisch. & Mey.	×	.	×	.	.	.	.	.	.	.
Trigonella capitata Boiss.	×	.	×	×	.	.	.	.	.	.
Trigonella coerulescens (Bieb.) Hal.	×	×	.	.	.	.	.	.	.	.
Trigonella crassipes Boiss.	×	×	.	.	.	.	.	×	.	.
Trigonella cf. *fischeriana* Ser.	×	×	.	.	.	.	.	.	.	.
Trigonella monantha C.A. Meyer	×	×	.	.	.	.	.	.	.	.
Trigonella cf. *orthoceras* Kar. & Kir.	×	×	.	.	.	.	.	.	.	.
Vicia sp.	.	.	×	.	.	.	.	.	.	.
Fagaceae: *Quercus cerris* L.	.	.	.	.	.	.	.	.	.	.
Quercus pubescens Willd.	.	.	.	.	.	.	.	.	.	.
Frankeniaceae: *Frankenia hirsuta* L.	.	.	.	.	.	.	.	.	.	.
Geraniaceae: *Erodium cicutarium* (L.) L'Hérit.	×	×	×	×	.	.	.	×	.	.
Geranium lucidum L./*rotundifolium* L.	.	.	×	.	.	.	.	.	.	.
Geranium pusillum Boiss.	.	.	×	.	.	.	.	.	.	.
Globurlariaceae: *Globularia orientalis* L.	.	.	.	.	.	.	.	.	.	.
Hypericaceae: *Hypericum origanifolium* Willd.	.	.	.	.	.	.	.	.	.	.
cf. *Hypericum scabroides* Robson et Poulter	.	.	.	.	.	.	.	.	.	×

Table C2A cont'd.

Genus/nm #	Tumulus MM	City Mound	Roadside, tracks	Irrigated field edge	Ditch	Gardens	Sakarya bank	Sakarya plain	Grazed	South ridge
Illecebraceae: *Herniaria incana* Lam.	X	X	X	.	.	.	.	.	.	.
Paronychia (nm 1699)	.	.	.	.	.	.	.	.	.	X
Scleranthus cf. *annuus* L.	X	X	.	.	.	.	.	.	X	.
Juglandaceae: *Juglans regia* L.	.	.	.	.	.	cultiv.	.	.	.	.
Juncaceae: *Juncus effusus* L.	.	.	.	.	.	.	.	.	.	.
Juncus cf. *gerardi* Loisel.	.	.	.	.	.	.	.	X	.	.
Juncus cf. *inflexus* L.	.	.	.	.	.	.	X	.	.	.
Lamiaceae: *Acinos rotundifolius* Pers.	X	X	.	.	.	.	.	.	.	.
Ajuga chamaepitys L. Schreber	X	.	.	.	.	.	.	.	X	.
Lamium amplexicaule L.	.	X	.	.	.	.	.	.	.	.
Marrubium parviflorum Fisch. et Mey.	X	.	.	.	.	.	.	.	X	X
Marrumbium vulgare L.	X	.	.	X	.	X	.	X	.	.
Marrubium, various	.	.	X	.	.	.	.	.	.	.
Mentha aquatica L.	.	.	.	.	.	.	X	.	.	.
Molucella laevis L.	.	.	.	.	.	.	.	X	.	.
Nepeta congesta Fisch. & Mey.	.	.	.	.	.	.	.	.	.	.
Nepeta stricta (Banks & Sol.) Hedge & Lamond	.	.	.	.	.	.	.	.	.	.
Phlomis cf. *armenaica* Willd.	.	.	.	.	.	.	.	.	.	.
Phlomis pungens Willd.	.	.	.	X	X	.	.	X	X	.
Salvia ceratophylla L.	X	.	.	.	.	.	.	.	.	.
Salvia cryptantha Montbret & Aucher ex Bentham	.	.	.	.	.	.	.	.	.	X
Salvia (nm 2227)	.	.	.	.	.	.	.	.	.	.
Scutellaria orientalis L. subsp. *pinnatifida*	X	.	.	.	.	.	.	.	X	X
Sideritis montana L.	.	.	.	.	.	.	.	.	.	.
Stachys cretica L.	.	.	.	.	.	.	.	X	.	.

Table C2A cont'd.

Genus/nm #	Tumulus MM	City Mound	Roadside, tracks	Irrigated field edge	Ditch	Gardens	Sakarya bank	Sakarya plain	Grazed	South ridge
cf. *Stachys* (nm 2552)	.	.	.	.	.	.	.	.	.	.
Teucrium polium L.	x	.	.	.	.	.	.	.	x	x
Thymus, various	x	.	.	.	.	.	.	.	.	x
Wiedemannia orientalis Fisch. & Mey.	x	x	.	x	.	.	.	.	.	.
Ziziphora taurica Bieb.	x	.	.	.	.	.	.	.	x	x
Lamiaceae (nm 1970)	.	.	.	.	.	.	.	.	.	.
Liliaceae: *Allium rotundum* L.	x	.	x	.	.	.	.	.	.	.
Allium (nm 2380)	x	.	.	.	.	.	.	.	.	x
cf. *Asparagus tenuifolius* Lam.	.	.	.	.	.	x	x	.	.	.
cf. *Bellevalia* sp.	x	.	.	.	.	.	.	.	.	.
Linaceae: *Linum bienne* Mill.	x	.	.	.	.	.	.	.	.	x
Lythraceae: *Lythrum salicaria* L.	.	.	.	.	.	.	x	.	.	.
Malvaceae: *Hibiscus trionum* L.	.	.	.	.	.	x	.	.	.	.
cf. *Lavatera bryonifolia* Miller	.	.	.	.	.	.	x	.	.	.
Malva moschata L.	.	.	.	.	.	.	.	.	.	.
Malva sp.	.	.	.	.	.	.	.	.	.	.
Orobanchaceae: *Orobanche* sp.	x	.	.	.	.	.	.	.	.	.
Papaveraceae: *Fumaria* cf. *vaillantii* Lois.	.	x	.	.	.	x	.	.	.	.
Glaucium corniculatum (L.) Rud.	x	.	.	x	.	.	.	.	.	.
Glaucium hausknechtii Bornm. & Fedde ?	.	.	.	.	.	.	.	.	.	.
Hypecoum imberbe Sibth. & Sm.	x	x	x	.	.	.	.	.	.	.
Hypecoum pendulum L.	x	.	.	.	.	.	.	.	.	.
Papaver cf. *dubium* L.	x	x	.	.	.	.	.	.	.	.
Papaver hybridum L.	x	x	.	x	.	.	.	.	.	.
Papaver rhoeas L.	.	.	x	x	.	.	.	.	.	.

Table C2A cont'd.

Genus/nm #	Tumulus MM	City Mound	Roadside, tracks	Irrigated field edge	Ditch	Gardens	Sakarya bank	Sakarya plain	Grazed	South ridge
Papaver sp.	x	.	.	.	.	.	.	.	.	.
Roemeria hybrida (L.) DC subsp. hybrida	.	x	x	.	.	.	.	.	.	.
Pinaceae: *Cedrus libani* A. Rich	.	.	.	.	.	x	.	.	.	.
Pinus nigra Arn. subsp. *pallasiana* (Lamb.) Holmboe	.	.	.	.	.	x	.	.	.	.
Plantaginaceae: *Plantago lanceolata* L.	.	.	.	.	.	.	x	.	.	.
Plantago major L.	.	.	.	.	.	.	x	.	.	.
Plantago media L.	.	.	x	.	.	.	x	.	.	.
Plumbaginaceae: *Acantholimon* cf. *acerosum* (Willd.) Boiss.	x	.	.	.	.	.	.	.	.	x
Acantholimon sp.	.	.	.	.	.	.	.	.	.	.
Limonium lilacinum (Boiss. & Bal.) Wagenitz	.	.	x	x	x	x	.	.	.	.
Poaceae: *Aegilops bicornis* Jaub. & Spach	.	x	.	.	.	.	.	.	.	.
Aegilops cylindrica Host	x	.	x	x	x	.	.	x	.	.
Aegilops umbellulata Zhuk.	x	.	.	.	.	.	.	.	.	.
Aegilops triuncalis L./*geniculatum* Roth	x	.	.	.	.	.	.	.	.	.
Aeluropus littoralis (Gouan) Parl.	.	.	.	.	.	.	.	x	.	.
Aegilops (nm 2349)	x	.	.	.	.	.	.	.	.	.
Agropyron cristatum (L.) Gaertner subsp. *pectinatum*	x	.	.	x	.	.	.	.	x	x
cf. *Alopecurus myosuroides* Hudson	.	.	x	.	.	.	.	x	.	.
Amblyopyrum muticum (Boiss.) Eig/*Elymus* sp.	x	.	x	.	x	.	.	x	x	.
Avena sativa L.	.	.	.	x	.	.	.	.	.	.
Bothriochloa ischaemum (L.) Keng	.	.	.	.	.	.	.	.	.	.
Briza humilis M. Bieb.	x	.	.	.	.	.	.	.	.	.
Bromus cappadocicus Boiss.	x	.	.	.	.	.	.	.	x	x

Table C2A cont'd.

Genus/nm #	Tumulus MM	City Mound	Roadside, tracks	Irrigated field edge	Ditch	Gardens	Sakarya bank	Sakarya plain	Grazed	South ridge
Ceratocephalus	x	.	.	.	.	.	.	.	.	.
Consolida raveyi (Boiss.) Schröd .	x	.	.	.	.	.	.	.	.	.
Consolida orientalis (Gay) Schröd.	x	x	.	x	.	.	.	.	.	.
Consolida cf. *saccata* (Huth) Davis	x	x	x	x	.	.	x	.	.	.
Delphinium cf. *cinereum* Boiss.	.	.	.	.	.	.	.	.	x	.
Nigella arvensis L.	x	x	.	x	.	.	.	x	x	x
Nigella nigellastrum (L.) Willd.	.	.	.	.	.	.	.	.	.	x
Nigella segetalis Bieb.	.	.	.	x	x	.	.	.	.	.
Ranunculus cornutus DC.	.	.	.	.	x	.	.	.	.	.
Ranunculus muricatus L.	.	.	.	.	x	.	.	.	.	.
Resedaceae: *Reseda lutea* L.	x	x	x	x	.	.	x	x	.	x
Reseda microcarpa Müller	.	x	x	.	.	.	x	.	.	.
Rhamnaceae: *Paliurus spina-christi* Miller	.	.	.	.	.	.	.	.	.	.
Rosaceae: *Crataegus* sp.	.	.	.	x	.	.	.	.	.	.
Malus sylvestris Miller	.	.	.	.	.	cultiv.	.	.	.	.
Potentilla erecta (L.) Räuschel	.	.	.	.	.	.	.	.	.	.
Potentilla reptans L.	.	.	.	.	.	.	x	.	.	.
Prunus amygdalus [*A. communis* L.]	.	.	.	.	.	cultiv.	.	.	.	.
Prunus divaricata Ledeb.	.	.	.	.	.	.	.	.	.	.
Prunus orientalis (Miller) Koehne	.	.	.	.	.	.	.	.	.	.
Prunus (nm 1446)	.	.	.	.	.	cultiv.	.	.	.	.
Pyrus amygdaliformis Vill.	.	.	.	.	.	cultiv.	.	.	.	.
Pyrus elaegnifolia Pallas subsp. *elaeagnifolia*	.	.	.	.	.	.	.	.	.	.
Pyrus (nm 1962)	.	.	.	.	.	cultiv.	.	.	.	.

Table C2A cont'd.

Genus/nm #	Tumulus MM	City Mound	Roadside, tracks	Irrigated field edge	Ditch	Gardens	Sakarya bank	Sakarya plain	Grazed	South ridge
Rubiaceae: *Asperula*, various	x	.	.	.	.	.	.	.	.	x
Galium verum L.	x	x	x	.	x	.	Eski	.	.	.
Galium (nm 1242)	x	x	.	x	.	.	.	.	.	.
Galium (nm 2657)	x	.	.	.	.	.	.	.	.	.
Salicaceae: *Salix* sp.	.	.	.	x	.	cultiv.	x	.	.	.
Populus sp.	.	.	.	.	.	cultiv.	.	.	.	.
Scrophulariaceae: *Bungea trifida* (Vahl) C.A. Meyer	.	.	.	.	.	.	.	.	.	x
Linaria kurdica Boiss. & Hohen.	.	.	.	.	x	.	x	.	.	.
Linaria simplex (Willd.) DC.	x	.	.	.	.	.	.	.	.	.
Verbascum sp.	.	.	x	x	.	.	.	.	.	.
Veronica multifida L.	x	x	.	.	.	.	.	.	.	.
Solanaceae: *Datura stramonium* L.	.	.	.	x	.	.	.	.	.	.
Solanum dulcamara L.	.	.	.	.	.	.	x	.	.	.
Tamaricaceae: *Tamarix* sp.	.	.	x	x	.	cultiv.	Eski	x	.	.
Thymelaeaceae: *Thymelaea passerina* (L.) Cosson & Germ.	x	.	.	.	.	.	.	.	x	x
Typhaceae: *Typha* cf. *domingensis* Pers.	.	.	.	.	x	.	.	.	.	.
Ulmaceae: *Celtis* cf. *glabrata* Steven ex Planchon	.	.	.	.	.	.	.	.	.	.
Ulmus glabra Huds.	.	.	.	x	.	.	.	.	.	.
Ulmus minor Miller	.	.	.	.	.	.	.	.	.	.
Zygophyllaceae: *Peganum harmala* L.	x	x	.	.	.	.	.	.	x	.
Tribulus terristris L.	.	.	.	x	.	x	.	.	.	.
Zygophyllum fabago L.	.	.	x	.	.	.	.	.	.	.

Table C2B

Genus/nm #	N ridge, Kızlarkayası	Conglomerate outcrop	Grassy steppe	Roman road	Çile Dağı	Ankara Çay	Juniper/ pine	Dig house
Acanthaceae: *Acanthus* sp.	.	.	.	x	.	.	.	.
Amaranthaceae: *Amaranthus albus/graecizans*	.	.	.	.	.	.	.	.
Amaranthus, misc.	.	.	.	.	.	.	.	.
Anacardiaceae: *Rhus coriaria* L.	.	.	.	.	.	.	pine	.
Pistacia terebinthus L.	.	.	.	.	.	x	.	.
Apiaceae: *Bifora radians* Bieb.	.	.	.	.	.	.	.	.
Bupleurum cf. *flavum* Forssk	x	.	x	.	.	.	juniper	.
Bupleurum cf. *turcicum* Snogerup	.	.	.	.	.	.	juniper	.
Caucalis platycarpos L.	.	.	.	.	.	.	.	.
Conium maculatum L.	.	.	.	.	.	.	.	x
Daucus carota L.	.	.	.	.	.	.	.	.
cf. *Echinophora* sp.	.	.	.	.	.	.	.	.
Eryngium campestre L.	.	x	.	.	x	.	.	x
Eryngium creticum Lam.	x	.	x	.	.	.	.	.
Falcaria vulgaris Bernh.	.	.	.	.	.	.	.	.
Malabaila carvifolia Boiss. & Mal./*Peucedanum palimboides* Boiss.	.	.	.	.	.	.	.	.
Malabaila cf. *secacul* Banks & Sol.	.	.	.	.	.	.	.	.
Malabaila sp.	.	.	.	.	.	.	.	.
Opoponax sp.	.	.	.	.	.	.	.	.
Scandix sp.	.	.	.	.	.	x	.	.
Torilis leptophylla (L.) Reichb.	.	x	.	.	.	.	.	x
Turgenia latifolia (L.) Hoffm.	.	.	.	.	.	.	.	x
Asclepiadaceae: *Cynanchum acutum* L. subsp. *acutum*	.	.	.	.	.	.	.	.

Table C2B cont'd.

Genus/nm #	N ridge, Kızlarkayası	Conglomerate outcrop	Grassy steppe	Roman road	Çile Dağı	Ankara Çay	Juniper/ pine	Dig house
Asteraceae: *Achillea* cf. *biebersteinii* Afan.								
Achillea tenuifolia Lam.			x					
Achillea wilhelmsii C. Koch								
Achillea perennial (like *A. teretifolia* Willd.)				x				
Achillea, various			x				pine	
Acroptilon repens (L.) D.C.								
Anthemis (nm 1232)								
Artemisia sp.	x	x						
Calendula/Pulicaria (nm 1977)						x		
Carduus nutans L.		x						
Carduus pycnocephalus L.								
Carduus (nm 2495)								
Centaurea calcitrapa L.								
Centaurea patula DC.								
Centaurea pseudoreflexa Hayek	x							
Centaurea pulchella Ledeb.		x						
Centaurea cf. *rigida* Banks & Sol.								
Centaurea solstitialis L. subsp. *solstitialis*								x
Centaurea virgata Lam.	x	x			x			
Centaurea (nm 1227; blue)		x	x					
Centaraurea, misc.			x		x			
Chondrilla juncea L.								
Cichorium intybus L.								x
Cirsium sp.								

Table C2B cont'd.

Genus/nm #	N ridge, Kızlarkayası	Conglomerate outcrop	Grassy steppe	Roman road	Çile Dağı	Ankara Çay	Juniper/ pine	Dig house
Cnicus benedictus L.	.	.	.	.	.	.	.	.
Cousinia halysensis Hub.-Mor.	.	.	.	.	.	.	.	.
Cymbolaena griffithii (A. Gray) Wagenitz	.	×	.	.	.	.	.	.
Echinops sp.	.	.	.	.	.	.	.	.
Gundelia tournefortii L.	×	.	.	.	.	.	.	.
Helichrisus sp.	.	.	.	×	.	.	.	.
Jurinea pontica Hausskn. & Freyn ex Hausskn	.	.	.	.	.	.	.	.
Koelpinea linearis Pallas	.	.	.	.	.	.	.	.
Lactuca serriola L.	.	×	.	.	.	.	.	×
cf. *Leucocyclus* sp.	×	.	.	.	.	.	.	.
Matricaria aurea (L.) Schultz Bip.	.	.	.	.	.	.	.	.
Matricaria cf. *macrotis* Rech. fil.	.	×	.	.	.	.	.	.
Onopordum anatolicum (Boiss.) Eig	.	.	.	.	.	.	.	.
Pulicaria vulgaris Gaertner	.	.	.	.	.	.	.	.
Scorzonera laciniata L.	.	.	.	.	.	.	.	.
Scorzonera/Crepis	.	.	.	.	.	.	.	.
Senecio cf. *vernalis* Waldst. & Kit.	.	.	.	.	.	.	.	.
Taraxacum sp.	.	.	.	.	.	.	.	.
Tragopogon dubius Scop.	.	×	.	.	.	.	.	×
Xanthium spinosum L.	.	.	.	.	.	.	.	.
Xanthium strumarium subsp. *cavanillesii* (Schouw) D. Löve & P. Dansereau	.	.	.	.	.	.	.	.
Xeranthemum inapertum (L.) Miller	×	×	.	×	.	.	.	.
Asteraceae, various	.	.	.	×	×	.	.	.

Table C2B cont'd.

Genus/nm #	N ridge, Kızlarkayası	Conglomerate outcrop	Grassy steppe	Roman road	Çile Dağı	Ankara Çay	Juniper/pine	Dig house
Berberidaceae:J *Berberis* cf. *vulgaris* L.							X	
Boraginaceae:J *Anchusa arvensis* (L.) Bieb.								
Asperugo procumbens L.								
Buglossoides arvensis (L.) Johnston								
Echium italicum L.		X			X			
Heliotropium spp.								X
Lappula patula (Lehm.) Aschers. ex Gürke								
Moltkia/Alkanna		X						
Nonea caspica (Willd.) G. Don		X						
Rochelia disperma (L.fil.) C.Koch								
Boraginaceae, various						Orta ağıl	Dölleme	
Brassicaceae: *Alyssum* cf. *linifolium* Steph. ex Willd.								
Alyssum sibiricum Willd.								
Alyssum (nm 1648)		X						X
Alyssum (nm 2393)								
Alyssum, various								
Boreava orientalis Jaub. & Spach								
Camelina rumelica Vel.		X						
Capsella bursa-pastoris (L.) Medik.								
Cardaria draba (L.) Desv.								X
Conringia orientalis (L.) Andrz.								
Descurainia sophia (L.) Webb & Berth. ex Prantl		X						X
Erysimum repandum L.								
Isatis cf. *cappadocica* Desv.								
Lepidium latifolium L.								

Table C2B cont'd.

Genus/nm #	N ridge, Kızlarkayası	Conglomerate outcrop	Grassy steppe	Roman road	Çile Dağı	Ankara Çay	Juniper/pine	Dig house
Lepidium perfoliatum L.								X
cf. *Lepidium* (nm 2682)								
Matthiola longipetala (Vent.) DC. subsp. *bicornis*								
Myagrum perfoliatum L.								
Rapistrum rugosum (L.) All.								
Sinapis arvensis L.								
Sisymbrium altissimum L.		X						
Sisymbrium sp. (nm 2482)		X						
Brassicaceae (nm 2526)								
Capparidaceae: *Capparis spinosa* L. var. *spinosa*	X							
Caryophyllaceae: *Bufonia* cf. *virgata* Boiss.								
Cerastium dichotomum L.								
Dianthus zonatus Fenzl								
Dianthus zonatus Fenzl/*floribundus* Boiss.	X							X
Gypsophila eriocalyx Boiss.	X		X					
Gypsophila cf. *lepidioides* Boiss.	X							
Gypsophila perfoliata L.								
Gypsophila pilosa Hudson								
Gypsophila viscosa Murray		X						
Gypsophila (nm 2193)		X						
Lychnis sp.								
Minuartia anatolica (Boiss.) Woron								
Minuartia hamata (Hausskn.) Mattf.		X						
Minuartia sp.		X						
Silene conica L.								

Table C2B cont'd.

Genus/nm #	N ridge, Kızlarkayası	Conglomerate outcrop	Grassy steppe	Roman road	Çile Dağı	Ankara Çay	Juniper/ pine	Dig house
Silene conoidea L.	.	.	.	.	.	.	.	.
Silene cf. *otites* (L.) Wibel	.	.	.	x	.	.	.	.
Silene cf. *subconica* Friv.	.	.	.	.	.	.	.	.
Silene supina Bieb.	.	.	.	.	.	.	.	.
Spergularia media (L.) C. Presl	.	.	.	.	.	.	.	.
Stellaria media (L.) Vill.	.	.	.	.	.	.	.	x
Velezia sp.	.	x	.	.	.	.	.	.
Caryophyllaceae (nm 2530)	.	.	.	.	.	.	.	.
Chenopodiaceae: *Atriplex laevis* C.A. Meyer	.	.	.	.	.	.	.	.
Atriplex cf. *leucoclada* Boiss.	.	.	.	.	.	.	.	.
Atriplex, various	.	.	.	.	.	.	.	.
Beta vulgaris L.	.	.	.	.	.	.	.	.
Camphorosma sp.	.	.	.	.	.	.	.	x
Chenopodium album L.	.	.	.	.	.	.	.	.
Chenopodium cf. *vulvaria* L.	.	.	.	.	.	.	.	x
Chenopodium, various	.	.	.	.	.	.	.	.
Kochia prostrata (L.) Schrad.	.	x	.	.	.	.	.	.
Krascheninnikovia ceratoides (L.) Güldenst.	.	.	x	.	.	.	.	.
Noaea mucronata (Forssk.) Aschers. & Schewinf.	.	.	x	.	.	.	.	x
Petrosimonia sp.	.	.	.	.	.	.	.	.
Salsola laricina Pallas	x	x	.	.	.	.	.	.
Suaeda altissima (L.) Pall.	.	.	.	.	.	.	.	.
cf. *Suaeda confusa* Iljin	.	.	.	.	.	Magara ağıl	Dölleme	.
Chenopodiaceae, various	.	.	.	.	.	Satılmış	.	x

Table C2B cont'd.

Genus/nm #	N ridge, Kızlarkayası	Conglomerate outcrop	Grassy steppe	Roman road	Çile Dağı	Ankara Çay	Juniper/ pine	Dig house
Cistaceae: *Cistus laurifolius* L.	.	.	.	.	.	.	pine	.
Fumana paphlagonica Bornm. & Janchen	x	.	.	.	.	.	.	.
Helianthemum salicifolium (L.) P. Mill.	x	x	.	.	.	.	pine	.
Convolvulaceae: *Convolvulus arvensis* L.	.	.	.	.	.	.	.	x
Convolvulus galaticus Rostan ex Choisy	.	.	.	.	.	.	.	.
Convolvulus scammonia L.	.	.	.	.	.	.	.	.
Convolvulus (nm 2395)	.	.	.	.	.	.	.	.
Convolvulus, various perennial	.	x	.	.	.	.	.	x
Cupressaceae: *Cupressus sempervirens* L.	.	.	.	.	.	.	.	.
Juniperus excelsa Bieb.	.	.	.	.	.	.	juniper	.
Juniperus oxycedrus L.	.	.	.	.	.	.	x	.
Cuscutaceae: *Cuscuta* sp.	.	.	.	.	.	.	.	.
Cyperaceaed: *Cyperus* (nm 1408)	.	.	.	.	.	.	.	.
Cyperus/Scirpus nm 2130	.	.	.	.	.	.	.	.
Eleocharis mitrocarpa Steudel/*palustris* (L.) Roemer & Schultes	.	.	.	.	.	.	.	.
Scirpoides holoschoenus (L.) Sojak	.	.	.	.	.	.	.	.
Dipsaceae: cf. *Dipsacus cephalarioides* Matthews & Kupicha	.	.	x	.	.	.	.	.
Dipsacus cf. *laciniatus* L.	.	.	.	.	.	.	.	.
Scabiosa cf. *rotata* Bieb.	.	.	.	.	.	.	.	.
Scabiosa cf. *argentea* L.	.	.	.	.	.	.	.	.
Scabiosa, various	.	.	.	.	.	.	.	.
Elaeagnaceae: *Elaeagnus angustifolia* L.	.	.	.	.	.	.	.	.
Equisetaceae: *Equisetum fluviatile* L.	.	.	.	.	.	.	.	.

Table C2B cont'd.

Genus/nm #	N ridge, Kızlarkayası	Conglomerate outcrop	Grassy steppe	Roman road	Çile Dağı	Ankara Çay	Juniper/ pine	Dig house
Euphorbiaceae: *Euphorbia macroclada* Boiss.	X	.	.	.	.	.	.	.
Euphorbia (nm 1681)	.	X	.	.	.	.	.	X
Euphorbia (nm 1772)	.	.	X	.	.	.	.	.
Fabaceae: *Alhagi pseudalhagi* (Bieb.) Desv.	.	.	.	.	.	.	.	.
Astracantha sp.	X	.	.	.	.	.	.	.
Astragalus asterias Stev. ex Ledeb.	.	.	.	.	.	.	.	.
Astragalus hamosus L.	.	X	.	.	.	.	.	X
Astragalus lycius Boiss.	.	X	.	.	.	.	.	.
Astragalus macrocephalus Willd.	.	.	.	X	.	.	juniper	.
Astragalus microcephalus Willd.	.	.	.	Çek	.	.	.	.
Astragalus odoratus Lam.	.	.	.	.	.	.	.	.
Astragalus triradiatus Bunge	.	X	.	.	.	.	.	.
Astragalus (nm 2347)	.	X	.	.	.	.	.	.
Cicer arietinum L.	.	.	.	.	cultiv.	.	.	.
Galega officinalis L.	.	.	.	.	.	.	.	.
Genista sessilifolia DC.	X	.	.	.	X	.	.	.
Glycyrrhiza echinata L.	.	.	.	.	.	.	.	.
Glycyrrhiza glabra L.	.	X	.	.	.	.	.	X
Hedysarum cappadocicum Boiss.	.	.	.	.	.	.	.	.
Hedysarum varium Willd.	.	.	.	.	.	.	.	.
Lotus corniculatus L.	.	.	.	.	.	.	.	.
cf. *Lotus* (nm 1845)	.	.	.	X	.	.	.	.
Medicago constricta Dur.	.	.	.	.	.	.	.	X
Medicago minima (L.) Bart.	.	X	.	.	.	.	.	.
Medicago sativa L.	.	.	.	.	.	.	.	.

Table C2B cont'd.

Genus/nm #	N ridge, Kızlarkayası	Conglomerate outcrop	Grassy steppe	Roman road	Çile Dağı	Ankara Çay	Juniper/pine	Dig house
Melilotus alba Desr.	.	.	.	.	.	.	.	.
Melilotus officinalis (L.) Desr.	.	.	.	.	.	.	.	.
Onobrychis armena Boiss. & Huet	x	.	.	.	x	.	.	.
Onobrychis tournefortii (Willd.) Desv.	x	.	.	.	.	.	.	.
Ononis spinosa L.	.	.	.	.	.	.	.	.
Trifolium resupinatum L.	.	.	.	.	.	.	.	.
Trifolium (nm 2452)	.	.	.	Çek	.	.	.	.
Trifolium sp.	.	.	.	.	.	.	.	.
Trigonella astroites Fisch. & Mey.	.	.	.	.	.	.	.	.
Trigonella capitata Boiss.	.	.	.	.	.	.	.	.
Trigonella coerulescens (Bieb.) Hal.	.	x	.	.	.	.	.	.
Trigonella crassipes Boiss.	.	.	.	.	.	.	.	.
Trigonella cf. *fischeriana* Ser.	.	x	.	.	.	.	.	.
Trigonella monantha C.A. Meyer	.	.	.	.	.	.	.	.
Trigonella cf. *orthoceras* Kar. & Kir.	.	.	.	.	.	.	.	.
Vicia sp.	.	.	.	.	.	.	.	.
Fagaceae: *Quercus cerris* L.	.	.	.	.	.	.	pine	.
Quercus pubescens Willd.	.	.	.	.	.	.	x	.
Frankeniaceae: *Frankenia hirsuta* L.	x	.	.	.	.	.	.	.
Geraniaceae: *Erodium cicutarium* (L.) L'Hérit.	.	x	.	.	.	.	.	x
Geranium lucidum L./*rotundifolium* L.	.	.	.	.	.	.	.	x
Geranium pusillum Boiss.	.	.	.	.	.	.	.	x
Globurlariaceae: *Globularia orientalis* L.	x	.	.	.	x	.	juniper	.
Hypericaceae: *Hypericum origanifolium* Willd.	.	x	.	.	.	.	.	.
cf. *Hypericum scabroides* Robson et Poulter	.	.	.	.	.	.	.	.

Table C2B cont'd.

Genus/nm #	N ridge, Kızlarkayası	Conglomerate outcrop	Grassy steppe	Roman road	Çile Dağı	Ankara Çay	Juniper/ pine	Dig house
Illecebraceae: *Herniaria incana* Lam.								
Paronychia (nm 1699)								
Scleranthus cf. *annuus* L.		X						
Juglandaceae: *Juglans regia* L.								
Juncaceae: *Juncus effusus* L.						X		
Juncus cf. *gerardi* Loisel.								
Juncus cf. *inflexus* L.								
Lamiaceae: *Acinos rotundifolius* Pers.								
Ajuga chamaepitys L. Schreber								
Lamium amplexicaule L.								
Marrubium parviflorum Fisch. et Mey.	X							
Marrumbium vulgare L.								
Marrubium, various						Orta ağıl	plateau	
Mentha aquatica L.								
Molucella laevis L.								
Nepeta congesta Fisch. & Mey.		X	X					
Nepeta stricta (Banks & Sol.) Hedge & Lamond		X	X					
Phlomis cf. *armenaica* Willd.					X			
Phlomis pungens Willd.					X			
Salvia ceratophylla L.		X						
Salvia cryptantha Montbret & Aucher ex Bentham								
Salvia (nm 2227)				X			Dölleme	
Scutellaria orientalis L. subsp. *pinnatifida*	X							
Sideritis montana L.	X							
Stachys cretica L.								

Table C2B cont'd.

Genus/nm #	N ridge, Kızlarkayası	Conglomerate outcrop	Grassy steppe	Roman road	Çile Dağı	Ankara Çay	Juniper/ pine	Dig house
cf. *Stachys* (nm 2552)	.	.	.	.	.	Satılmış	.	.
Teucrium polium L.	x	.	.	.	.	.	.	.
Thymus, various	.	x	.	.	x	.	.	.
Wiedemannia orientalis Fisch. & Mey.	.	.	.	.	.	.	.	.
Ziziphora taurica Bieb.	x	.	.	.	.	.	.	.
Lamiaceae (nm 1970)	.	.	.	.	x	.	.	.
Liliaceae: *Allium rotundum* L.	.	x	.	.	.	.	.	.
Allium (nm 2380)	.	.	.	.	.	.	.	.
cf. *Asparagus tenuifolius* Lam.	.	.	.	.	.	.	.	.
cf. *Bellevalia* sp.	.	.	.	.	.	.	.	.
Linaceae: *Linum bienne* Mill.	x	.	x	.	.	.	.	.
Lythraceae: *Lythrum salicaria* L.	.	.	.	.	.	.	.	.
Malvaceae: *Hibiscus trionum* L.	.	.	.	.	.	.	.	.
cf. *Lavatera bryonifolia* Miller	.	.	.	.	.	.	.	.
Malva moschata L.	.	.	.	Çek	.	Magara ağıl	.	.
Malva sp.	.	.	.	.	.	.	.	x
Orobanchaceae: *Orobanche* sp.	.	.	.	.	.	.	.	.
Papaveraceae: *Fumaria* cf. *vaillantii* Lois.	.	.	.	.	.	.	.	.
Glaucium corniculatum (L.) Rud.	.	x	x	.	.	.	.	x
Glaucium hausknechtii Bornm. & Fedde ?	x	.	.	.	.	.	.	.
Hypecoum imberbe Sibth. & Sm.	.	.	.	.	.	.	.	x
Hypecoum pendulum L.	.	x	.	.	.	.	.	.
Papaver cf. *dubium* L.	.	.	.	.	.	.	.	.
Papaver hybridum L.	.	x	.	.	.	.	.	.
Papaver rhoeas L.	.	.	.	.	.	.	.	.

Table C2B cont'd.

Genus/nm #	N ridge, Kızlarkayası	Conglomerate outcrop	Grassy steppe	Roman road	Çile Dağı	Ankara Çay	Juniper/ pine	Dig house
Papaver sp.	.	.	.	.	.	.	.	.
Roemeria hybrida (L.) DC subsp. hybrida	.	.	.	.	.	.	.	x
Pinaceae: *Cedrus libani* A. Rich	.	.	.	.	.	.	.	.
Pinus nigra Arn. subsp. *pallasiana* (Lamb.) Holmboe	.	.	.	.	.	.	pine	.
Plantaginaceae: *Plantago lanceolata* L.	.	.	.	.	.	.	.	.
Plantago major L.	.	.	.	.	.	.	.	.
Plantago media L.	.	.	.	.	.	.	.	.
Plumbaginaceae: *Acantholimon* cf. *acerosum* (Willd.) Boiss.	x	.	.	x	.	.	.	.
Acantholimon sp.	x	.	.	.	.	.	juniper	.
Limonium lilacinum (Boiss. & Bal.) Wagenitz	.	.	.	.	.	.	.	.
PoaceaeJ *Aegilops bicornis* Jaub. & Spach	.	.	.	.	.	.	.	.
Aegilops cylindrica Host	.	.	.	.	.	.	.	.
Aegilops umbellulata Zhuk.	.	.	.	.	.	.	.	.
Aegilops triuncalis L./*geniculatum* Roth	.	.	.	.	x	.	.	.
Aeluropus littoralis (Gouan) Parl.	.	.	.	.	.	.	.	.
Aegilops (nm 2349)	.	x	.	.	.	.	.	.
Agropyron cristatum (L.) Gaertner subsp. pectinatum	.	x	x	.	.	.	.	.
cf. *Alopecurus myosuroides* Hudson	.	.	.	.	.	.	.	.
Amblyopyrum muticum (Boiss.) Eig/*Elymus* sp.	.	.	x	.	.	.	.	.
Avena sativa L.	.	.	.	.	.	.	.	.
Bothriochloa ischaemum (L.) Keng	.	.	.	Çek	.	.	.	.
Briza humilis M. Bieb.	x	.	.	.	.	.	.	.
Bromus cappadocicus Boiss.	x	.	x	.	.	.	.	.

Table C2B cont'd.

Genus/nm #	N ridge, Kızlarkayası	Conglomerate outcrop	Grassy steppe	Roman road	Çile Dağı	Ankara Çay	Juniper/ pine	Dig house
Bromus japonicus Thunb. ssp. *anatolicus*	.	X	.	.	.	.	.	.
Bromus tectorum L.	.	X	.	.	.	.	.	.
Bromus tomentellus Boiss.	.	.	.	.	.	.	.	.
Calamagrostis epigejos (L.) Roth	.	.	.	.	.	.	.	.
Chrysopogon gryllus (L.) Trin.	.	X	.	X	X	.	pine	.
Cynodon dactylon (L.) Pers.	.	.	.	.	.	.	.	X
Echinaria capitata (L.) Desf.	.	.	.	.	.	.	.	.
Echinochloa crus-galli (L.) P. Beauv.	.	.	.	.	.	.	.	.
Elymus cf. *hispidus* (Opiz) Melderis	.	X	.	.	.	.	.	.
cf. *Eragrostis* sp.	.	.	.	.	.	.	.	.
Eremopyrum bonaepartis (Sprengel) Nevski	.	.	.	.	.	.	.	X
Eremopyrum triticeum (Gaertn.) Nevski	.	.	.	.	.	.	.	.
Festuca ovina L.	.	X	X	.	X	.	pine	.
Hordeum bulbosum L.	.	.	.	X	.	.	.	.
Hordeum distichum L.	.	.	.	.	.	.	.	.
Hordeum geniculatum All.	.	.	.	.	.	.	.	.
Hordeum murinum L.	.	X	.	.	.	.	.	.
Koeleria nitida-like (nm 1593)	.	X	.	.	.	.	.	.
Leymus sp.	.	.	.	.	.	.	.	.
Lolium sp.	.	.	.	.	.	.	.	.
Melica ciliata L.	.	.	.	X	X	.	.	.
Pennisetum orientale Rich.	.	.	.	Çek	.	.	.	.
Phalaris arundinacea L.	.	.	.	.	.	.	.	.
Phleum pratense L.	.	.	.	.	.	.	.	.
Phleum cf. *boissieri* Bornm.	.	X	X	.	.	.	.	.

Table C2B cont'd.

Genus/nm #	N ridge, Kızlarkayası	Conglomerate outcrop	Grassy steppe	Roman road	Çile Dağı	Ankara Çay	Juniper/ pine	Dig house
Phragmites australis (Cav.) Trin. ex Stendel	.	.	.	.	.	.	.	.
Poa bulbosa L.	.	.	.	.	.	.	.	.
Poa bulbosa L., proliferous	.	x	.	.	.	.	.	.
Polypogon monspeliensis (L.) Desf.	.	.	.	.	.	.	.	.
Sclerochloa sp.	.	.	.	.	.	.	.	.
Secale cereale L.	.	.	.	.	.	.	.	.
Setaria verticillata (L.) P. Beauv.	.	.	.	.	.	.	.	.
Stipa arabica Trin. et Rupr.	x	.	x	.	.	.	.	.
Stipa holoseriaca Trin.	x	x	.	.	.	.	.	.
Stipa lessingiana Trin. & Rupr.	.	.	.	.	.	.	x	.
Taeniatherum caput-medusae (L.) Nevski	.	x	x	.	.	.	.	.
Triticum aestivum L.	.	.	.	.	.	.	.	.
Triticum boeoticum Boiss. subs. *boeoticum*	.	.	.	.	x	.	.	.
Triticum durum Desf.	.	.	.	.	.	.	.	.
Triticum turgidum L.	.	.	.	.	.	.	.	.
Poaeceae, various	.	.	.	.	.	.	.	.
Polygonaceae: *Atraphaxis billardieri* Jaub. & Spach	.	.	.	.	.	Orta ağıl	.	.
Polygonum arenarium Waldst. & Kit.	.	.	.	.	.	.	.	.
Polygonum pulchellum Lois.	.	.	.	.	.	.	.	.
Rumex gracilescens Rech.	.	.	.	.	.	.	.	.
Rumex pulcher L.	.	.	.	.	.	.	.	.
Primulaceae: *Androsace maxima* L.	.	.	.	.	.	.	.	.
Ranunculaceae: *Aconitum nasutum* Fisch. ex Reichb.	.	.	.	.	.	.	.	.
Adonis, misc.	.	.	.	.	.	.	.	.

Table C2B cont'd.

Genus/nm #	N ridge, Kızlarkayası	Conglomerate outcrop	Grassy steppe	Roman road	Çile Dağı	Ankara Çay	Juniper/ pine	Dig house
Ceratocephalus	x	.	.	.	.	.	.	.
Consolida raveyi (Boiss.) Schröd .	x	x	.	.	.	.	.	.
Consolida orientalis (Gay) Schröd.	.	.	.	.	.	.	.	.
Consolida cf. *saccata* (Huth) Davis	.	.	x	.	.	.	.	x
Delphinium cf. *cinereum* Boiss.	.	.	.	.	.	.	.	.
Nigella arvensis L.	x	.	.	.	.	.	.	x
Nigella nigellastrum (L.) Willd.	.	.	.	.	.	.	.	.
Nigella segetalis Bieb.	.	.	.	.	.	.	.	.
Ranunculus cornutus DC.	.	.	.	.	.	.	.	.
Ranunculus muricatus L.	.	.	.	.	.	.	.	.
Resedaceae: *Reseda lutea* L.	.	x	.	.	.	.	.	.
Reseda microcarpa Müller	.	.	.	.	.	.	.	.
Rhamnaceae: *Paliurus spina-christi* Miller	.	.	.	.	.	.	juniper	.
Rosaceae: *Crataegus* sp.	.	.	.	.	x	.	pine	.
Malus sylvestris Miller	.	.	.	.	.	.	.	x
Potentilla erecta (L.) Räuschel	.	x	.	.	.	.	.	.
Potentilla reptans L.	.	.	.	.	.	.	.	.
Prunus amygdalus [*A. communis* L.]	.	.	.	.	.	.	.	x
Prunus divaricata Ledeb.	.	.	.	.	x	.	.	.
Prunus orientalis (Miller) Koehne	.	.	.	Çek	v	.	.	.
Prunus (nm 1446)	.	.	.	.	.	.	.	.
Pyrus amygdaliformis Vill.	.	.	.	.	.	.	.	x
Pyrus elaegnifolia Pallas subsp. *elaeagnifolia*	.	.	.	.	.	.	pine	.
Pyrus (nm 1962)	.	.	.	.	.	.	.	.

Table C2B cont'd.

Genus/nm #	N ridge, Kızlarkayası	Conglomerate outcrop	Grassy steppe	Roman road	Çile Dağı	Ankara Çay	Juniper/ pine	Dig house
Rubiaceae: *Asperula*, various	.	.	.	.	.	.	.	.
Galium verum L.	.	.	.	.	.	.	.	.
Galium (nm 1242)	.	.	.	.	.	.	.	.
Galium (nm 2657)	.	.	.	.	.	.	.	.
Salicaceae: *Salix* sp.	.	.	.	.	.	.	.	.
Populus sp.	.	.	.	.	.	.	.	.
Scrophulariaceae: *Bungea trifica* (Vahl) C.A. Meyer	.	.	.	.	.	Orta ağıl	juniper	.
Linaria kurdica Boiss. & Hohen.	.	.	.	.	.	.	.	.
Linaria simplex (Willd.) DC.	.	X	.	.	.	.	.	.
Verbascum sp.	.	.	.	.	.	.	.	.
Veronica multifida L.	.	.	.	.	.	.	juniper	X
Solanaceae: *Datura stramonium* L.	.	.	.	.	.	.	.	.
Solanum dulcamara L.	.	.	.	.	.	.	.	.
Tamaricaceae: *Tamarix* sp.	.	.	.	.	.	.	.	.
Thymelaeaceae: *Thymelaea passerina* (L.) Cosson & Germ.	.	.	.	.	.	.	.	.
Typhaceae: *Typha* cf. *domingensis* Pers.	.	.	.	.	.	.	.	.
Ulmaceae: *Celtis* cf. *glabrata* Steven ex Planchon	.	.	.	.	.	.	juniper	.
Ulmus glabra Huds.	.	.	.	.	.	.	.	.
Ulmus minor Miller	.	.	.	.	X	.	.	.
Zygophyllaceae: *Peganum harmala* L.	.	X	.	.	.	.	.	.
Tribulus terristris L.	.	.	.	.	.	.	.	.
Zygophyllum fabago L.	.	.	.	.	.	.	.	.

<h1 style="text-align:center">Appendix D</h1>

<h1 style="text-align:center">Wild and Weedy Taxa: Seed Identification
and Ecological Information</h1>

Description of the Taxa

The taxa of wild and weedy plants found in the Gordion flotation samples are listed below in alphabetical order by family. The identifications of the ancient material are based on illustrations in seed atlases and archaeobotanical reports (especially W. van Zeist's many publications in the journal *Palaeohistoria*) and the comparative collection housed in the ethnobotanical laboratory at the University of Pennsylvania Museum, which includes many types collected in the spring and summer in the environs of Gordion. General comments based on personal observation around Gordion and published information about the taxa follow the seed descriptions.

Apiaceae (Umbelliferae–carrot family)

In fresh specimens, members of the Apiaceae are distinguished by general morphology, specific variations in shape, and surface. Charring destroys or distorts many features, so many of the seeds cannot be determined even to genus. Generally, members of this large, diverse family are plants of open ground.

cf. *Anthriscus*. Long drop-shaped seed; surface eroded (YH 28338: L 3.5 mm, B 1.8 mm, T 1.4 mm). Not seen growing today.

Artedia. [Fig. D1a] Distinctive flat seed in only one sample. Not seen growing today.

Bifora. Distinctive round seed with heart-shaped hilum. *Bifora radians* seen at the edge of an irrigated wheat field.

Bupleurum. Size, shape, and surface texture (rugose, but with no hint of spines) consistent with *Bupleurum*. YH-Apiaceae 6 may be *Bupleurum* based

on general size and shape, but seed coat absent. *Bupleurum turcicum* and *B. flavum* seen in uncultivated steppe.

cf. *Daucus*. [Fig. D1d] Size and shape (roughly parallel sides, hint of spines) consistent with *Daucus carota,* which is common along roadsides, irrigated field edges, and the banks of the Sakarya.

Eryngium. [Fig. D1c] Five distinctive flat seeds occur in a single sample. Both *Eryngium campestre* and *E. creticum* grow in the area today. Their leaves and inflorescences have spiny tips, and they grow in overgrazed pasture as well as on Tumulus MM.

Torilis. [Fig. D1e]; *Torilis* cf. *leptophylla*. [Fig. D1f] Similar to cf. *Daucus,* but in examples where spines have been abraded away, wavy longitudinal ridges visible (YH 30664: L 6.1 mm, B 2.0 mm, T 1.0 mm).

cf. *Turgenia*. [Fig. D1b] Seed wider and thicker at base than at apex; base of some spines visible. Seen on Tumulus MM.

Apiaceae, various

Several members of the family occur throughout the sequence in small numbers. Given the inherent variability of charred seeds, I have not described and illustrated all these types; the reader would be advised to lump them as miscellaneous Apiaceae, which tend to be plants of open ground.

YH-Apiaceae 2. [Fig. D2a; Table D1] A small seed, ridged; may not have spines. Seven measurable seeds from four samples average length 2.3 mm, breadth 1.1 mm, and thickness 0.9 mm.

YH-Apiaceae 3. A small seed, relatively flat and ridges not prominent (YH 22096: L 1.9 mm, B 1.2

mm, T 0.7 mm). The single examplar has remains of fine spines.

YH-Apiaceae 4/8 [Fig. D2b; Table D2] A plump seed; surface texture may be shiny or dull, with or without bumps indicative of spines. Designation based primarily on length and plumpness. There were 13 measurable seeds from 13 different samples, with length averaging 2.8 mm, length:breadth about 1.65, and thickness:breadth about 0.90.

YH-Apiaceae 6. May be *Bupleurum* without its seedcoat.

YH-Apiaceae 7. [Fig. D2c] See seed illustration.

YH-Apiaceae 9. A whole fruit (i.e., two attached carpels) in one sample.

YH-Apiaceae 10/Unknown 31. [Fig. D2d, e; Table D3] Similar to YH-Apiaceae 4/8, but larger and longer. Surface texture may be shiny or dull, with or without bumps indicative of spines. Identification based primarily on length and plumpness, with length being about 3.5 mm, length:breadth about 1.87, and thickness:breadth 0.88.

Asteraceae
(Compositae–daisy family)

The Asteraceae is one of the largest plant families in Turkey, with diverse genera that tend to prefer open ground. In both ancient seed samples and modern vegetation survey, they can be difficult to identify. I have been unable to collect seeds of several common genera because they ripen in the late summer or fall (notably *Cousinia halysensis, Xeranthemum inapertum*); perhaps some of the unidentified seeds belong to these uncollected genera.

cf. *Anthemis/Matricaria*. [Fig. D3a]. Differs from YH-Asteraceae 1 because it is slightly bigger. Both of these genera are common in lightly grazed steppe.

Artemisia. Along with wild thyme (*Thymus* sp.) *Artemisia* sp. is one of the most common shrubs in the overgrazed environs of Gordion.

Carthamus. [Fig. D4a]. *Carthamus* is a large seed; one seed (YH 26472) measures 4.3 x 2.5 mm. Seen growing today in ruderal habitats.

Table D1. YH-Apiaceae 2 measurements

N = 7	Range (mm)	Mean (mm)	Standard Deviation
L	1.6–2.7	2.3	0.4
B	0.8–1.3	1.1	0.2
T	0.6–1.1	0.9	0.2
L/B	2.00–2.25	2.06	0.09
T/B	0.73–0.92	0.82	0.17

Table D2. YH-Apiaceae 4/8 measurements

N = 13	Range (mm)	Mean (mm)	Standard Deviation
L	2.3–3.2	2.8	0.3
B	1.3–2.2	1.7	0.3
T	1.3–2.1	1.5	0.3
L/B	1.38–1.94	1.65	0.21
T/B	0.86–1.00	0.90	0.04

Table D3. YH-Apiaceae 10/unknown 13 measurements

N = 20	Range (mm)	Mean (mm)	Standard Deviation
L	3.1–4.3	3.5	0.3
B	1.5–2.3	1.9	0.2
T	1.3–2.2	1.7	0.3
L/B	1.60–2. 25	1.87	0.19
T/B	0.76–0.96	0.88	0.06

Centaurea. [Fig. D4b]. *Centaurea* achenes in the Gordion assemblage are very variable in size. One sample had a *Centaurea* head (capitulum). In and around Gordion today I have seen many different species of *Centaurea*. (In addition to *C. calcitrapa, C. patula, C. pseudoreflexa, C. pulchella, C. solstitialis, C. virgata*, there are six that I have been unable to determine to species.) Some have spiny leaves and calyces and some not, and seed size is quite variable. For that reason, it was not possible to categorize the seeds by size or shape, with the exception of a particularly large one with an oval hilum that is similar to *C. cyanus* or *C. depressa*.

Cirsium. [Fig. D3b] The single *Cirsium* is smooth and relatively flat with an umbo (raised part of achene apex, characteristic of the Cardueae (Davis, vol. V,

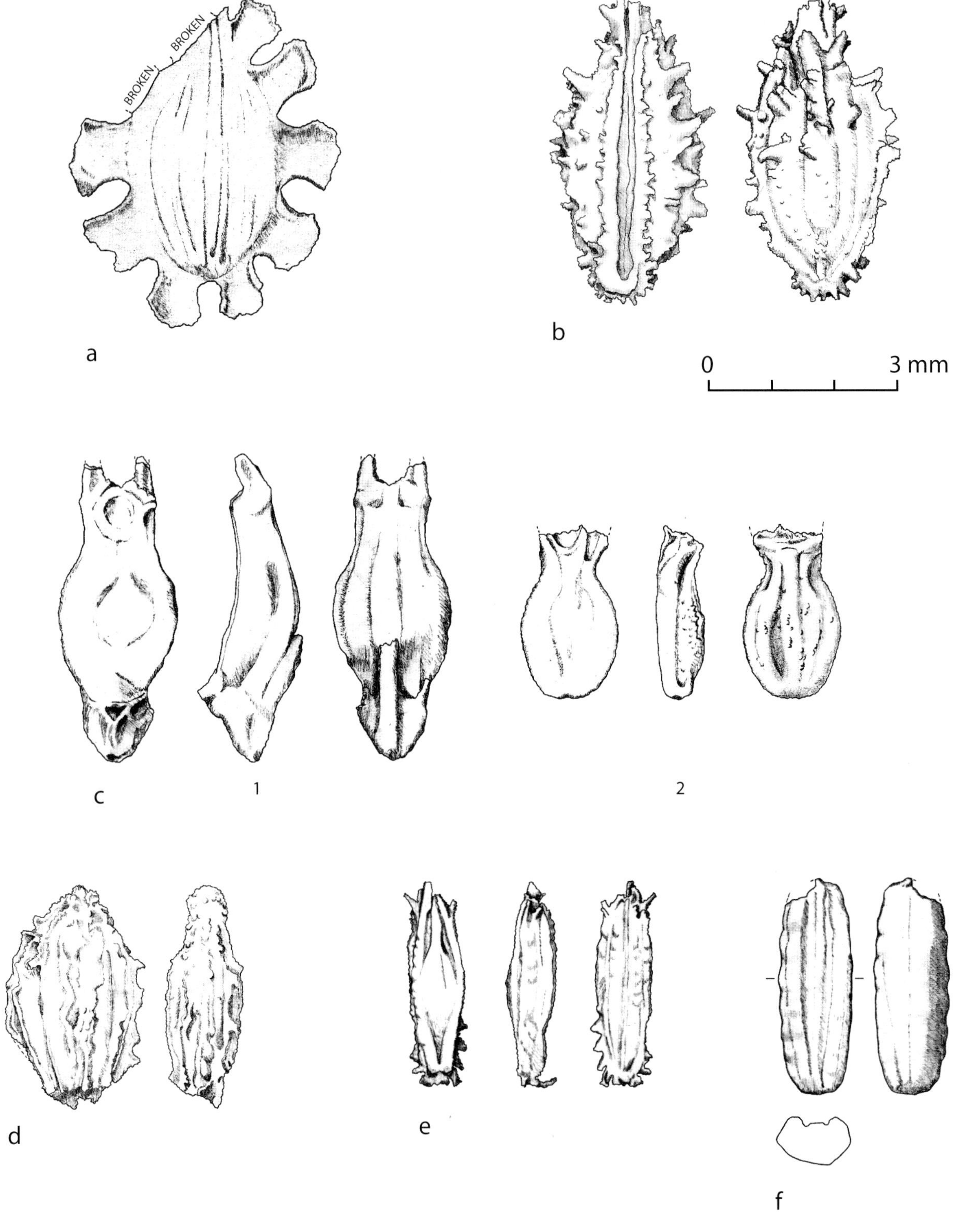

Fig. D1. a. *Artedia*; b. cf. *Turgenia*; c. *Eryngium*; d. cf. *Daucus carota*; e. *Torilis*; f. *Torilis leptophylla*.

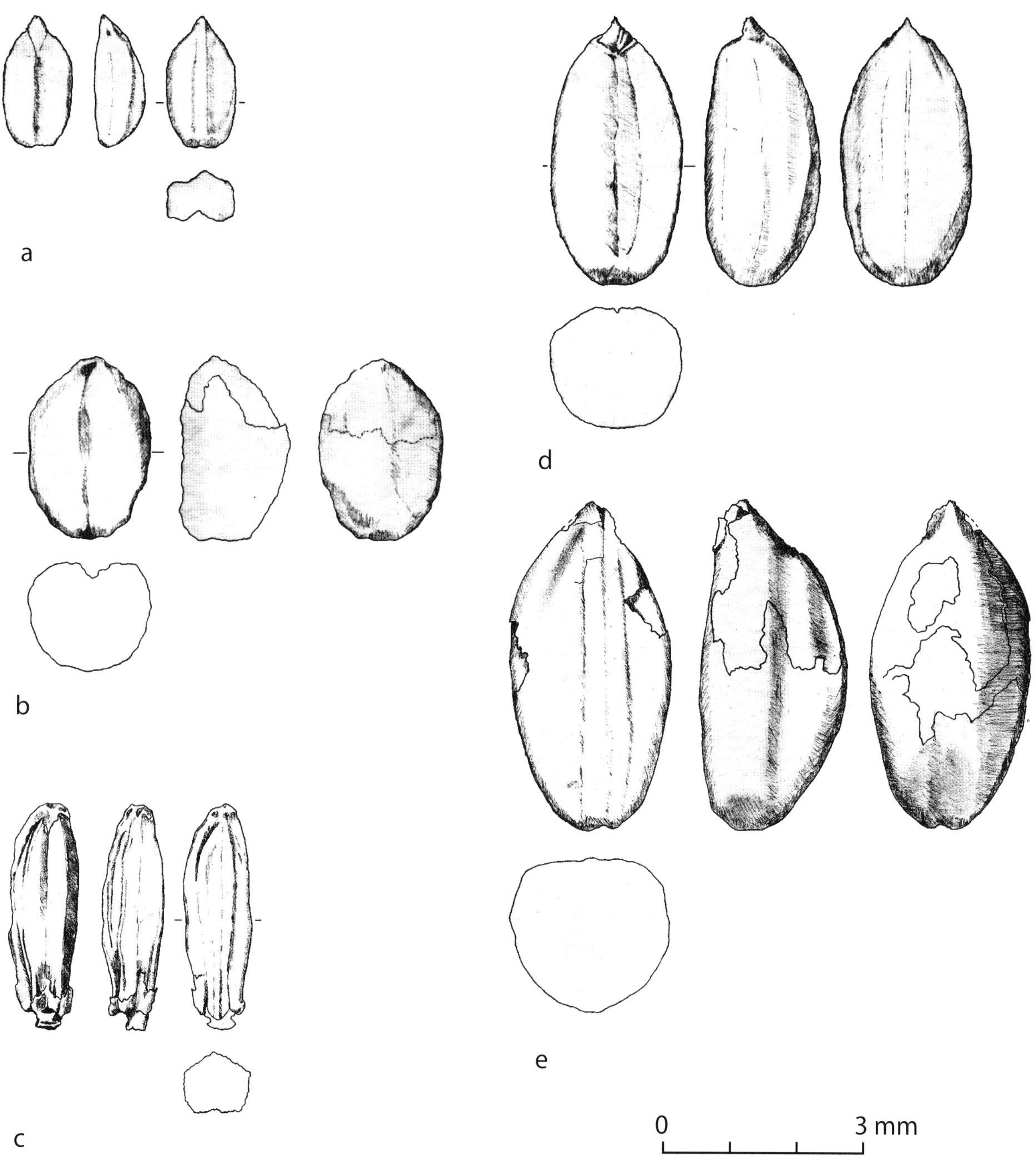

Fig. D2. a. YH-Apiaceae 2; b. YH-Apiaceae 4/8; c. YH Apiaceae 7; d, e. YH-Apiaceae 10/Unknown 31.

p. 5). *Cirsium* sp. is seen at roadsides and other disturbed ground.

cf. *Koelpinia*. [Fig. D3k] Curved with the bases of stiff bristles on the outer side. *Koelpinia linearis* is seen on Tumulus MM.

Onopordum. [Fig. D3l] A large achene; a separable ring of connate (fused) pappus hairs (see Davis, vol. 5, p. 356) is also encountered in some samples. *Onopordum anatolicum* is a prominent thistle seen on Tumulus MM and also in some poorly drained terrain.

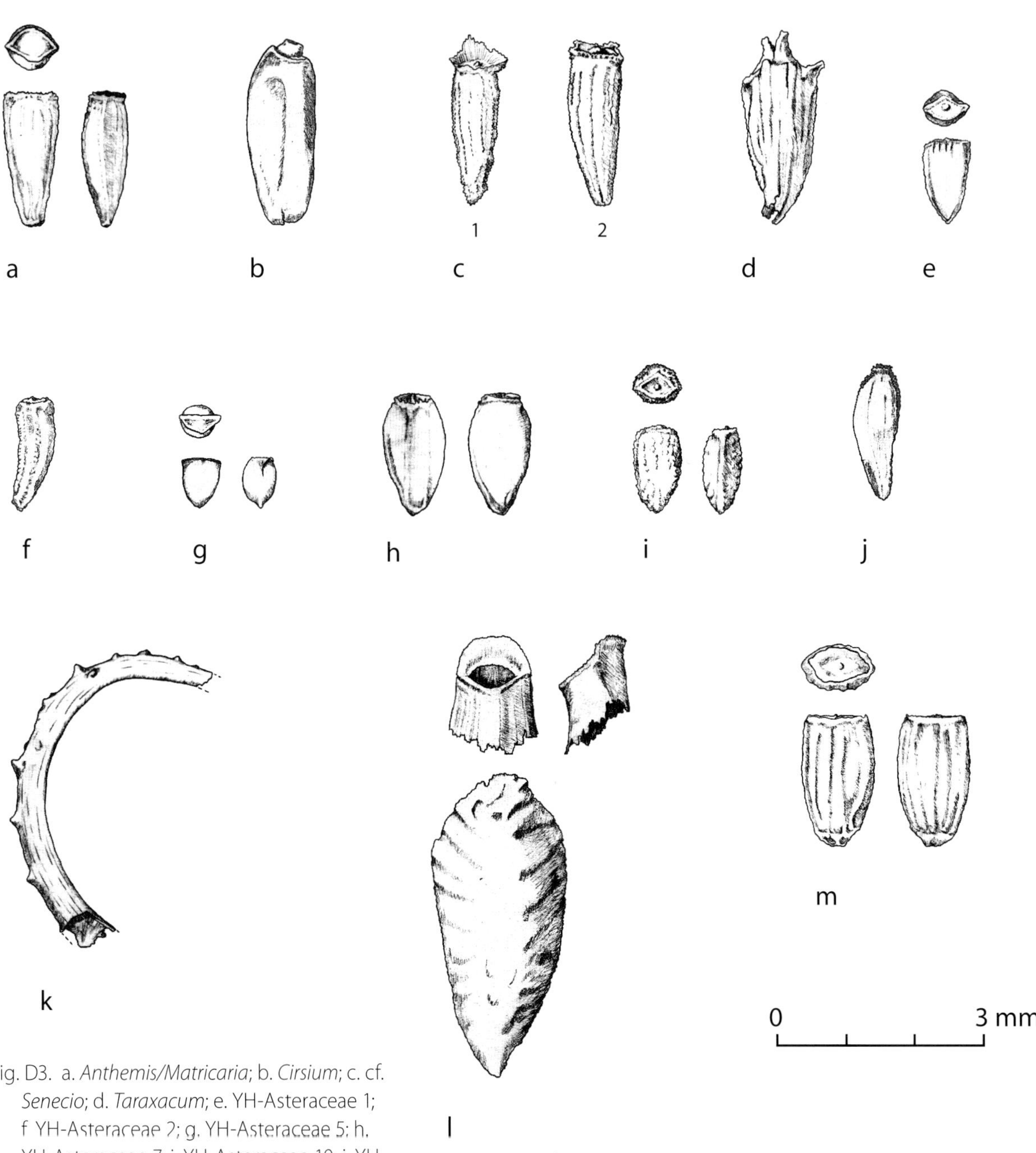

Fig. D3. a. *Anthemis/Matricaria*; b. *Cirsium*; c. cf.
Senecio; d. *Taraxacum*; e. YH-Asteraceae 1;
f. YH-Asteraceae 2; g. YH-Asteraceae 5; h.
YH-Asteraceae 7; i. YH-Asteraceae 10; j. YH-
Asteraceae 11; k. *Koelpinea*; l. *Onopordum*;
m. YH-Asteraceae 13.

cf. *Senecio*. [Fig. D3c] *Senecio* sp. is seen on Tumulus MM.

Taraxacum. [Fig. D3d] A single *Taraxacum*
achene has been identified. *Taraxacum* sp. (dandelion)
has been noticed on the Gordion Citadel Mound.

Asteraceae, various

Several members of the family occur throughout
the sequence, some in large numbers. In addition to
the seeds (actually, achenes), other parts of the flower
head have been seen: Cardueae involucres and phyllaries and several forms of receptacles. Some seeds are
illustrated and described. Given the inherent variability of charred seeds, I have not described and illustrated all these types; many could be lumped as
miscellaneous Asteraceae, which tend to be plants of

open ground. In view of the complexity and wide distribution of this family, no further interpretations are provided.

YH-Asteraceae 1. [Fig. D3e] A small, smooth achene; narrow ridge follows the edge of the seed, and it has a slightly rounded cross-section. It may be an *Achillea*. There are several *Achillea* species growing near Gordion (most common in overgrazed pasture is a small perennial, *A. wilhelmsii*, but on Tumulus MM and lightly grazed areas several bushier perennial *Achillea* species are seen).

YH-Asteraceae 2. [Fig. D3f] Unlike YH-Asteraceae 1, YH-Asteraceae 2 has small bumps arranged in longitudinal ribs and a rounder cross-section.

YH-Asteraceae 5. [Fig. D3g] A tiny seed, most probably a member of the family based on shape and apparent apex.

YH-Asteraceae 7. [Fig. D3h] Slight ridge on anterior side.

YH-Asteraceae 9. Similar to YH-Asteraceae 3. Tubercles more pronounced in general, especially on slightly ridged anterior side.

YH-Asteraceae 10. [Fig. D3i] Tubercles on longitudinal ribs, rounded cross-section.

YH-Asteraceae 11. [Fig. D3j] Only one distinctive exemplar, reminiscent of *Taraxacum* or *Sonchus*.

YH-Asteraceae 12. Only 3 examples; low ribs, smooth, rounded cross-section, about 1.5 mm long and about 0.6 mm in diameter.

YH-Asteraceae 13. [Fig. D3m] A flattish Asteraceae with shallow ribs, with only 6 designated in 3 samples.

Phyllaries (136) and tips (142) in one sample, YH 27718, that also has 14 *Centaurea* seeds (achenes), 1 *Centaurea* capitulum, 16 *Onopordum* seeds, 4 fragments of an *Onopordum* capitulum, 24 *Onopordum* "connate ring" of pappus hairs [Fig. D14]. *Centaurea* and *Onopordum* are both in the tribe Cardueae, and members of both genera could have spiny phyllaries like the ones in this sample.

Receptacles. One sample, YH 23307, has two receptacles. YH-Asteraceae plant part 1 is about 5 mm in diameter with ephemeral paleas; it is reminiscent of *Matricaria*. YH-Asteraceae plant part 2 is conical, about 2 mm in diameter and 4 mm long, with paleas "cuneate" at the base; it is consistent with *Anthemis*/*Matricaria*. Other capitula occur occasionally.

Boraginaceae (borage family)

It is now well known that many members of the Boraginaceae preserve well in uncarbonized form, and when they do char, they sometimes turn white or gray rather than black. Distinguishing modern from ancient examples presents problems. I have incorporated some gray and black seeds in the main analysis (reported with other charred seeds), and am assuming that white and tan ones, if not modern, arrived uncharred in the samples (reported with mineralized and uncharred seeds). All types identified here are herbaceous plants.

Anchusa cf. *azurea*. A single charred seed, YH 33246, YHSS 620; consistent with this species as shown in Davis, vol. 6, p. 247, fig. 8b.

Arnebia/*Lithospermum*. All but 5 of the seeds classified as *Arnebia* or *Lithospermum* are uncharred (white, gray, tan).

Asperugo. There is no reason to think these tan seeds are ancient; *Asperugo procumbens* has been seen growing profusely within the excavated area of the Citadel Mound.

Heliotropium. Only dark gray, charred seeds are included in charred seed data tables. *Heliotropium* is a common ruderal (plant of disturbed ground) near Gordion today.

cf. *Buglossoides*. Some of the nutlets are clearly charred, and some are white or tan; only the dark gray ones are included in charred seed data tables. *Buglossoides arvensis* has been seen at Gordion.

Moltkia. A few uncharred seeds of this type were encountered. *Moltkia coerulea* has been seen in disturbed steppe.

Nonea. A single uncharred seed of this type was seen. *Nonea caspica* grows on Tumulus MM and on the Citadel Mound.

Brassicaceae (Cruciferae–mustard family)

Seeds of members of the mustard family are distinguished by general morphology. Some have relatively distinctive shape and surface texture, but more often than one would prefer, one must be satisfied to identifcation at the family level. Some of the Brassicaceae siliques (seed pods) in the assemblage are distinctive. Generally, members of this large, diverse family are plants of open ground.

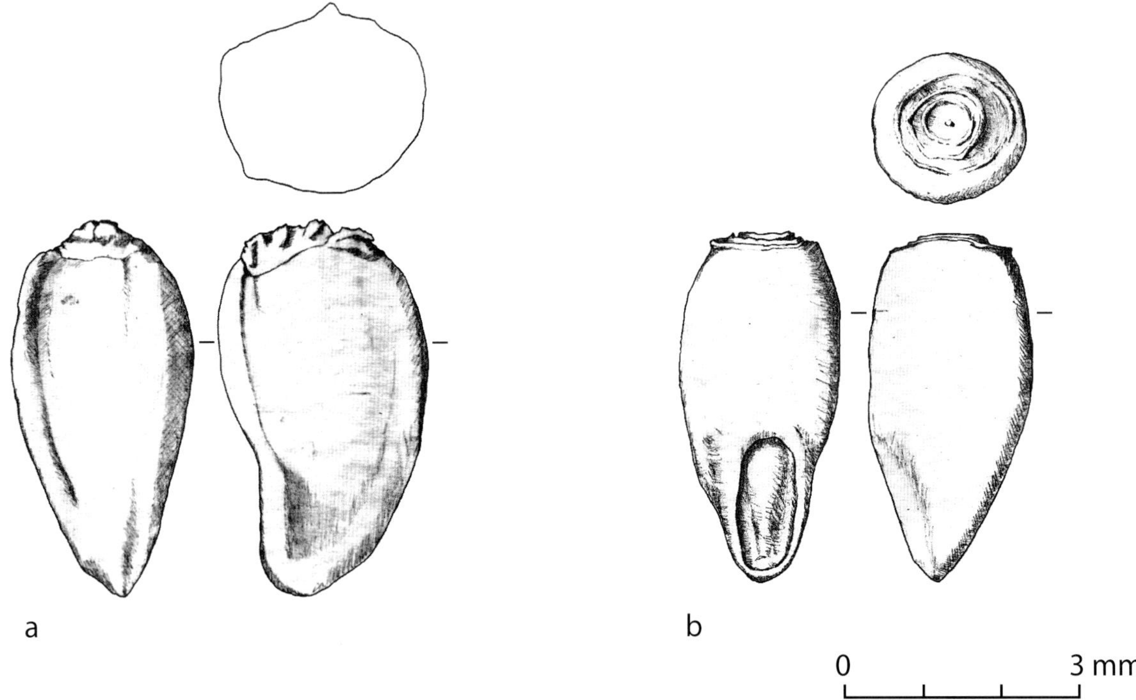

Fig. D4. a. *Carthamus*; b. *Centaurea*.

cf. *Alyssum*. A few tentatively identified *Alyssum* seeds occur in the samples. The plant *Alyssum*, however, is quite widespread on Tumulus MM as well as in waste areas. At least four species have been recognized, though not identified, growing in the area today.

Boreava orientalis. [Fig. D5o] Several siliques of this species occur in the samples. The fruit is almost spherical with a wavy ridged margin on two sides. It has some surface texture and a flat beak-like projection. The wavy margin distinguishes it from the other Turkish species, *B. aptera*. *B. orientalis* has been seen in irrigated fields near the site.

cf. *Camelina rumelica* (was YH-Brassicaceae 14). [Fig. D5a]. Three seeds from Early Iron Age context, YHSS 7, compare well to *C. rumelica* collected at Gordion in size, shape (the boundary between the radicle and the rest of the seed is pronounced), and overall surface distribution of small tubercles. Though not common today, the plant has been seen on Tumulus MM, on the Citadel Mound, and uncultivated field edges.

cf. *Camelina sativa*. The four seeds identified as cf. *Camelina sativa* are somewhat bigger than those of *C. rumelica*.

Cardaria draba. [Fig. D5b] A single example of the distinctive silique (flat inverted heart shape) occurs in YH 22192, which also has a lot of YH-Brassicaceae 3/5. *Cardaria* grows on Tumulus MM, but is widespread in ruderal habitats. *C. draba* is the only *Cardaria* species in Turkey.

Conringia (was YH-Brassicaceae 9). [Fig. D5c] *Conringia* seeds are a bit more common earlier in the sequence (Early Iron Age 8 and 7). They have a distinctive surface texture that compares well with seeds of *Conringia orientalis*, which was collected in irrigated fields near Gordion.

Euclidium syriacum. [Fig. D5k] A silique type identified as *Euclidium syriacum* makes a sporadic appearance. There is only one species in Turkey, but I have not seen it in the area.

cf. *Lepidium*. [Fig. D5d] One hundred of the 107 tentatively identified *Lepidium* seeds come from a single sample (YH 27461, YHSS 705). A typical one is about 1.8 mm long and 0.8 mm wide, lies flat with radicle to one side, and radicle curves along the edge of the seed.

Sisymbrium altissimum-type. [Fig. D5e] This is a

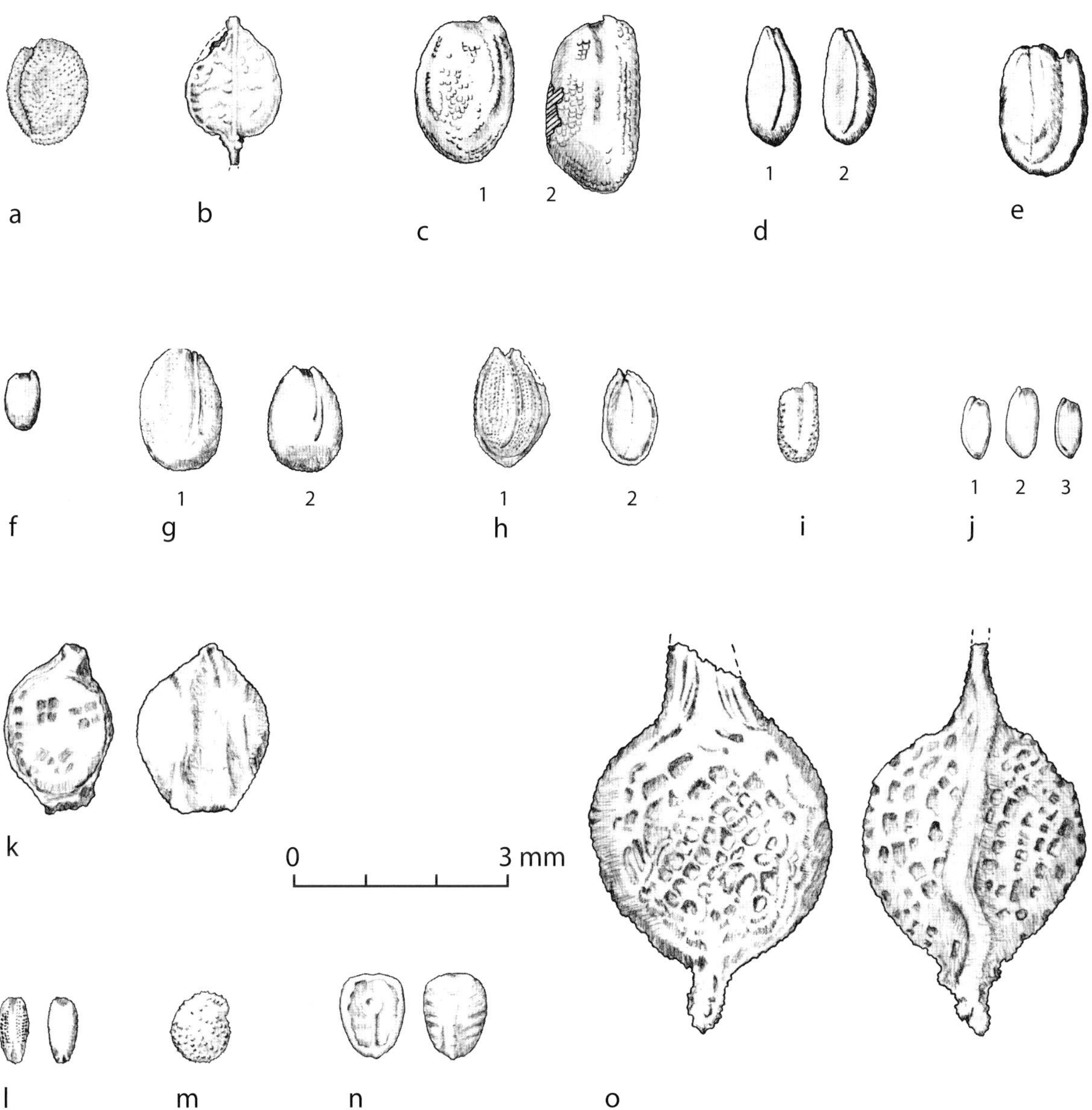

Fig. D5. a. cf. *Camelina rumelica*; b. *Cardaria draba* silique; c. *Conringia*; d. cf. *Lepidium*; e. *Sisymbrium altissimum*-type; f. YH-Brassicaceae 2; g. YH-Brassicaceae 3/5; h. YH-Brassicaceae 7; i. YH-Brassicaceae 10; j. YH-Brassicaceae 11; k. *Euclidium syriacum* silique; l. *Bufonia*; m. cf. *Cerastium*; n. YH-Caryophyllaceae 1; o. *Boreava* silique.

rather blocky seed. The radicle is pronounced. These specimens compare most closely to *S. altissimum* type in the comparative collection. *S. altissimum* is a very common plant of disturbed ground.

Thlaspi. [Fig. D7c] This type was found in a jar of flaxseed (Destruction Level), along with other Brassicaceae (e.g., Fig. D7d).

Brassicaeae, various

A variety of Brassicaceae seeds and some silique fragments have been separated out. Identifications for some are suggested, but at this point it would be better to be more cautious.

YH-Brassicaceae 2. [Fig. D5f] This small,

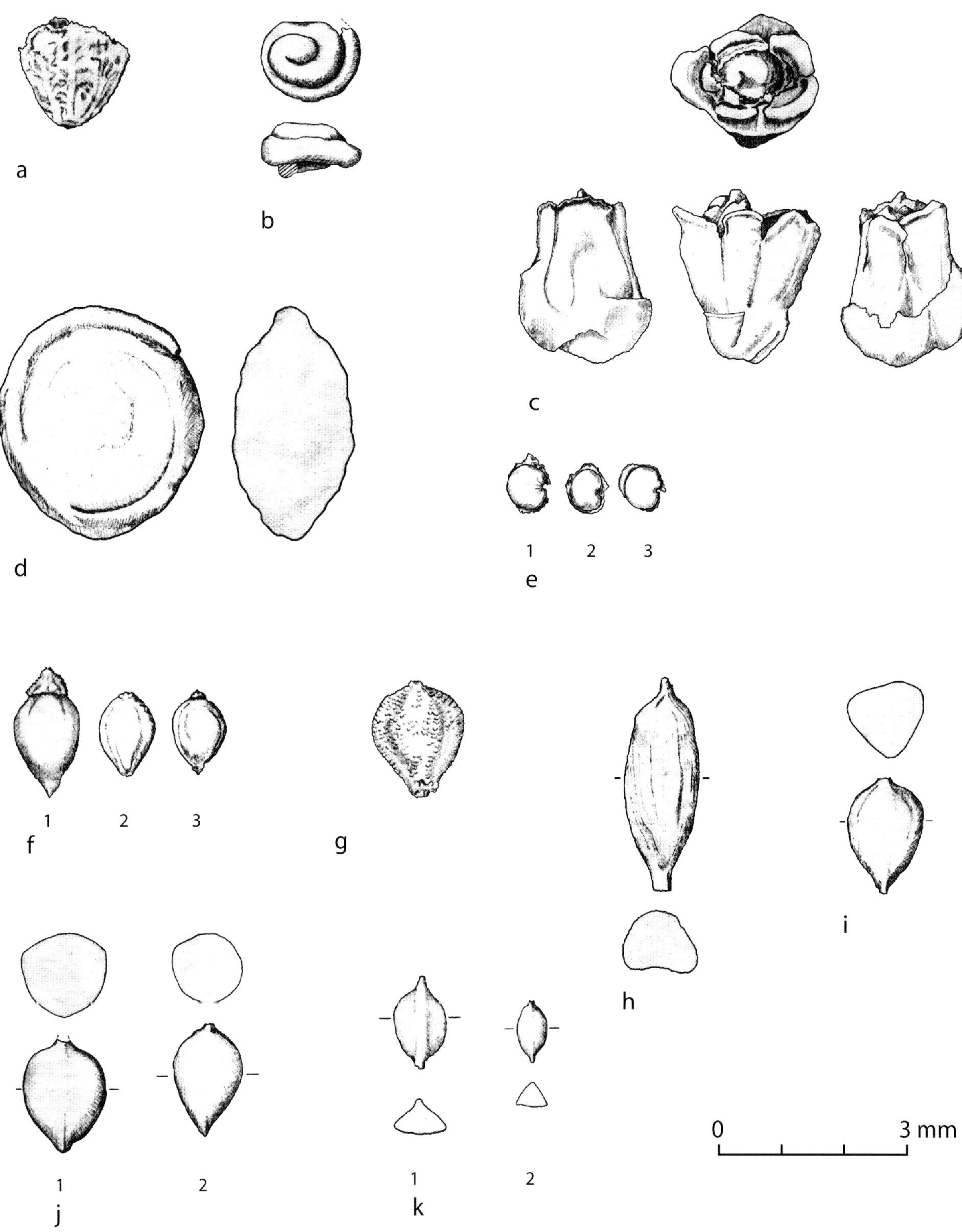

Fig. D6. a. *Atriplex* bract; b. *Salsola kali*-type; c. *Salsola/Kochia* fruit; d. cf. *Atriplex*; e. YH-Chenopodiaceae 2; f. *Eleocharis*; g. *Fimbristylis*; h. YH-Cyperaceae 3; i. YH-Cyperaceae 5; j. Polygonum (was YH-Cyperaceae 6); k. YH-Cyperaceae 7.

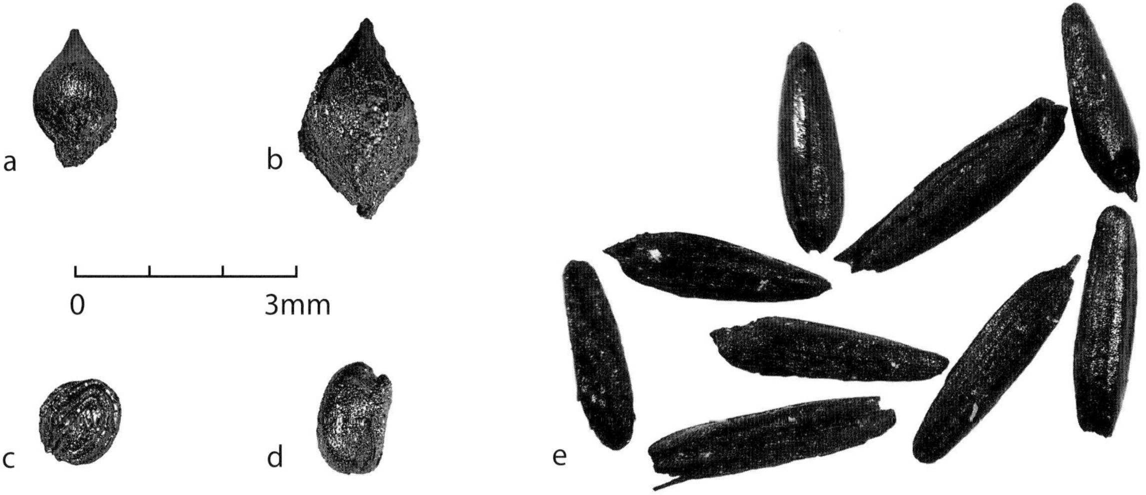

Fig. D7. a, b. *Polygonum*; c. *Thlaspi*; d. YH-Brassicaceae unspecified; e. *Eremopyrum*.

blocky Brassicaceae (about 1 mm long) is a morphological category that might include more than one genus.

YH-Brassicaceae 3/5. [Fig. D5g] This seed type is fairly numerous. Fine cell structure is visible at 30x magnification. The shape and size is consistent with *Cardaria draba,* and it is perhaps not an accident that the one sample with a *C. draba* silique also has a lot of this seed type.

YH-Brassicaceae 7. [Fig. D5h] This type compares well with *Lepidium perfoliatum* in the comparative collection. There is a flat rim around the edge; fine cell structure is visible at 30x magnification; the seed is about 1.4 to 1.6 mm long, and is flatter than cf. *Lepidium* above.

YH-Brassicaceae 10. [Fig. D5i] Another blocky Brassicaceae with fairly large tubercles.

YH-Brassicaceae 11. [Fig. D5j] Narrower than YH-Brassicaceae 2, this numerous type is a morphological category that might include more than one genus.

YH-Brassicaceae 12. Not illustrated, this type has the same general shape as YH-Brassicaceae 11, but is well under 1 mm in length. It is a morphological category that might include more than one genus.

YH-Brassicaceae silique 3. This appears to be the pedicel of a completely dehisced silique.

YH-Brassicaceae silique 4. This is consistent with *Sinapis arvensis.*

Caryophyllaceae (pink family)

Bufonia. [Fig. D5l] The seeds are a bit over 1 mm in length. The hilum is on one of the narrow sides of the relatively flat oval seed, which has nearly linear arrangement of tubercles following the perimeter. *Bufonia virgata,* a small (ca. 10 cm) delicate plant, has been seen on Tumulus MM as well as in unprotected areas.

Cerastium? [Fig. D5m] A seed type tentatively identified as *Cerastium* based on size and relatively sparse (compared to other Caryophyllaceae) distribution of tubercles. *Cerastium dichotomum* has been seen in fields around Gordion.

Gypsophila. One of the more common genera (415 seeds), *Gypsophila* is found throughout the sequence. At least five species grow in the area today: *G. eriocalyx,* a small steppe shrub that is abundant on Tumulus MM and also in unprotected steppe; two other perennials—cf. *G. lepidioides,* similar to *G. eriocalyx,* and *G. perfoliata*; and two annuals, *G. viscosa,* common on tumulus MM, and *G. pilosa,* which has been seen in fields.

Silene. Seeds identified as *Silene* are scattered throughout the sequence. I have seen at least three kinds of *Silene—Silene conoidea* in an irrigated field, *Silene subconica* on Tumulus MM, and *Silene supina,* a small shrub, also on Tumulus MM. The genus is sufficiently varied that one cannot specify its requirements and habits.

Vaccaria pyramidata. Seeds of this type are scattered throughout the sequence. The seed is spherical with small bumps; the archaeological specimens are split open on the equatorial plane. *V. pyramidata* is a field weed, and the only species of this genus in Turkey.

Caryophyllaceae, various

Many in this indeterminate category include seeds that are likely to be *Gypsophila* (beaked) or *Silene* (unbeaked). In addition, the form of several unknowns are small, flattish seeds with hilum on concave side; convex side may have parallel ridges (YH-Unknowns 14, 16, 29) or ridges not noticeably parallel (YH-Unknown 38); genera that have been considered include *Dianthus* in the Caryophyllaceae and *Veronica* in the Scrophulariaceae.

YH-Caryophyllaceae 1. [Fig. D5n] A smooth, flattish seed with hilum on concave side; *Tunica* is a possible match, though that genus is not present today.

Chenopodiaceae (goosefoot family)

Members of the Chenopodiaceae are an important component of the vegetation of the central Anatolian steppe; many are salt-tolerant. Others are common weeds of irrigated fields and gardens.

Atriplex. [bract, Fig. D6a] *Atriplex* seeds, recognized by the embryo curled around the perimeter, occur in a few samples. In addition, the distinctive bract-enclosed fruit has been recognized. *Atriplex* cf. *leucoclada* grows on the unprotected part of the Citadel Mound and it is an early colonizer of the steep baulks within the fenced part. *A. laevis* was seen in an irrigated field.

Beta. [inflorescence, Fig. D15d]. A single example occurs in a Medieval deposit. In this late period, either the wild or cultivated type could be present.

Chenopodium. The seeds designated *Chenopodium* compare well in size and surface texture to modern specimens. *Chenopodium album* has been seen on waste areas and in gardens and irrigated fields. Complete ancient charred seeds are sometimes difficult to distinguish from the modern black ones.

Salsola. Two types of *Salsola* seeds and two types of inflorescences have been recognized. Seeds that compare well with *Salsola kali* (was YH-Chenopodiaceae 1) occur in the early and later parts of the sequence. The curled embryo is visible in the seeds (Fig. D6d). *Salsola soda*-type. [Fig. D6b] In contrast to *Salsola kali*-type seeds, the embryos of *S. soda*-type seeds can be coiled (Fig. D6b).

Salsola sp. *S. kali*-type seeds are relatively flat, and *S. soda*-type look like little coiled mounds; intermediate forms (or incomplete seeds) that could be either are designated *Salsola* sp. *Salsola laricina* is a common plant in the vegetation today, but I have not been able to collect its seeds.

Salsola/Kochia inflorescence. [Fig. D6c] This inflorescence compares well with those of *S. kali* and *S. salsola,* as well as *Kochia. Kochia* seeds and enclosing bracts are more elongated than those of *Salsola.* The specimens here are pentamerous, but bilaterally symmetrical, which would suggest *Kochia.* Seeds of *Kochia* have not been identified, however, and other Chenopodiaceous fruits may be similar, too.

Salsola, formerly YH-Unknown 10. [Fig. D13m] Twenty-five of this type occur in only one Hellenistic sample (YH 21719, YHSS 380.18). It is most probably the inflorescence, with the delicate leafy bracts burned off (Mac Marston, personal communication, October 14, 2009).

Suaeda. Suaeda is relatively easy to identify (see illustration, van Zeist and Bakker-Heeres 1985:fig. 4.1). It has been seen growing as a weed in gardens and irrigated fields around Gordion. One modern specimen is *Suaeda altissima.*

Chenopodiaceae, various

The many small, lenticular seeds (ca. 1 mm diam.) listed under the family taxon have not been determined further.

YH-Chenopodiaceae 2. [Fig. D6e] This seed looks like a tiny *Chenopodium.* It has a tendency to burst on an equatorial plane.

Cistaceae (rock-rose family)

Helianthemum. Thirteen *Helianthemum* seeds from Gordion are similar in shape to those from, e.g., Sweyhat (Miller 1997:fig. 6.1a). *Helianthemum salicifolium* is a small annual herb that is very common on the lower, drier slopes of Tumulus MM.

Convolvulaceae (morning glory family)

Convolvulus. Eleven *Convolvulus* seeds have been identified. The most common species in the area today is *C. arvensis* (bindweed), an invasive perennial plant of disturbed ground, but there are at least three other perennial species that have been seen at Gordion: *C. galaticus* and *C. scammonia,* near the river, and an as yet unidentified one that compares well with *C.* cf. *aucheri* on Tumulus MM.

Cyperaceae (sedge family)

Sedges occur mainly as seeds, but a few stem fragments, recognized by a triangular cross-section, are also encountered. Sedges are underrepresented in my modern botanical collections from around Gordion.

cf. *Carex.* Seeds identified as *Carex* are among the more numerous sedges (1154). They are relatively flat, about 1.5 mm long, and the linear cell structure is commonly visible at low magnification.

YH-*Carex* 3, of which only 11 exemplars were seen, is relatively flat and has the surface texture of *Carex,* but is about 2 mm long and 1 mm wide.

Eleocharis. [Fig. D6f] The seed of *Eleocharis* was identified by comparison with fresh examples. It has a flat side and a rounded side; the rounded side has two shallow furrows. Some specimens have a cap-like structure at the apex. Some are charred black, but a greater number are gray or white. *Eleocharis mitrocarpa/palustris* has been seen in a ditch on the valley bottom.

Fimbristylis. [Fig. D6g] The seed has a distinctive surface texture. All identified specimens come from a single sample dated to YHSS 1 (YH 21728).

Cyperaceae, various

YH-Cyperaceae 1. The most numerous identified sedge (1478) is most probably *Scirpus/Cyperus* (as illustrated by van Zeist and Bakker-Heeres [1982:fig. 24.1] and others). Instead of *Scirpus maritimus* L., some archaeobotanists use the synonym *Bolboschoenus maritimus* Palla.

YH-Cyperaceae 2. This type is now listed as *Polygonum.*

YH-Cyperaceae 3. [Fig. D6h] This type has a rounded triangular cross-section (i.e., it is trigonous).

Reticulate surface texture is visible at low magnification; it compares with, for example, *Cyperus fuscus,* illustrated in Schoch et al. (1988:77).

YH-Cyperaceae 4. [see YH-Unknown 26 for description]

YH-Cyperaceae 5. [Fig. D6i] This type, about 1.8 mm long, has a rounded triangular cross-section. The seed is reminiscent of *Carex flava.*

YH-Cyperaceae 6. [Fig. D6j] This type is now listed as *Polygonum.*

YH-Cyperaceae 7. [Fig. D6k] This type is a small triquetrous seed.

YH-Cyperaceae 8. This type compares well with *Fimbristylis bisumbellata* illustrated in Townsend and Guest (1985:pl. 84); like *Fimbristylis,* all in YHSS 1.

Dipsacaceae (teasel family)

cf. *Cephalaria.* Some seeds identified as *Cephalaria* are spindle-shaped. Sometimes, the outer surface, similar to that of *Dipsacus,* is preserved, but the point that extends beyond the (outer surface) is more obvious.

cf. *Dipsacus* (was YH-Unknown 9, 9.1). [Fig. D8a, b] Most of the seeds in this category occur in a single Medieval sample, and all but one of the remainder occur in Hellenistic or Medieval contexts. The seed is roughly four-sided, with ribs at the corners and middles of each side.

Scabiosa. [Fig. D8c] A small number of this distinctive seed were seen. Two or three types grow on Tumulus MM, at the edges of the old Gordion excavations, and on the hills north and south of Yassıhöyük.

Euphorbiaceae (spurge family)

Euphorbia cf. *falcata.* [Fig. D8d] The seed in sample YH 22491 is similar to *E. falcata.* Several *Euphorbia* species grow in the area today, in a variety of habitats.

Fabaceae (Leguminosae–pea family)

Alhagi (camelthorn). [Fig. D8f, g, h] The samples contain both seeds and pod fragments. *Alhagi* is most prevalent in deposits from YHSS phases 4 and 3, and is common but less numerous in the Medieval deposits. Camelthorn is a plant of the steppe, but

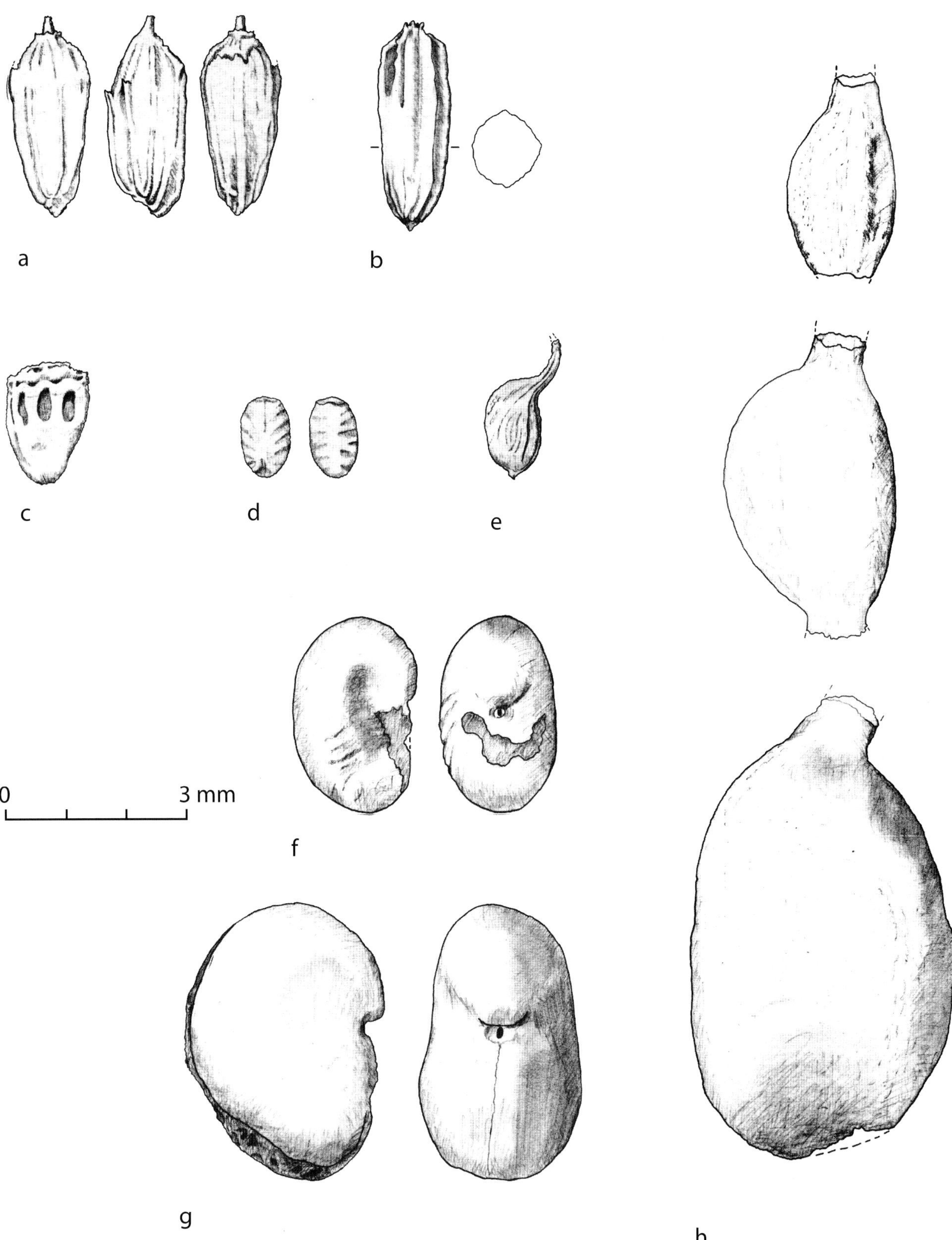

Fig. D8. a, b. cf. *Dipsacus*; c. *Scabiosa*; d. *Euphorbia*; e. *Trigonella capitata* pod; f, g. *Alhagi* seeds; h. *Alhagi* pod fragments.

particularly of disturbed soil. It has a deep taproot that is not destroyed by plowing. Today it grows primarily out in the middle of the plain and in fallow fields; it also may be regarded as an indicator of degraded pasture. Animals avoid eating it because of its spine-tipped branches, though ethnographically it is known as a fuel.

Astragalus. One of the most widespread and varied genera in the Middle East, *Astragalus* grows in many different habitats, from steppe to cultivated fields, so no ecological generalizations can be made. Some are perennial and some annual, and spiny types which have been removed from the genus, now called *Astracantha*, also grow in the area. In addition to *Astragalus hamosus, A. lycius, A. odoratus, A. triaradiatus,* at least five or six other species have been seen.

Coronilla. Only two of this small cylindrical seed were seen. Sometimes charred exemplars have a bump at the area around the hilum.

Medicago. Small kidney-shaped legume seeds are identified as *Medicago.* In the area today, *Medicago constricta* is common in recently disturbed ungrazed areas, and *M. minima* has been seen on the conglomerate outcrop across the Sakarya from Gordion.

Medicago radiata. The distinctive seed of identified as *Medicago radiata* compares well with the type illustrated in van Zeist and Bakker-Heeres (1982[1985]).

Onobrychis. When the seed can be seen through the reticulate pod it is easy to identify. Other seeds have been only tentatively assigned to *Onobrychis.*

Trifolium/Melilotus. This small, rounded legume has not been further determined. Modern members of the two genera, clover and melilot, have been seen growing in relatively moist habitats near river banks, ditches, and irrigated gardens and fields.

Trigonella. Trigonella is the single most numerous genus of wild plant seed at Gordion. It is one of the endemics of the central Anatolian steppe, and today is common on Tumulus MM; in 1988 it was also common within the fenced area of the Citadel Mound. At least 7 species have been seen growing around Gordion, including: *Trigonella astroites, T. capitata, T. coerulescens, T. crassipes, T.* cf. *fischeriana, T. monantha,* and *T.* cf. *orthoceras.*

Trigonella cf. *astroites. Trigonella* seeds with tuberculate surface are assigned a morphological species, though other species cannot be excluded.

Trigonella capitata. [Fig. D8e] This type is recognized by its pod, which compares well with that of *T. capitata.* With only one seed per pod, it appears in the tables under "seed" rather than "plant part."

Vicia. Three spherical members of the Fabaceae have been identified as *Vicia.* Other wild vetches may be part of the category "unidentified pulse."

Fabaceae, various

There are a number of seeds of small- and medium-sized legumes.

Geraniaceae (geranium family)

cf. *Erodium.* [Fig. D9a] Today, *Erodium cicutarium* is a common plant on waste areas and at the base of Tumulus MM.

cf. *Geranium.* [Fig. D9b] A seed type tentatively identified as *Geranium* occurs almost exclusively in one Hellenistic sample, YH 28338. Two species of *Geranium* have been seen at the Gordion dighouse, *G. lucidum/rotundifolium* and *G.* cf. *pusillum.*

Hypericaceae (Guttiferae–St. John's wort family)

Hypericum. A single *Hypericum* seed occurs in a Medieval sample. *H. origanifolium* has been seen on the conglomerate outcrop.

Juncaceae (rush family)

cf. *Juncus.* [Fig. D9c] A capsule filled with seeds compares well with *Juncus.* Not only is it the right size and shape, but the tiny seeds with longitudinal striations also fit. Today, *Juncus* grows in some of the poorly drained parts of the valley.

Lamiaceae (Labiatae–mint family)

Ajuga chamaepitys. [Fig. D9d] A single seed identified to species occurs in a Late Bronze Age sample, YH 31836. It has been seen growing on Tumulus MM.

cf. *Lamium amplexicaule.* Five seeds from a Late Phrygian deposit (YH 22109) are consistent with *L. amplexicaule,* seen on the Citadel Mound of Gordion.

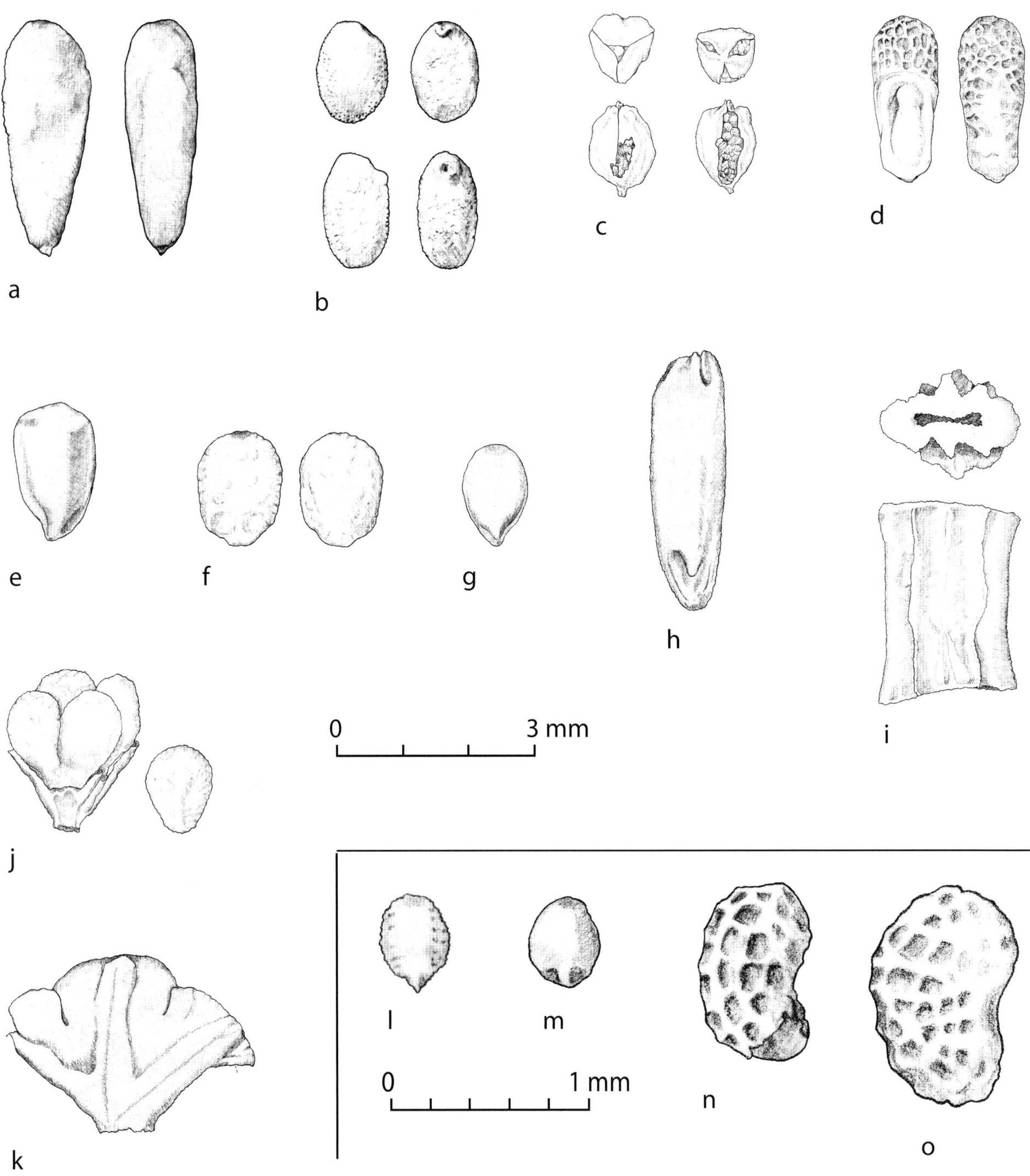

Fig. D9. a. cf. *Erodium*; b. *Geranium*; c. *Juncus* capsules with seeds; d. *Ajuga chamaepitys*; e. YH-Lamiaceae 1; f. YH-Lamiaceae 3; g. YH-Lamiaceae 5; h. YH-Lamiaceae 4; i. *Hypecoum* fruit segment; j. YH-Lamiaceae 7; k. *Papaver* disk fragment; l. *Mentha*; m. YH-Lamiaceae 6; n, o. *Papaver*.

Mentha. [Fig. D9l] Two tiny mints with cell structure visible at low magnification have been identified as *Mentha*. They come from Late Phrygian levels. *Mentha aquatica* has been seen on the banks of the Sakarya.

Teucrium. A distinctive type. Sparsely distributed in the flotation samples. It is fairly common on the overgrazed steppe around Gordion. *Teucrium polium* grows on Tumulus MM.

Ziziphora. Ziziphora is the most numerous member of the mint family in the samples. Today it is a component of the steppe vegetation, and it is quite common on the lower east and south slopes of Tumulus MM.

Lamiaceae, various

Many mints have tiny seeds that are difficult to distinguish.

YH-Lamiaceae 1. [Fig. D9e] The distal end is relatively flat.

YH-Lamiaceae 2. Small mint seeds, some of which were enclosed in the calyces (4 per flower). They measure about 0.7 mm long and 0.3 mm wide. The seeds are angular with flat sides; the distal end is flat.

YH-Lamiaceae 3. [Fig. D9f] Based on size and surface, this type is consistent with *Nepeta congesta*, which grows on undisturbed steppe in the region.

YH-Lamiaceae 4. [Fig. D9h] A long seed. The surface is distinctive, but seems to flake off. Nearly all of this type come from an Early Iron Age sample, YHSS 7 (YH27461).

YH-Lamiaceae 5. [Fig. D9g] *Marrubium* was considered but discarded as a possible identification for this seed. Unlike YH-Lamiaceae 1, the distal end is rounded. Cell structure is visible at low magnification, and sometimes the seed is encrusted with a white substance.

YH-Lamiaceae 6. [Fig. D9m] There are only a few of these small rounded seeds.

YH-Lamiaceae 7. [Fig. D9j] There are only a few of these seeds, but their surface is distinctive and 4 are still attached to each other.

Liliaceae (lily family)

A few seeds considered to be in this family are thin walled and have a hole at one end.

Linaceae (flax family)

Linum. Two seeds tentatively identified as wild flax were seen. *Linum bienne* has been seen on the north upper slopes of Tumulus MM.

Malvaceae (mallow family)

cf. *Malva*. The wedge-shaped seed of *Malva* is similar to other members of the family (e.g., *Alcea*, *Lavatera*). *Lavatera bryonifolia* grows by the river.

Papaveraceae (poppy family)

Glaucium. *Glaucium* seeds have a distinctive reticulate surface and in contrast to reniform *Papaver*, the hilum area is straight. *Glaucium corniculatum* and/or *G. hausknechtii* grow on the drier slopes of Tumulus MM, the south ridge, the conglomerate outcrop, and field edges.

Hypecoum. [Fig. D9i] The scimitar-shaped fruit of *Hypecoum* breaks cleanly into segments when dry. The species seen growing in Yassıhöyük are *H. imberbe* and *H. pendulum*.

Papaver. [Fig. D9n, o] The poppy seeds are small, presumably from uncultivated plants. The seeds occur charred, but gray and white mineralized examples are fairly common. In addition, one sample had part of the disk that tops the poppy capsule [Fig. D9k]. I cannot distinguish *Papaver* seeds from those of *Roemeria* (though unlike poppy, the fruit is an elongated capsule, more like *Glaucium*). Several poppy species grow today in uncultivated steppe as well as fields and field edges: *Papaver rhoeas*, which has a large prominent flower, and the smaller *P. hybridum* and *P. lateritium/dubium*. Much less common, *Roemeria hybrida* has been seen in protected places on overgrazed land.

Fumaria. Fumaria, sometimes put in a separate family, has a very distinctive seed: it is small, lens-shaped, with sharply defined circumference, irregular surface texture, and the hilum is a double circular depression (one on either side of the circumference). *Fumaria vaillantii* has been seen in the protected excavation area on the Citadel Mound, as well as in the watered garden at the dighouse.

Plantaginaceae (plantain family)

Plantago. *Plantago* is not particularly common in these samples. In Europe, *Plantago* pollen is consid-

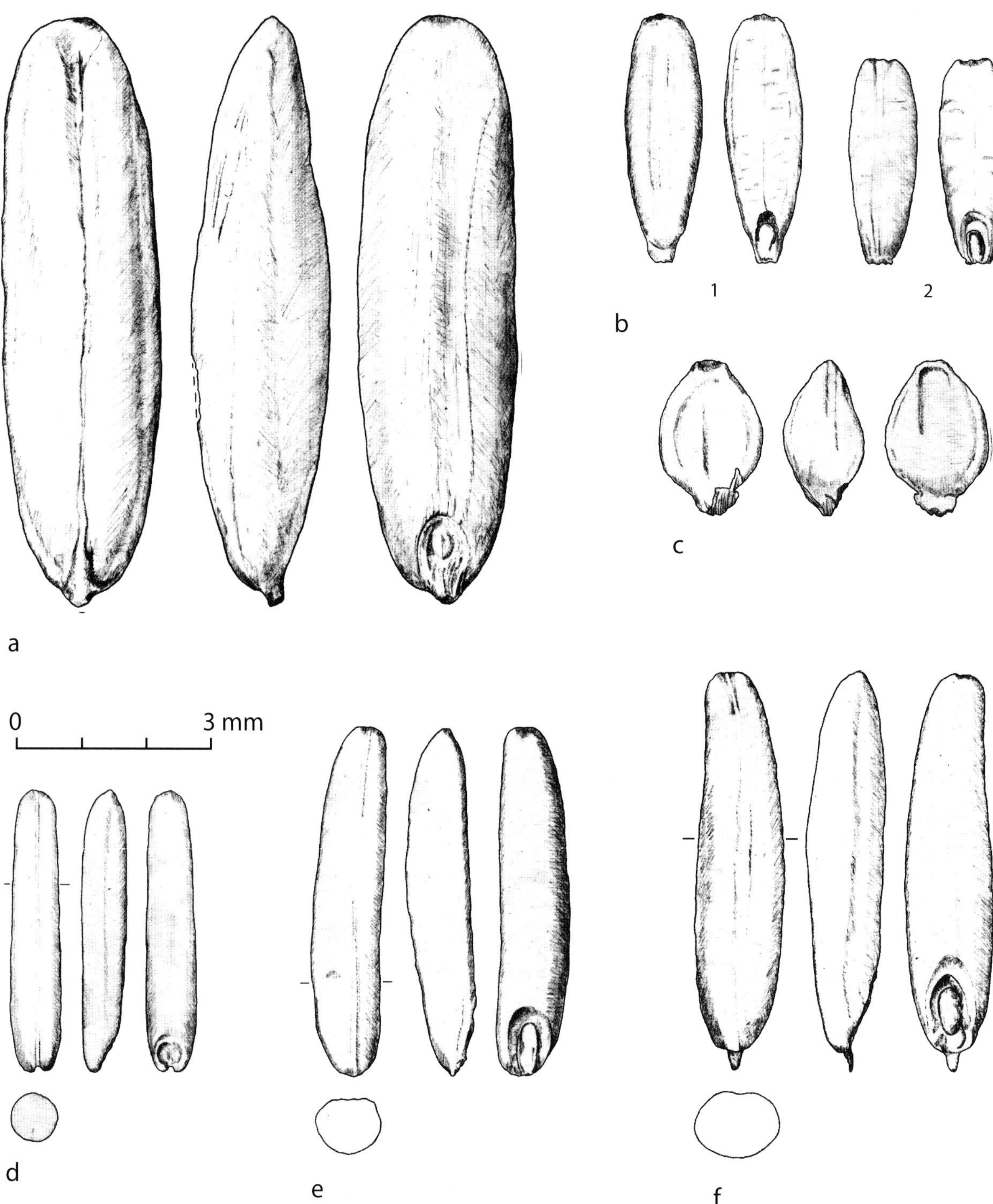

Fig. D10. a. *Avena*; b. *Hordeum murinum*-type; c. *Setaria*; d. *Stipa* (small); e. *Stipa* (medium); f. *Stipa* (large).

ered an indicator of agricultural disturbance. At Gordion, it is more likely to indicate relatively moist, grazed conditions: *Plantago lanceolata, P. major,* and *P. media* have all been seen growing along the Sakarya.

Poaceae (Gramineae—grass family)

Aegilops. Aegilops is a minor component that occurs both as seed and glume base. *Aegilops* cf. *truncialis/geniculatum* grows at the base of Tumulus MM, as does a second type, *Aegilops umbellullata,* and *Aegilops cylindrica.* The latter is common near ditches and irrigated fields.

Avena. [Fig. D10a] There are only a few oat grains. Like today, oat was probably growing as a weed in grain fields.

Bromus. Bromus is one of the most numerous identified grass genera; there are at least two morphological types in the assemblage: a long, narrow one, and a short, broad one. *Bromus* fragments are frequently recognizable, however, but cannot be distinguished further. The annuals *Bromus tectorum* and *B. japonicus* grow both in steppe and waste areas; the perennials *Bromus cappadocicus* and *B. tomentellus* are common on Tumulus MM, but are also seen on overgrazed edges of the valley bottom. The seeds of *B. cappadocicus* are long and relatively broad; on phytogeographical grounds it seems likely that some of the *Bromus* are of this type.

Bromus cf. *japonicus.* This morphological type, equivalent to van Zeist's *Bromus danthoniae* type, is relatively broad and flat.

Bromus cf. *tectorum.* This morphological type, equivalent to van Zeist's *Bromus sterilis,* is long and thin.

Eremopyrum. [Fig. D7e] *Eremopyrum* is one of the most common identified grasses in the samples. It is particularly plentiful in the grassy steppe and in steppe vegetation at the edges of fields, as well as on Tumulus MM and in the fenced excavation area of Gordion.

Hordeum. Seeds of wild barley have been identified. A few could be underdeveloped *H. vulgare.* In addition, a few rachises of wild barley (i.e., with smooth dehiscence scar) occur in some of the samples. Today, the ubiquitous annual weedy *Hordeum murinum* and less common perennial *Hordeum bulbosum* both grow in the area, so there is no reason to doubt the presence of wild barley in the samples.

Hordeum cf. *murinum.* [Fig. D10b] Archaeological examples assigned to this type are much smaller than the domesticated type. Nowadays, *Hordeum murinum,* with its spiky-awned seed-dispersal unit, is one of the most common plants on roadsides and overgrazed areas. Once the awns form, herbivores avoid eating it.

Hordeum cf. *spontaneum.* Examples of a large-seeded wild barley resemble *H. spontaneum.* They are more likely to be undeveloped *H. vulgare* than the locally available large-seeded wild barley, the perennial *H. bulbosum,* which tends to be quite flat on the distal half.

cf. *Lolium. Lolium* seeds have distinctive glume folds; identification is tentative because the plant is

Table D.4. Archaeological *Stipa* measurements

a. Treated as a single population:

Stipa	N	Width (mm) mean (range)	Length (mm) mean (range)	L/W mean (range)
Total	58	1.0 (0.6–1.4) SD = 0.2	4.8 (3.2–5.9) SD = 0.5	4.8 (3.7–6.7) SD = 0.7

b. Treated as three populations, widths 0.6 –0.9, 1.0, and 1.1–1.4 mm:

Stipa	N	Width (mm) mean (range)	Length (mm) mean (range)	L/W mean (range)
Total	18	0.8 (0.6–0.9) SD = 0.1	4.4 (3.2–5.5) SD = 0.1	5.49 (4.50–6.71) SD = 0.6

Stipa	N	Width (mm) mean (range)	Length (mm) mean (range)	L/W mean (range)
Total	9	1.0 (1.0) SD = 0	5.0 (4.5–5.5) SD = 0.35	5.0 (4.5–5.5) SD = 0.35

Stipa	N	Width (mm) mean (range)	Length (mm) mean (range)	L/W mean (range)
Total	31	1.2 (1.1–1.4) SD = 0.1	5.0 (4.2–5.9) SD = 0.4	4.4 (3.75–5.39) SD = 0.4

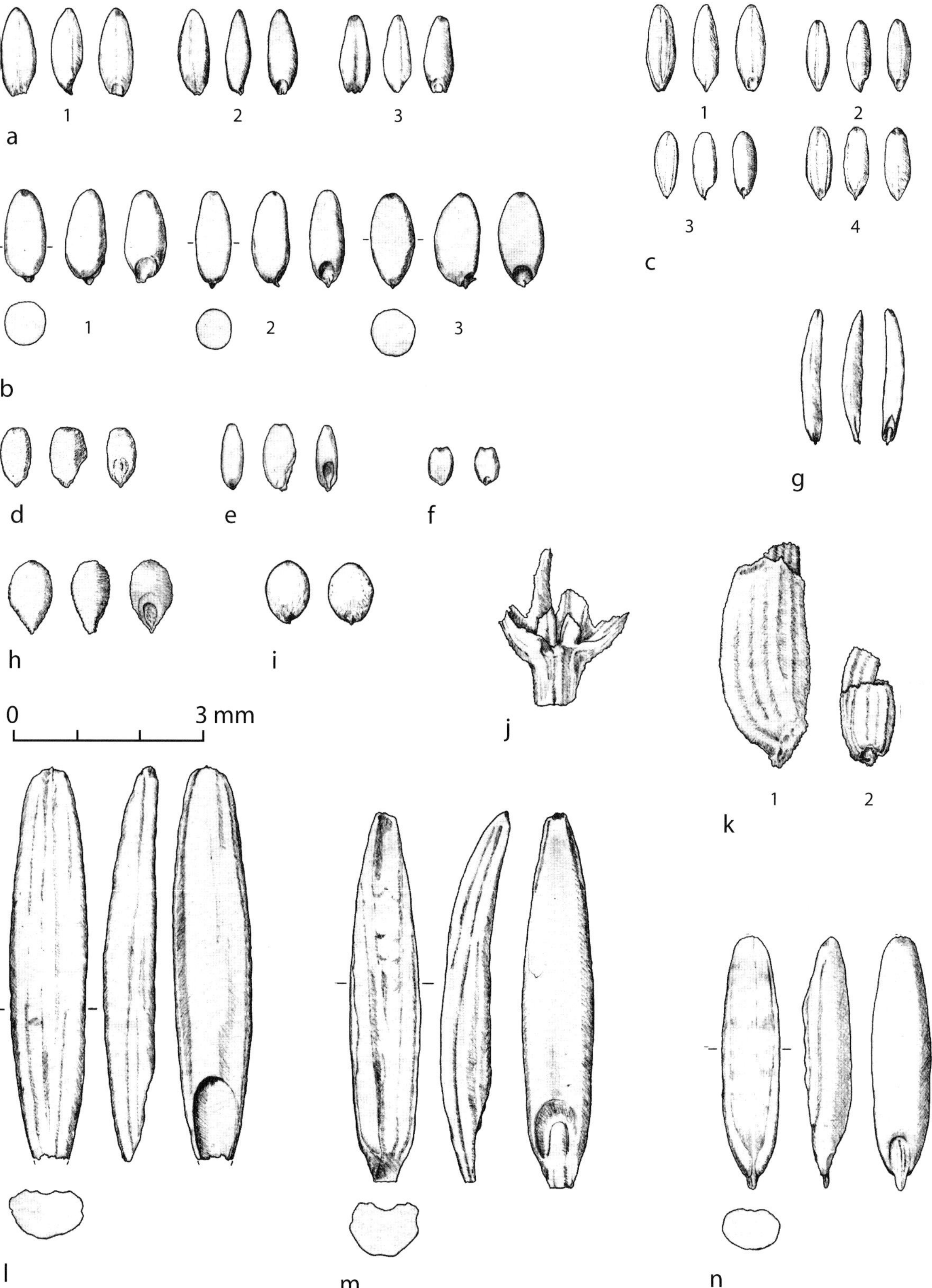

Fig. D11. a. YH-Poaceae 2; b. YH-Poaceae 13; c. YH-Poaceae 5; d. YH-Poaceae 4; e. YH-Poaceae 8; f. YH-Poaceae 10/15; g. YH-Poaceae 3; h. cf. *Phleum pratense*; i. YH-Poaceae 1; j. *Taeniatherum caput-medusae* glume base; k. *Poa bulbosa*; l, m. cf. Taeniatherum (was YH-Poaceae 7); n. YH-Poaceae 7.1.

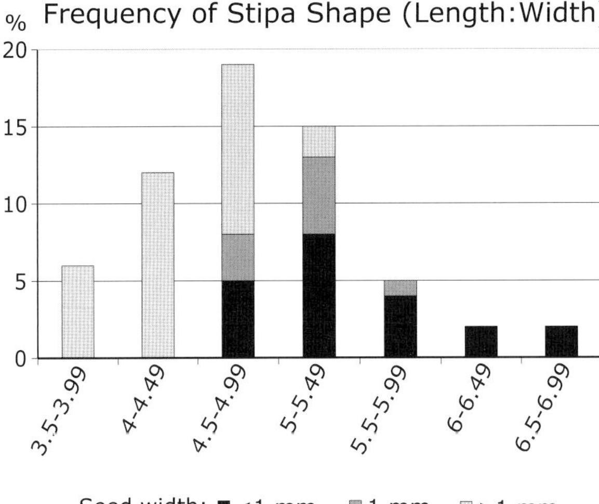

Fig. D12. Frequency of widths of *Stipa* seeds according to shape.

not that common today and there are few archaeological examples.

cf. *Phalaris.* Like *Lolium, Phalaris* is usually distinctive; identification is tentative because the plant is not that common today and there are few archaeological examples.

cf. *Phleum pratense.* [Fig. D11h] Seven of the 8 drop-shaped seeds that compare well with modern examples of *Phleum pratense* come from a Hellenistic deposit. The seeds have a wrinkled appearance. Today, this species grows in and near irrigated fields.

Phragmites. One Hellenistic sample (YH 20825, YHSS 320) had some grass stem fragments of such great diameter that they are most probably from *Phragmites,* a plant which today grows along the river, the former riverbed, and in irrigated fields near the river.

cf. *Poa bulbosa.* [Fig. D11k] No complete (i.e., unbroken) one has been seen. If these are the bulblets, the parallel ridges support an identification as a monocot, and the general size and shape are consistent with *P. bulbosa,* which may exhibit vivipary. Technically not a seed, the bulblets are formed in the inflorescence.

Setaria. [Fig. D10c] The Medieval level at Gordion has cultivated *Setaria italica. S. italica* and smaller wild *Setaria* are scattered through most of the sequence.

Stipa. [Fig. D10d, e, f; Fig. D12; Table D4] *Stipa* seeds are distinctively round in cross-section, with parallel sides. There is a tendency in the archaeological specimens for length and width to be roughly correlated (i.e., longer seeds tend to be broader, too), but the larger seeds tend to have a smaller length to breadth ratio than the smaller ones (i.e., their overall appearance is less slender). A range of shapes is illustrated here: small slender (Fig. D10d), medium (Fig. D10e), and large broad (Fig. D10f). Further study may enable us to confirm two or three overlapping morphological types. Today, *Stipa* is an important component of Tumulus MM vegetation, and is also present in overgrazed steppe. There are at least two species that grow on Tumulus MM. The seeds are about 1 mm wide, but are generally longer than the archaeological specimens: *Stipa arabica* (seeds smaller, more slender) and *S. holosericea* (seeds larger, broader). On the south ridge, *Stipa lessingiana,* with even larger seeds, has been seen.

Taeniatherum (was YH-Poaceae-7). [Fig. D11j; see also D11l, m] *Taeniatherum* is a relatively large seed. In cross-section, the glumes create a fold on either side and towards the ventral side. The ventral side flattens out towards the tip. In addition to seeds, a few rachis fragments have also been seen. *Taeniatherum caput-medusae* is an annual grass characteristic of relatively undisturbed areas (along the railroad tracks, on the conglomerate outcrop).

"Triticoid." There are a few seeds that are reminiscent of *Triticum* but quite a bit smaller.

Poaceae (small grasses), various

Grasses are generally plants of open ground, and at least when green tend to be fodder plants. They are difficult to identify, and even the numbered types mentioned below may not be true scientific taxa. If a type is consistent with a known taxon growing in the area, I mention it.

YH-Poaceae 1, cf *Eragrostis.* [Fig. D11i] This is a small round seed (commonly less than 1 mm). It is numerous and widely distributed in the samples.

YH-Poaceae 2. [Fig. D11a] Similar to YH-Poaceae 5.

YH-Poaceae 3. [Fig. D11g] Longer than YH-Poaceae 2 and 5.

YH-Poaceae 4. [Fig. D11d] Germ takes up about one-half the length of the seed, which is relatively broad.

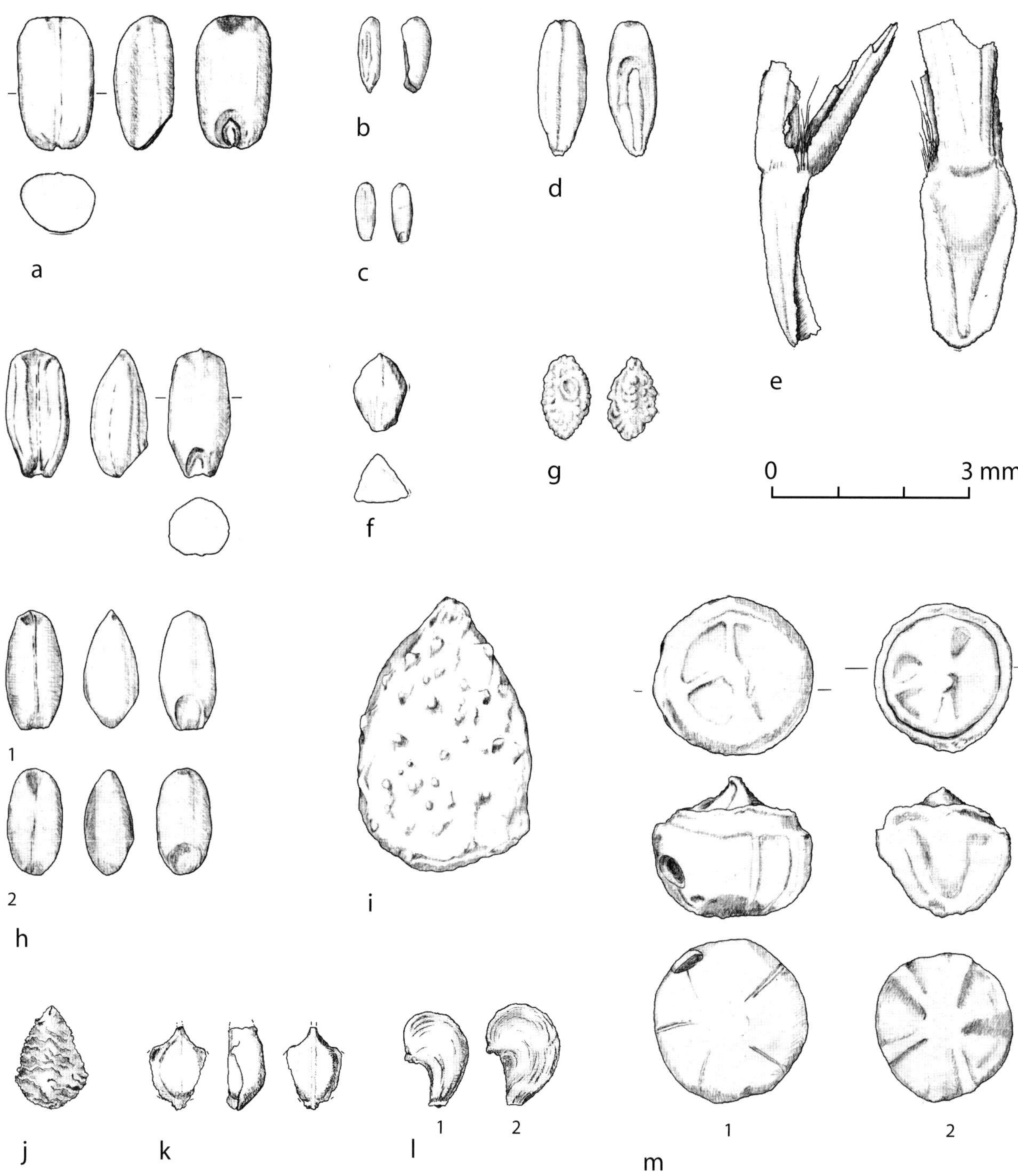

Fig. D13. a. YH-Poaceae 12; b. YH-Poaceae 14; c. YH-Poaceae 16; d. YH-Poaceae 19; e. YH-Poaceae rachis 1; f. YH-Primulaceae 1; g. YH-Primulaceae 2; h. YH-Poaceae 17/18; i. *Ranunculus arvensis*-type; j. *Aconitum/Consolida*; k. *Ceratocephalus*; l. *Ranunculus*; m. *Salsola* (was YH-Unknown 10).

YH-Poaceae 5. [Fig. D11c] See seed illustration; similar to YH-Poaceae 2.

YH-Poaceae 7.1. [Fig. D11n] Nearly all of these seeds come from a single Hellenistic sample, YH 28338 (59 of 62). They are reminiscent of *Taeniatherum caput-medusae.*

YH-Poaceae 8. [Fig. D11e] Germ is not quite one-half the length of the seed, and is thinner than YH-Poaceae 4. It occurs throughout the sequence. It is consistent with *Cynodon dactylon,* which is an aggressive perennial that propagates by deep underground runners and is not particularly good forage.

YH-Poaceae 10/15. [Fig. D11f] The glumes of this tiny (less than 0.4 mm long) seed are sometimes visible as a fold along the edge. By size and shape, the seed is consistent with *Aeluropus littoralis,* which grows on the poorly drained plain.

YH-Poaceae 11. No illustration available.

YH-Poaceae 12. [Fig. D13a] Only a single seed of this distinctive large, plump grass was seen.

YH-Poaceae 13. [Fig. D11b] These seeds compare well in size and position of embryo (i.e., dorsal side relatively flat, ventral side bulged toward base) with *Phleum paniculatum,* which, though not collected at Gordion, is in its range (Davis, vol. 9).

YH-Poaceae 14. [Fig. D13b] This is a distinctive seed. If it is a grass, the embryo extends almost the entire length of the seed.

YH-Poaceae 16. [Fig. D13] A small nondescript grass; see illustration.

YH-Poaceae 17/18. [Fig. D13] Some of these seeds retain the impression of the glume on the ventral side. They are plump, but flatten towards the tip.

YH-Poaceae 19. [Fig. D13] The germ, which extends about 3/4 of the length of the seed, is similar to the seed of *Bothriochloa ischaemum,* a perennial grass that has been seen near Çekerdeksiz.

YH-Poaceae 20/21. Most of these come from a single sample, YH 29312 (YHSS 705). Similar in size to YH-Poaceae 17/18, they do not have a fold on the ventral side. YH-Poaceae 19 is slightly concave at the tip and bulges at the base of the ventral side, whereas YH-Poacae 20 is slightly concave toward the base and bulges at the tip. They appear to come from a two-seeded floret, with the growth of the first one constraining that of the second.

YH-Poaceae rachis 1. [Fig. D13e]

Polygonaceae (knotweed family)

Polygonum. Polygonum may be underrepresented. As this manuscript went to press, Mac Marston (personal communication, October 13, 2009), demonstrated that two types classified as Cyperaceae compare very well with *Polygonum pulchella* seeds collected near Gordion. A seed first designated YH-Cyperaceae 2 is round in cross-section except for three seams at one end, where the seed has a tendency to burst. A seed first designated YH-Cyperaceae 6 (Fig. D7a, b) is about 1.5 mm long and 1.2 mm wide, is round in cross-section for most of its length. One end has 3 seams; the seeds are unburst examples that first type. Other *Polygonum* seeds may also have been misidentified as sedge. Seeds tentatively identified as *Polygonum* are wider towards the base than the tip. Some seeds with a rounded triangular cross-section that could not be oriented (i.e., it is impossible to tell whether the seed is wider at the base or the end) are designated *Polygonum*/Cyperaceae. Unlike sedge, the cell structure of *Polygonum* is not clearly visible at low magnification. I have seen *Polygonum arenarium* growing along ditches and roadsides, and *P.* cf. *pulchellum* in gardens, irrigated fields, and along the Sakarya near Gordion.

Rumex. The tetrahedral shape with sharp edges make *Rumex* easily recognized. It occurs in low numbers throughout the sequence. I have seen *Rumex gracilescens* and *R. pulcher* growing on the edges of irrigated fields near Gordion.

Portulaceae (purslane family)

Portulaca. It is not always easy to determine whether a *Portulaca* seed is ancient or modern, since both are black. I have seen *Portulaca* in gardens and moist areas near Gordion.

Primulaceae (primrose family)

Androsace. This type is roughly triangular in cross-section, and the surface has broad shallow ridges perpendicular to the side edges. Sometime a hilum is visible at the center of one edge. *Androsace maxima,* a small (ca. 10 cm high) annual, is fairly common in uncultivated steppe around Gordion, including Tumulus MM.

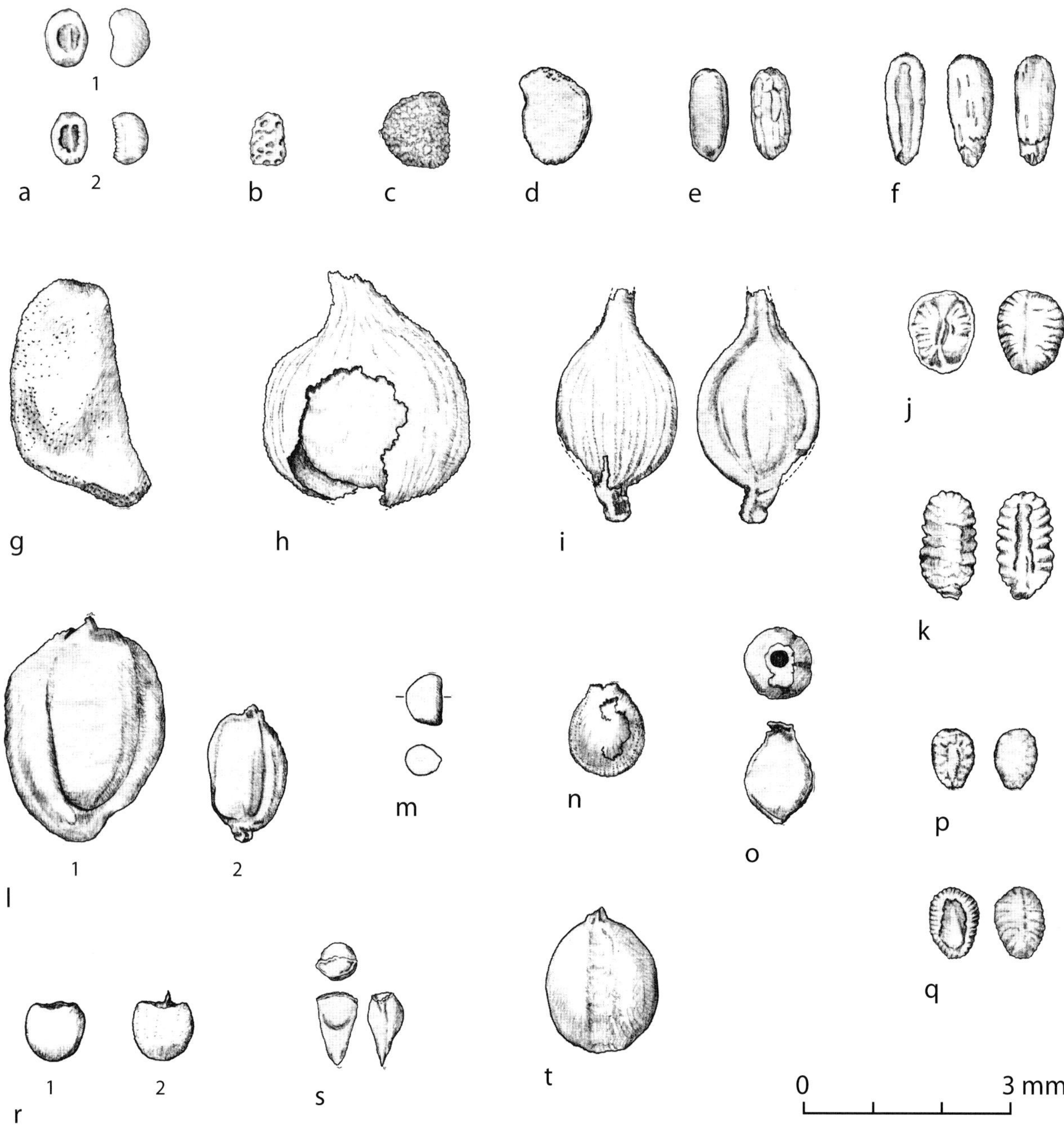

Fig. D14. a. YH-Rubiaceae 1; b. Scrophulariaceae 1; c. *Hyoscyamus*; d. *Solanum*; e. *Verbena officinalis*; f. YH-Unknown 7 (cf. *Sherardia*); g. *Zygophyllum*; h. YH-Unknown 12; i. YH-Unknown 26 (listed in Table F2 as YH-Cyperaceae 4); j. YH-Unknown 14; k. YH-Unknown 16; l. YH-Unknown 27; m. YH-Unknown 23; n. YH-Unknown 32; o. YH-Unknown 37; p. YH-Unknown 28; q. YH-Unknown 29; r. YH-Unknown 21; s. YH-Unknown 30; t. YH-Unknown 35.

Primulaceae, various

YH-Primulaceae 1. [Fig. D13f] This seed type has been tentatively assigned to the Primulaceae because of its triangular cross section. In contrast to *Androsace*, the surface is relatively smooth.

YH-Primulaceae 2. [Fig. D13] This seed type has been tentatively assigned to the Primulaceae because the hilum is on the center of one side.

Ranunculaceae (buttercup family)

cf. *Aconitum/Consolida.* [Fig. D13j] This seed is quite distinctive, pointed at one end with thin wavy ridges perpendicular to the sides. It most resembles two Ranunculaceae genera, both of which grow on uncultivated ground in the area today. *Consolida orientalis* is common; *C. raveyi,* and *C.* cf. *saccata,* and *Aconitum* cf. *nasutum* grow mostly in areas protected from grazing.

Adonis. The distinctive seeds of *Adonis* occur in low numbers throughout the sequence. *Adonis* is fairly ubiquitous and is seen on low-lying overgrazed areas.

Ceratocephalus. [Fig. D13k] A few seeds of *Ceratocephalus* occur; identification is based on comparison with fresh *C. falcatus.* The plant is inconspicuous (about 10 cm high), seen growing on Tumulus MM.

Ranunculus. [Fig. D13l] This flat asymmetrical seed has a shape reminiscent of *Ranunculus repens,* but the surface is characterized by low irregular ridges roughly parallel to the edges. The two *Ranunculus* species seen at Gordion today, both of which grow along ditches, are not under consideration; *R. cornutus* has large (ca. 2 mm) seeds with bumps, and *R. muricatus* seeds have spines on the surface, similar to *R. arvensis* (which was not noticed).

Ranunculus arvensis-type. [Fig. D13i] Nine large (> 2 mm) seeds with remnants of the spines were seen. *Ranunculus muricatus,* which has been seen along ditches, has spiny seeds, too.

Resedaceae (mignonette family)

Reseda. Today, two types of *Reseda* have been seen: the seeds of *R. lutea* are more than 1 mm long (typically between 1.2 and 1.5 mm), and those of *R. microcarpa* are under 1 mm. The ancient seeds are probably *R. lutea. R. lutea* grows on Tumulus MM and lightly grazed areas; *R. microcarpa* has been seen in low areas and roadside ditches.

Rosaceae (rose family)

The Rosaceae family is well represented in the wood charcoal by *Pyrus/Crataegus* (whose wood cannot be distinguished; both types occur in the area) and *Prunus* (various wild almonds, plums, cherries, and other stone fruits). The seeds of these woody types are less common: cherry and nutshell of wild almond have been seen in the samples from the 1988/89 excavation.

Potentilla. Of four seeds identified as *Potentilla* in YH22074 (YHSS 1), one measurable one is 0.9 mm long and 0.6 mm wide, asymmetrically drop-shaped with low-ridged relief similar to *Potentilla erecta.* Today, *P. erecta* has been seen on the conglomerate ridge and *P. reptans* grows along the Sakarya.

Rubiaceae (bedstraw family)

cf. *Asperula* (was YH-Rubiaceae 2). Following Riehl 1999:108, hollow globose seeds internally divided by a septum—visible because there is a hole at the hilum—are assigned to the genus *Asperula.* It resembles *Galium,* except the latter is undivided inside. Today, yellow-flowered *A. stricta* subsp. *grandiflora* grows on grazed and ungrazed steppe, as does a white-flowered type.

Galium. Galium is one of the most common and numerous seeds in the assemblage. When charred, the globose seed has an undivided hollow interior visible through a hole at the hilum. On fresh material, the same pedicel may have a tiny seed and a large one. The genus has many species with a wide range of habitats, both disturbed and undisturbed, and may have a perennial or annual habit. *Galium verum,* though not common, grows in roadside ditches and in the poorly drained Roman road excavation just inside the Tumulus MM fence.

YH-Rubiaceae 1. [Fig. D14a] This type is most likely a member of the Rubiaceae. Unlike the more globular *Galium* and *Asperula,* it is longer than it is wide. The central depression is divided, like *Asperula,* but the charred seed is not hollow. Note that species of both *Galium* and *Asperula* are not necessarily globose, and not all are hollow. In other words, YH-Rubiaceae 1 may belong to one of these genera.

Scrophulariaceae (figwort family)

Veronica. Only three seeds identified as *Veronica* were encountered. One of them (YH 23637) compares well with *V. persica,* with ridges on the dorsal side; that seed is about 1.4 mm long. One species of this genus, *Veronica* cf. *multifida,* has been seen on Tumulus MM and in juniper scrub near Ahırozu. See also discussion of unknowns in Caryophyllaceae.

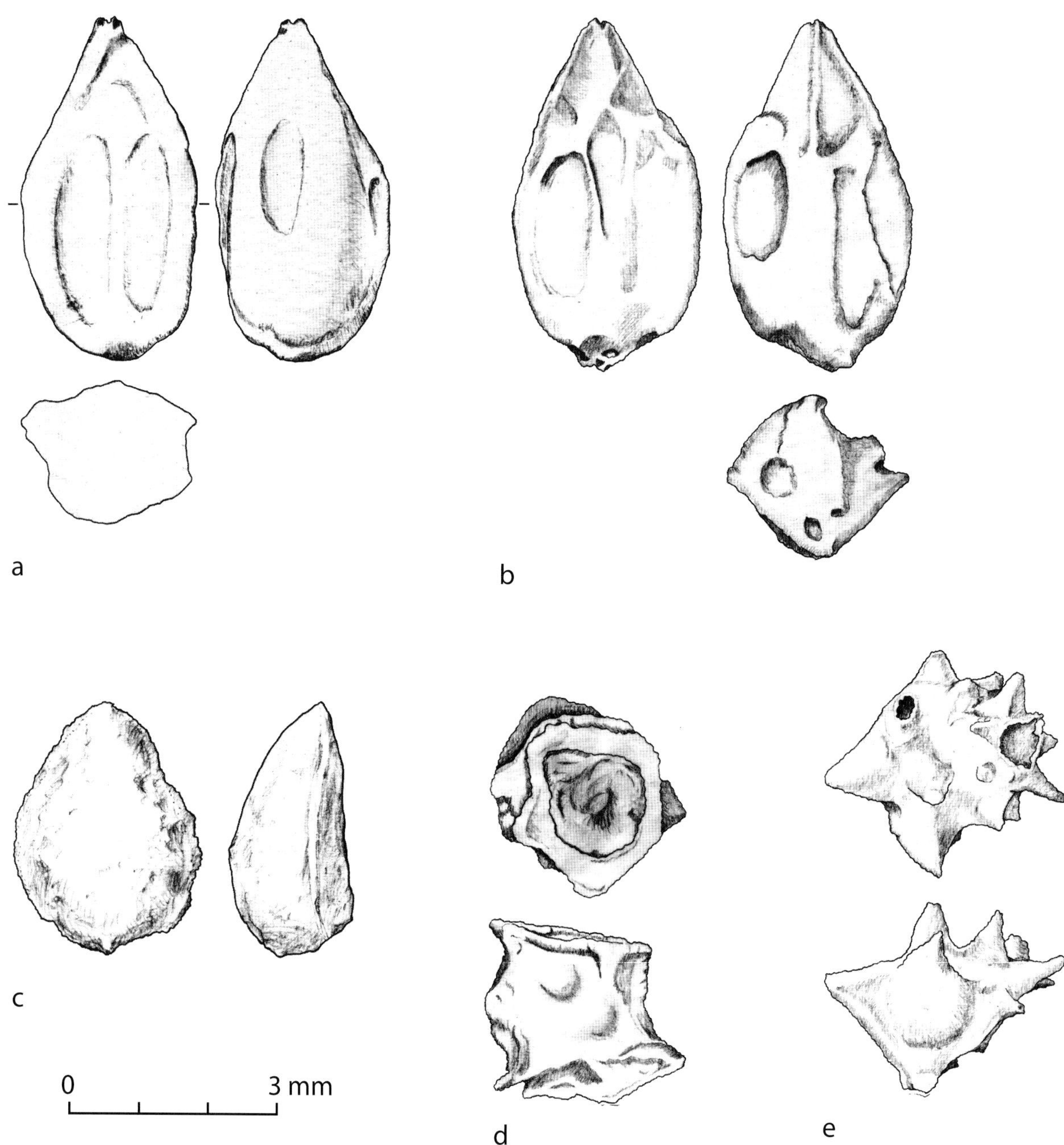

Fig. D15. a. YH-Unknown 17; b. YH-Unknown 18; c. YH-Unknown 19; d. *Beta* (was YH-plant part 16); e. YH-plant part 25.

Verbascum. Only three seeds identified as *Verbascum* were encountered. In contrast to YH-Scrophulariaceae 1 (see next entry), the depressions are very sharply delineated. Otherwise, they are similar in size and shape.

YH-Scrophulariaceae 1. [Fig. D.14b] This seed is very similar to *Verbascum* and *Scrophularia,* with irregular blocky shape and short horizontal depressions aligned in vertical rows along the length. The seed is typically about 0.8 mm in length. *Verbascum,* with its candelabra-shaped inflorescences, is prominent in the summer landscape along roadsides around Gordion and towards Polatlı, and in Turkey generally, and seems a likely identification for this seed.

Solanaceae

Hyoscyamus (was YH-Solanaceae 1). [Fig. D14c] The surface of the seed has the wavy-edged ridges typical of the Solanaceae, and the irregular shape, flattish but thicker at one end, supports an identification of *Hyoscyamus*. I have not seen *Hyoscyamus* growing, but there is no reason members of this widespread genus could not have been part of the local flora.

Solanum (was YH-Solanaceae 3). [Fig. D14d] A single flat Solanaceous seed is designated *Solanum*. *Solanum dulcamara* grows along the Sakarya today.

YH-Solanaceae 2. One flat seed is relatively smooth on surface, except low sculpting is visible on the rim at high magnification; the identification as Solanaceae is a best guess.

Thymelaeaceae (daphne family)

Thymelaea. The distinctive seed of Thymelaea is pointed at one end and rounded at the other (van Zeist and Bakker-Heeres 1984:fig. 7.12 for illustration). *Thymelaea passerina*, though fairly common on Tumulus MM, is actually quite inconspicuous thanks to its very slender green branches.

Valerianaceae (valerian family)

Valerianella. In several publications, Willem van Zeist has distinguished three morphologically distinct types of *Valerianella*, all of which occur in the Gordion samples: *V. coronata, V. dentata,* and *V. vesicaria*. At Gordion, *Valerianella coronata* is the most numerous type (105 seeds), followed by *V. vesicaria* (14) and *V. dentata* (3), with 18 indeterminate *Valerianella*. I have not recognized this genus growing today.

Verbenaceae (verbena family)

Verbena officinalis. [Fig. D14e] Two *Verbena officinalis* occur in Late Phrygian contexts. I have not seen it growing, but there is no reason this plant could not have been part of the local flora. There is only one other *Verbena* species listed in the *Flora of Turkey*.

Zygophyllaceae (caltrop family)

Peganum harmala. Despite the presence of only a few intact seeds, a number of seeds with traces of the surface and a shiny endosperm are identified as wild rue. A native shrubby steppe plant, wild rue has hallucinogenic alkaloids that make it unpalatable to livestock. With overgrazing, it would tend to increase through time. Today, it is most common on the uncultivated tumuli and archaeological mounds in the valley. It is not common on the degraded steppe areas where thyme predominates. Although nowadays people do not take advantage of its psychotropic properties, belief in its magical effects is widespread in the Middle East; in some places, its seeds are tossed into fires for general good fortune, and even today women make charms (*nazarlık*) against the evil eye from its seed pods.

Zygophyllum fabago (was YH-Unknown 11/13). [Fig. D12g] This asymmetrical seed sometimes looks mineralized rather than charred. Well-preserved exemplars look like *Zygophyllum fabago* (a ruderal that has been seen along the railroad tracks near Gordion; the only other Turkish *Zygophyllum, Z. album,* grows on dunes and salt flats [Davis, vol. 2, p. 492]).

Recognizable Unknowns

Many seeds remain unidentified because they are poorly preserved, or are non-descript (for example, the ubiquitous category "small round seed"), or occur in such small numbers or fragmentary state that there is no point trying to describe them. Sometimes, a seed type is so distinctive or occurs in sufficient quantity so that it is possible to get a sense of the range of variation that some day, when the comparative collection is extensive enough or a fellow archaeobotanist passes by the laboratory, it will be identified.

YH-Unknown 7. [Fig. D14f] This seed is likely to be in the Rubiaceae. The linear white flecks are similar to those seen on fresh *Sherardia arvensis*.

YH-Unknown 12. [see Fig. D14h] The outer coat of this unidentified seed/fruit is thin; see also *Carex divisa*.

YH-Unknown 14. [Fig. D14j] This seed is similar to members of the Caryophyllaceae in form, as well as *Veronica* (Scrophulariaceae), but did not match illustrations or seeds in the comparative collection. The Caryophyllaceae with similar seeds that have been collected near Gordion are *Dianthus floribundus* or *D. zonatus,* which have smooth seeds, as does *Veronica*.

YH-Unknown 16. [Fig. D14k] Only two of this type occur. The central ridge on the ventral side is reminiscent of *Plantago,* but the ridges are not.

YH-Unknown 17. [Fig. D15a] This large, irregular seed is reminiscent of *Crataegus* (hawthorn).

YH-Unknown 18. [Fig. D15b] This large, irregular seed remains unidentified; may be the same as YH-Unknown 17.

YH-Unknown 19 [Fig. D15c] A large seed with a fine cell structure visible at high magnification and warty surface (small rounded bumps) was found in YH33270.

YH-Unknown 21 [Fig. D14r] This small item is probably the endosperm of a flat round seed.

YH-Unknown 23. [Fig. D14m] A small wedge-shaped seed with rounded edges.

YH-Unknown 26. [Fig. D14i] [listed as YH-Cyperaceae 4 on Table F2] A few of these sedge-like seeds occur. The type has a series of ridges on the rounded side. A possible identification not yet ruled out is *Carex divisa* (or *C. muricata*). Despite superficial similarity in the drawings, this is unlikely to be the same as YH-Unknown 12 (Fig. D14h), which is considerably larger.

YH-Unknown 27. [Fig. D14l] Irregularly shaped and very variable in size, this form remains unidentified; it may not be a seed. Brassicaceae (*Myagrum* in particular) has been considered, but the distal end is not smooth. Some kind of small animal fecal pellet has also been considered, but at high magnification some cell structure is visible, and one broken specimen had a 0.2 mm thick seed coat.

YH-Unknown 28. [Fig. D14p] Two of this type occur in a single Hellenistic sample (YH28338).

YH-Unknown 29. [Fig. D14q] This seed is similar to YH-Unknowns 14 and 16 (Figs. D14j, k). The central rise is wider than that of 14 and the ridges are narrower than those of 16.

YH-Unknown 30. [Fig. D14s] This small item, if it is a seed, looks a bit like a member of the Asteraceae.

YH-Unknown 32. [Fig. D14n] Small lens-shaped seed with tiny tubercles visible around the edge. At first was thought to be in the Brassicaceae.

YH-Unknown 35. [Fig. D14t] The surface of this rounded seed is smooth and shiny.

YH-Unknown 36. Fairly large (> 1 mm diameter) globose seed with low tubercles. Brassicaceae is under consideration.

YH-Unknown 37. [Fig. D14o] A unique, distinctive item, this may be a seed capsule rather than a seed.

YH-Unknown 38. Similar to YH-Unknowns 14,

16, and 29, but not ridged; entire sample (YH22706) sent to P. I. Kuniholm for radiocarbon dating.

Plant Parts

YH-Plant part 16. [Fig. D15d] As this volume went to press, this plant part was recognized as *Beta*.

YH-Unknown 25. [Fig. D15e]

Samples from which the Illustrated Seeds Come

YH no. of illustrated types: D1a–b: 23307; D1c: 22109; D1d: 22780; D1e: 22109; D1f: 21526; D2a: 22706; D2b: 28856; D2c: 30664; D2d: 28856; D2e: 22096; D3a: 28838; D3b: 30990; D3c: 26899, 28774; D3d: 22337; D3e–f: 31053; D3g: 23200; D3h: 32818; D3i: 25652; D3j: 30664; D3k: 29965; D3l: 28338; D3m: 23208; D4a: 26472; D4b: 26870; D5a: 29312; D5b: 22109; D5c: 29965; D5d: 27461; D5e: 23633; D5f: 22343; D5g: 22192; D5h: 21728; D5i: 22343; D5j: 23633; D5k: 22466; D5l: 22305; D5m: 22337; D5n: 22491; D5o: 22441; D6a: 21728; D6b: 22780; D6c: 26789; D6d: 22075; D6e: 23637; D6f: 27461; D6g: 21728; D6h: 28338; D6i: 25652; D6j: 22276, 25652; D6k: 28629, 22281; D;7a–d: 33555; D7e: 26482; D8a: 21719; D8b: 21728; D8c: 22496; D8d: 28522; D8e: 28338; D8f–h: 22096; D9a: 28774; D9b: 28338; D9c: 33335; D9d: 31836; D9e: 23307; D9f: 21585; D9g: 32590; D9h: 26398; D9i: 22307; D9j: 30191; D9k: 22307; D9l: 26789; D9m: 23393; D9n: 31560; D9o: 22407; D10a: n/a; D10b–c: 21728; D10d–f: 31666; D11a: 22337; D11b: 21728; D11c: 22343; D11d: 22780; D11e: 32486; D11f: 21728; D11g: 31666; D11h: 28338; D11i: 23307; D11j: 27718; D11k: 28338; D11l: 23571; D11m: 29965; D11n: 21301; D13a: 26870; D13b: 28338; D13c: 23393; D13d: 31053; D13e: 23307; D13f: 23637; D13g: 21585; D13h: 30990; D13i: 22691; D13j: 22023; D13k: 22276; D13l: 22491; D13m: 30664; D14a: 31560; D14b: 21207; D14c: 30191; D14d: 21585; D14e: 29295; D14f: 22780; D14g: 29295; D14h: 21526; D14i: 29003; D14j: 22343; D14k: 23633; D14l: 28338; D14m: 22281; D14n: 25248; D14o: 33270; D14p: 28338; D14q: 22343; D14r: 25945; D14s: 31053; D14t: 33165; D15a: 21526; D15b: 33280; D15c: n/a; D15d: 22343; D15e: 29295. Cultigen figures: 5.2: 22075, 23307; 5.5a,b: 26899; 5.5c,d: 28338; 5.5e, f: 26899; 5.6a: 33554; 5.6b: 33379; 5.6c: 33335; 5.6d: e,f: 33243; 5.9: 28838; 5.10: 33555.

Appendix E

Charcoal Samples

The tables in Appendix E include the inventory of samples analyzed and their contents. In the digital files accompanying this volume (YH E1 and YH E2), samples are numbered 1 to 209 in Column A or Row A in rough order by period and locus number. Following those samples are the ones from the floor deposits of the three burned buildings: 210 to 232 (Hellenistic "Abandoned Village," YHSS 320, 330, 350); 233 to 260 (Early Phrygian Terrace Building

2A, YHSS 610, 620); 261 to 282 (Early Iron Age "BRH"=Burnt Reed House, YHSS 725); 283–to 403 (other "carbon" samples taken in 1988 and 1989)

Table E1.
Inventory of hand-picked charcoal samples

Table E2.
Data from hand-picked charcoal samples

Table E1. Inventory of charcoal samples

E1a. Ordinary debris samples analyzed

	YH	Op	Locus	Lot	Stratum	Description	# Id'd.	Wt. Id'd.	Total Wt.
1	25745	89.07	9	21	100	plow zone	6	1.75	1.75
2	26092	89.02	42	58	100	plow furrows?	2	0.39	0.43
3	27094	89.07	10	53	110.01	pit	4	1.50	4.87
4	27397	89.07	10	68	110.01	pit	3	0.76	0.95
5	20510	88.02	17	23	110.02	pit	7	3.71	3.91
6	21089	88.02	17	23	110.02	pit	6	1.24	1.42
7	26262	89.02	42	77	110.04	pit	10	2.09	3.10
8	26478	89.02	42	88	110.04	pit	15	4.94	4.96
9	26506	89.07	8	23	110.07	pit	10	2.29	4.70
10	20500	88.01	11	21	120	pit	2	0.43	0.43
11	20781	88.01	12	37	140	trash, collapse	1	0.62	0.62
12	20946	88.01	24	47	150	wall collapse	1	0.66	0.66
13	20833	88.02	21	36	300	collapse/floor deposit	15	11.16	21.44
14	21086	88.02	21	45	300	collapse w/ rodent burrows	10	12.84	15.48
15	22696	88.01	90	179	315	pit in robber trench	1	1.15	1.15
16	26114	89.01	13.2	42	315	pit	1	3.11	3.11
17	22077	88.01	28	123	345.01	pit & storage jar	10	1.92	4.75
18	23580	88.02	27	60	355	floor deposit	1	0.95	0.95
19	26540	89.07	13	31	360	mixed in excavation	10	4.53	15.94
20	28066	89.07	24	94	360.03	wall	3	12.75	13.33
21	25712	89.07	7	17	360.05	ash lens	14	1.35	2.75

Table E1a cont'd.

	YH	Op	Locus	Lot	Stratum	Description	# Id'd.	Wt. Id'd.	Total Wt.
22	26534	89.07	8	29	360.05	pit	10	1.08	2.98
23	25731	89.07	8	20	360.06	floor deposits	9	1.93	2.03
24	26204	89.07	13	33	360.06	floor deposit	7	1.04	1.52
25	27055	89.07	13	47	360.06	floor matrix	10	4.48	10.44
26	26225	89.07	13	38	360.09	basin hearth	3	0.45	0.69
27	26230	89.07	15	40	360.10	pyrotechnic feature, floor	6	1.97	2.87
28	25748	89.07	8	22	360.13	floor	9	2.10	2.26
29	26502	89.07	8	22	360.13	floor	3	2.41	2.68
30	26525	89.07	8	26	360.13	floor	2	1.06	1.06
31	26526	89.07	8	26	360.13	floor	5	1.03	1.33
32	26209	89.07	12	34	365	trash	4	0.81	0.87
33	26529	89.07	12	28	365	trash	10	2.39	3.22
34	27070	89.07	20	50	370	exterior surface	10	5.55	16.09
35	27372	89.07	20	57	370	exterior surface	10	3.57	4.35
36	28306	89.07	29	102	370.02	brick-lined bin/pit	10	11.36	46.36
37	23397	88.07	37	56	370.03	pit	1	2.36	2.36
38	28657	89.07	30	110	370.05	pyrotechnic feature	4	0.63	0.63
39	31276	89.07	45	219	370.13	pit	10	2.55	2.55
40	23603	88.07	29	41	375.02	floor deposits	1	2.73	2.73
41	22021	88.02	42	114	380.01	basin	5	10.40	12.89
42	22358	88.02	47	121	380.06	posthole	6	1.21	2.04
43	21971	88.01	38	73	380.15	hearth	10	2.37	5.04
44	21586	88.01	50	93	380.17	pit	5	1.66	1.98
45	22059	88.01	62	119	380.19	shallow pit	8	0.91	0.93
46	22364	88.02	46	120	380.24	pit	2	0.62	0.64
47	23120	88.02	71	169	380.26	basin pit	6	0.77	0.88
48	25613	89.01	10	23	390	surfaces	1	1.36	1.36
49	22852	88.02	64	155	400	pits	3	1.54	1.54
50	25226	89.01	5	7	400	mixed	4	2.27	2.27
51	25515	89.01	5	7	400	mixed	1	0.47	0.47
52	26900	89.02	55	119	400	test trench	1	2.57	2.57
53	27226	89.02	57	134	400	pit?	10	1.34	4.06
54	27900	89.07	26	91	400	mixed in excavation	5	0.61	1.14
55	28542	89.01	45	120	400	pits	5	4.08	4.08
56	29251	89.07	26	136	400	mixed	1	0.66	0.66
57	29261	89.07	26	127	400	?	10	3.93	9.96
58	29423	89.07	26	93	400		10	1.83	4.51
59	30141	89.01	53	156	400	robber trench	1	7.24	7.24
60	30503	89.07	36	178	400	mixed in excavation	1	1.64	1.64
61	31045	89.07	39	209	400	mixed in excavation	1	0.10	0.10
62	31259	89.07	39	214	400	mixed in excavation	7	1.62	1.62
63	31260	89.07	41	213	400	mixed in excavation	4	0.83	0.83
64	31290	89.07	46	221	400	floors	7	4.68	4.72
65	31547	89.07	53	242	400	surfaces	10	1.85	2.23
66	31960	89.07	54	244	400	mixed in excavation	15	8.18	10.00
67	22406	88.01	71	139	410.07	pit	4	1.08	1.24

Table E1a cont'd.

	YH	Op	Locus	Lot	Stratum	Description	# Id'd.	Wt. Id'd.	Total Wt.
68	22895	88.02	65	166	410.15	pit	10	1.59	2.22
69	26966	89.01	15	44	410.17	pit	5	4.80	5.46
70	25544	89.01	7	14	410.18	pit, metallurgical	1	1.20	1.20
71	25661	89.01	7	14	410.18	pit, metallurgical	10	5.82	6.94
72	25679	89.01	7	16	410.18	pit, metallurgical	3	0.64	0.64
73	30522	89.07	38	177	415	bricky collapse	10	3.09	4.71
74	30824	89.07	39	191	415	pit?	1	0.14	0.14
75	30846	89.07	38	200	415	exterior surfaces	7	1.57	1.61
76	31003	89.07	39	199	415	lensed collapse & trash, hearth	8	0.93	0.95
77	31040	89.07	43	210	415	robber trench	4	1.41	2.67
78	31305	89.07	47	225	415	collapse	10	3.84	4.40
79	31330	89.07	48	232	415	collapse	15	6.75	13.14
80	31340	89.07	39	230	415	lensed collapse & trash	10	1.94	3.74
81	31342	89.07	39	227	415	mixed in excavation	5	0.07	0.11
82	31995	89.07	57	251	415	mixed in excavation?	10	3.38	5.23
83	31998	89.07	58	258	415	mixed in excavation	11	5.29	5.95
84	29737	89.07	31	160	415.03	pit	5	0.77	1.10
85	31546	89.07	51	240	415.04	pit	15	18.82	25.07
86	31966	89.07	55	245	415.05	pit	6	15.18	15.18
87	31339	89.07	49	233	415.08	pit	10	3.75	11.39
88	31523	89.07	50	237	415.10	pit	10	1.33	1.71
89	22440	88.01	75	149	420	ash lens	20	12.53	33.64
90	23138	88.02	73	173	420	trash	10	2.31	2.71
91	28934	89.01	51	129	420.01	pit	2	2.87	2.87
92	23043	88.01	93	187	420.02	pit	10	3.86	4.20
93	25660	89.01	8	15	420.03	pit	3	0.58	0.64
94	25664	89.01	8	15	420.03	pit	10	6.10	9.32
95	22507	88.02	55	129	430	pits mixed in excavation	1	2.23	2.23
96	22894	88.02	69	165	430	pits mixed in excavation	2	4.96	2.38
97	22987	88.02	67	162	430	trash	2	0.56	0.56
98	23118	88.02	58	168	430	trash, pit?	10	4.74	6.10
99	23305	88.02	69	176	430	pits mixed in excavation	10	3.42	5.12
100	23311	88.02	64	175	430	pits?	15	4.27	6.37
101	23505	88.02	55	180	430	pits mixed in excavation	8	2.35	3.25
102	27510	89.01	25	75	430	trash, collapse	1	1.95	1.95
103	28429	89.02	64	177	430	pit	10	2.39	7.55
104	26781	89.01	23	62	430.02	yellow clay	1	1.05	1.05
105	27624	89.01	2.3	62	430.02	yellow clay	1	0.77	0.77
106	25522	89.01	5	10	430.03	furnace?	2	0.74	0.77
107	25524	89.01	5	10	430.03	furnace?	1	4.99	4.99
108	26624	89.02	53	96	430.12	pit	10	4.18	7.00
109	25588	89.02	28	36	430.15	pit	20	7.88	10.35
110	25282	89.02	10	14	430.17	pit w/ clay column	10	12.01	53.86
111	26185	89.02	48	70	435	mixed in excavation	7	2.78	3.27
112	26802	89.02	6	118	435	exterior surfaces	7	0.86	0.89
113	27432	89.02	60	146	435.05	pit	3	2.43	2.43

Table E1a cont'd.

	YH	Op	Locus	Lot	Stratum	Description	# Id'd.	Wt. Id'd.	Total Wt.
114	25247	89.01	3	6	450.01	pit/cellar collapse	7	2.99	3.64
115	23459	88.01	97	204	460	pit in robber trench	15	3.21	5.05
116	26646	89.02	30	98	470	collapse	2	1.98	2.15
117	27820	89.01	33	98	470	trash	7	2.61	2.61
118	28769	89.02	11	182	470	lensed trash	1	1.30	1.30
119	29529	89.01	52	133	495.06	floor deposit	1	2.13	2.13
120	31938	89.12	33	46	500	test trench	1	1.20	1.20
121	30694	89.02	78	243	510.05	pit	10	5.44	6.48
122	31438	89.02	88	287	515	pit?	10	2.17	6.29
123	32620	89.12	25	56	540.03	pit	1	10.82	10.82
124	32634	89.12	38	60	550.01	pit	1	2.97	2.97
125	32636	89.12	39	61	550.02	jar contents	10	5.30	13.65
126	32683	89.12	42	71	550.04	jar contents	20	25.52	131.31
127	32801	89.02	87	330	570.11	pit in cellar floor	1	0.31	0.31
128	20083	88.04	10	16	640	courtyard fill	4	1.40	1.76
129	20098	88.06	2	1	640	courtyard fill	10	2.94	3.71
130	25123	89.08	2	3	640	courtyard fill/chipping debris	2	0.32	0.32
131	20233	88.06	10	12	660	occup. debris & exterior surfaces	2	0.39	0.75
132	20342	88.05	10	10	660	occup. debris & exterior surfaces	7	2.34	2.54
133	25067	89.11	4	9	660	occup. debris & exterior surfaces	10	2.78	3.35
134	25069	89.11	4	9	660	occup. debris & exterior surfaces	1	5.23	5.23
135	25087	89.11	4	16	660	occup. debris & exterior surfaces	1	0.60	0.60
136	25111	89.08	1	2	660	occup. debris & exterior surfaces	10	1.54	2.80
137	25137	89.08	3	7	660	occup. debris & exterior surfaces	6	1.87	2.03
138	25171	89.09	2	6	660	occup. debris & exterior surfaces	2	2.06	2.06
139	25858	89.08	3	12	660	occup. debris & exterior surfaces	7	1.35	1.74
140	25865	89.08	3	14	660	occup. debris & exterior surfaces	2	0.29	0.29
141	25129	89.08	2	6	670	courtyard surface/constr. debris	8	2.59	2.78
142	25178	89.09	2	8	670	courtyard surface	4	2.87	2.90
143	25879	89.08	5	17	670	courtyard surface/constr. debris	10	5.48	9.66
144	20577	88.05	17	19	700	mixed in excavation	4	1.15	1.15
145	21280	88.05	26	30	700	mixed in excavation	1	1.03	1.03
146	27189	89.11	15	45	700	mixed in excavation	1	0.78	0.78
147	28118	89.09	21	73	700	mixed in excavation	10	0.27	0.81
148	28725	89.11	21	67	700	mixed in excavation	1	0.23	0.50
149	28745	89.11	21	73	700	mixed in excavation	1	1.21	1.21
150	28967	89.08	10	91	700	mixed in excavation	1	0.66	0.66
151	29236	89.11	20	83	700	mixed in excavation	6	1.85	1.93
152	29339	89.09	23	103	700	mixed in excavation?	10	7.49	16.94
153	31118	89.10	11	21	700	mixed in excavation	1	0.90	0.90
154	31777	89.10	3	37	700	mixed in excavation	5	3.26	4.13
155	31897	89.10	25	54	700	mixed in excavation	10	41.29	148.14
156	29815	89.08	10	100	700.01	depression/pit top	5	0.77	1.53
157	20865	88.04	18	33	705	exterior surface, pit tops	1	1.94	1.94
158	25909	89.09	5	19	705	exterior surface	16	5.72	5.72
159	28252	89.11	16	47	705	pit?	1	1.97	1.97

Table E1a cont'd.

	YH	Op	Locus	Lot	Stratum	Description	# Id'd.	Wt. Id'd.	Total Wt.
160	28459	89.14	31	60	705	pit	5	2.76	3.45
161	26701	89.09	12	39	705.01	pit	3	0.73	1.59
162	26703	89.09	12	39	705.01	pit	8	1.79	2.50
163	26719	89.09	12	42	705.01	pit	1	5.92	5.92
164	26720	89.09	12	42	705.01	pit	4	1.96	3.28
165	28733	89.11	14	70	705.01	bldg. collapse	5	0.70	0.70
166	30580	89.10	4	11	705.05	pit/hearth	2	1.66	1.66
167	27039	89.08	9.4	44	705.10	pit	2	0.82	1.19
168	28982	89.08	10	95	705.12	pit	1	1.00	1.00
169	31465	89.08	30	149	705.12	pit	6	11.00	11.13
170	30495	89.11	31	119	705.23	pit	3	3.81	3.81
171	27474	89.09	16	63	720	exterior surface	1	1.33	1.33
172	27481	89.09	16	62	720	exterior surface	10	12.84	16.88
173	29064	89.09	21	89	720	exterior surface	1	2.42	2.42
174	29334	89.09	23	101	720	exterior surface, burned bldg. collapse	10	4.91	8.69
175	29348	89.09	21	102	720	exterior surface	5	6.51	6.87
176	33321	89.10	25	76	720	burned bldg. collapse	10	18.92	55.58
177	22756	88.03	45	81	730.02	pit	3	1.40	1.40
178	23242	88.05	39	73	730.04	pit	12	4.46	7.48
179	21279	88.03	23	45	735	exterior surface	1	0.95	0.95
180	31796	89.10	16	41	740	bldg. collapse	4	0.96	0.96
181	31876	89.10	16	51	740	collapse/floor deposit	1	1.23	1.23
182	32487	89.10	29	67	745	floor deposit	11	3.00	8.08
183	28710	89.11	19	58	750.02	bldg. collapse	1	0.27	0.27
184	28734	89.11	20	71	750.02	bldg. collapse	1	0.97	0.97
185	29216	89.11	21	77	755.02	floor deposit	5	2.16	2.55
186	30460	89.11	14	110	755.03	bin	1	1.33	1.33
187	30395	89.11	25	105	755.05	bin	10	1.22	1.50
188	27033	89.08	9	42	760	bldg. collapse	4	0.60	0.84
189	27979	89.08	9	53	760	floor deposit/collapse	3	1.59	2.06
190	28952	89.08	9	84	760	wall collapse	2	0.46	1.07
191	28154	89.14	19	44	770	bldg. collapse	10	3.42	4.43
192	21482	88.06	27	39	795	wall collapse & floor deposit?	10	2.28	4.79
193	22704	88.06	30	46	795.02	floor deposit	5	0.85	3.49
194	21545	88.03	23	50	798	exterior surface	5	0.50	0.77
195	29202	89.11	21	73	798	floor deposit?	1	1.07	1.07
196	22162	88.03	37	62	800	trash	1	3.42	3.42
197	32736	89.11	44	184	840	wall collapse	2	5.31	5.31
198	23151	88.03	40	91	850	floor deposit	1	0.78	0.78
199	23178	88.03	40	96	850	floor deposit	10	8.73	12.77
200	23199	88.03	40	98	850	floor deposit	10	8.33	22.82
201	23213	88.03	40	100	850	floor deposit	10	16.21	23.67
202	23220	88.03	40	101	850	floor deposit	2	1.05	1.93
203	33156	89.11	46	189	850.01	hearth	4	7.51	7.52
204	29978	89.14	39	85	870.01	pit	7	4.74	5.08

Table E1a cont'd.

	YH	Op	Locus	Lot	Stratum	Description	# Id'd.	Wt. Id'd.	Total Wt.
205	33143	89.14	60	167	900	mixed in excavation	10	8.82	22.66
206	33252	89.14	60	168	900	mixed in excavation	10	6.65	7.93
207	28836	89.14	10	72	970	lensed trash	1	0.22	0.22
208	29371	89.14	10	75	970	lensed trash	2	3.84	3.85
209	33263	89.14	61	170	1030	erosion surface	1	3.22	3.22

E1b. Charcoal from burnt buildings

	YH	Op	Locus	Lot	Stratum	Description	# Id'd.	Wt. Id'd.	Total Wt.
210	20831	88.02	21	37	320	collapse & floor deposit	15	15.55	26.75
211	20839	88.02	21	37	320	collapse & floor deposit	8	2.83	2.86
212	21085	88.02	21	44	320	wall collapse	9	3.50	3.55
213	21126	88.02	21	50	320	wall collapse	10	10.43	15.53
214	23578	88.02	21	59	320	wall collapse	1	3.38	4.83
215	21094	88.02	21	46	330	roof collapse	8	548.44	548.44
216	21096	88.02	21	46	330	roof collapse	10	7.02	31.43
217	21122	88.02	28	52	330	roof collapse	10	7.14	12.71
218	22723	88.07	17	22	330	collapse	1	0.48	0.48
219	23294	88.07	23	38	330	collapse	10	3.70	4.61
220	23588	88.02	21	35	330	collapse	1	0.42	0.42
221	25724	89.07	7	19	330	wall collapse	10	4.49	6.32
222	20836	88.02	21	38	350	floor deposit	10	5.89	6.96
223	21056	88.02	21	40	350	floor deposit	10	9.50	17.11
224	21060	88.02	21	42	350	floor deposit	20	48.98	143.40
225	21143	88.02	21	55	350	floor deposit	10	17.91	25.78
226	21188	88.02	21	61	350	floor deposit	10	9.23	12.06
227	21190	88.02	21	61	350	floor deposit	20	55.70	176.90
228	21692	88.02	21	87	350	floor deposit	10	18.84	78.29
229	21831	88.02	21	94	350	floor deposit	21	16.11	80.54
230	23581	88.02	24	54	350	floor deposit	1	0.50	0.50
231	23590	88.02	21	42	350	floor deposit	1	2.20	2.20
232	21814	88.02	26	93	350.07	floor	1	1.72	1.72
233	31594	89.01	97	186	610	bldg. collapse/erosion	5	60.29	60.29
234	32130	89.01	97	191	610	bldg. collapse/erosion	10	37.11	61.35
235	32152	89.01	97	186	610	bldg. collapse/erosion	7	185.06	326.17
236	32188	89.01	97	189	610	bldg. collapse/erosion	6	95.55	200.00
237	32102	89.01	100	192	620	floor deposit	1	1100.00	1100.00
238	32132	89.01	100	192	620	floor deposit	20	239.18	399.61
239	32166	89.01	100	188	620	floor deposit	1	23.23	332.44
240	32182	89.01	100	190	620	floor deposit	5	75.71	231.38
241	32186	89.01	100	190	620	floor deposit	1	223.03	223.03
242	32956	89.01	100	194	620	floor deposit	1	793.59	793.59
243	32966	89.01	100	198	620	floor deposit	1	0.00	335.60
244	32972	89.01	100	198	620	floor deposit	1	0.00	127.83
245	33073	89.01	100	198	620	floor deposit	1	338.58	338.58
246	33074	89.01	100	198	620	floor deposit	10	50.48	200.60

Table E1b cont'd.

	YH	Op	Locus	Lot	Stratum	Description	# Id'd.	Wt. Id'd.	Total Wt.
247	33092	89.01	100	202	620	floor deposit	10	355.93	355.93
248	33214	89.01	100	203	620	floor deposit	6	109.59	140.23
249	33225	89.01	100	203	620	floor deposit	10	29.79	113.30
250	33531	89.01	100	192	620	floor deposit	1	112.77	112.77
251	33532	89.01	100	192	620	floor deposit	2	82.85	82.85
252	33565	89.01	100	205	620	floor deposit	1	125.13	125.13
253	33584	89.01	100	205	620	floor deposit	10	23.56	79.32
254	33630	89.01	100	209	620	floor deposit	2	10.41	10.41
255	33645	89.01	100	211	620	floor deposit	1	69.89	137.66
256	33661	89.01	100	211	620	floor deposit	5	27.51	76.06
257	33689	89.01	100	211	620	floor deposit	20	87.02	183.44
258	33754	89.01	100	216	620	floor deposit	1	201.89	201.89
259	33755	89.01	100	216	620	floor deposit	1	155.16	155.16
260	33756	89.01	100	216	620	floor deposit	1	157.03	157.03
261	28137	89.09	23	78	725	burned bldg. collapse, interior	10	4.47	9.02
262	28147	89.09	23	80	725	burned bldg. collapse, interior	1	6.29	6.29
263	28565	89.09	23	83	725	burned bldg. collapse	10	18.88	22.53
264	28584	89.09	23	83	725	burned bldg. collapse	2	71.45	71.45
265	28594	89.09	23	88	725	burned bldg. collapse	10	9.22	15.48
266	29085	89.09	23	93	725	burned bldg. collapse	10	6.85	7.47
267	29095	89.09	23	95	725	burned bldg. collapse	6	5.59	6.16
268	29486	89.09	23	106	725	burned bldg. coll/floor deposit	1	9.79	9.79
269	29497	89.09	23	109	725	burned bldg. coll/floor deposit	10	5.93	11.02
270	29904	89.09	23	110	725	burned bldg. coll/floor deposit	1	3.43	3.43
271	29915	89.09	23	111	725	floor deposit	10	3.83	18.81
272	29920	89.09	23	111	725	floor deposit	10	8.40	12.50
273	29921	89.09	23	111	725	floor deposit	5	3.19	8.93
274	30419	89.09	23	116	725	floor deposit	5	13.40	282.85
275	32466	89.10	25	57	725	floor deposit	20	99.59	190.12
276	33332	89.10	25	80	725	floor deposit	10	16.06	22.39
277	33336	89.10	25	80	725	floor deposit	17	54.48	100.98
278	33416	89.10	25	93	725	cleaning	10	14.38	29.46
279	33442	89.10	25	105	725	floor deposit	1	1.96	1.96
280	29906	89.09	23	112	725.05	burned bldg. collapse	10	7.07	21.62
281	29916	89.09	23	112	725.05	burned bldg. collapse	10	10.04	13.21
282	29326	89.09	23	99	725.06	burned bldg. collapse	reeds	0.00	2.90

Table E2. Weight and count of charcoal from occupation debris organized by time period

E2a. Debris (weight, g)

	YH	Op	Locus	Lot	Stratum	Quercus	Pinus	Juniperus	Conifer	Fraxinus	Populus/Salix	Rhamnus	Morus	Ulmus	Pyrus/Crataegus	Prunus	Unk. 3	Unk. 4 Tamarix?	Unk. 1	Indet.
1	25745	89.07	9	21	100	0.95	0.30	0.42			0.08									
2	26092	89.02	42	58	100		0.39													
3	27094	89.07	10	53	110.01	1.50														
4	27397	89.07	10	68	110.01		0.08							0.68						
5	20510	88.02	17	23	110.02	1.35	2.36													
6	21089	88.02	17	23	110.02	0.44	0.80													
7	26262	89.02	42	77	110.04		1.65							0.22						0.22
8	26478	89.02	42	88	110.04	2.20	0.33				0.12			1.80	0.28					0.21
9	26506	89.07	8	23	110.07	0.39	1.90													
10	20500	88.01	11	21	120		0.33		0.10											
11	20781	88.01	12	37	140		0.62													
12	20946	88.01	24	47	150		0.66													
13	20833	88.02	21	36	300	3.89	7.00													0.27
14	21086	88.02	21	45	300		12.84													
15	22696	88.01	90	179	315		1.15													
16	26114	89.01	13.2	42	315		3.11													
17	22077	88.01	28	123	345.01	1.92														
18	23580	88.02	27	60	355		0.95													
19	26540	89.07	13	31	360	4.09					0.44									
20	28066	89.07	24	94	360.03	12.75														
21	25712	89.07	7	17	360.05	0.36	0.10	0.17			0.72									
22	26534	89.07	8	29	360.05	0.12					0.96									
23	25731	89.07	8	20	360.06	0.22	0.82				0.59				0.18					0.12
24	26204	89.07	13	33	360.06	0.31	0.59							0.14						
25	27055	89.07	13	47	360.06	4.48														
26	26225	89.07	13	38	360.09	0.45														
27	26230	89.07	15	40	360.10	1.63					0.34									
28	25748	89.07	8	22	360.13	0.35	0.80	0.28										0.67		
29	26502	89.07	8	22	360.13	0.80	0.04													1.57

Table E2a cont'd.

	YH	Op	Locus	Lot	Stratum	Quercus	Pinus	Juniperus	Conifer	Fraxinus	Populus/Salix	Rhamnus	Morus	Ulmus	Pyrus/Crataegus	Prunus	Unk. 3	Unk. 4 Tamarix?	Unk. 1	Indet.
30	26525	89.07	8	26	360.13	.	0.73	.	.	.	0.33	.	.	.	.	.	.	.	.	.
31	26526	89.07	8	26	360.13	0.26	0.64	.	.	.	.	.	.	0.13	.	.	.	.	.	.
32	26209	89.07	12	34	365	0.29	.	.	0.36	.	.	.	.	.	.	.	.	.	0.16	.
33	26529	89.07	12	28	365	.	.	.	.	.	.	.	.	.	2.39	.	.	.	.	.
34	27070	89.07	20	50	370	2.19	3.36	.	.	.	.	.	.	.	.	.	.	.	.	.
35	27372	89.07	20	57	370	3.09	0.48	.	.	.	.	.	.	.	.	.	.	.	.	.
36	28306	89.07	29	102	370.02	10.83	.	.	.	.	0.53	.	.	.	.	.	.	.	.	.
37	23397	88.07	37	56	370.03	.	2.36	.	.	.	.	.	.	.	.	.	.	.	.	.
38	28657	89.07	30	110	370.05	0.32	0.31	.	.	.	.	.	.	.	.	.	.	.	.	.
39	31276	89.07	45	219	370.13	1.91	0.64	.	.	.	.	.	.	.	.	.	.	.	.	.
40	23603	88.07	29	41	375.02	2.73	.	.	.	.	.	.	.	.	.	.	.	.	.	.
41	22021	88.02	42	114	380.01	10.40	.	.	.	.	.	.	.	.	.	.	.	.	.	.
42	22358	88.02	47	121	380.06	.	0.51	.	.	.	.	.	.	.	.	.	.	.	.	0.70
43	21971	88.01	38	73	380.15	2.37	.	.	.	.	.	.	.	.	.	.	.	.	.	.
44	21586	88.01	50	93	380.17	.	1.62	.	.	.	.	.	.	.	.	.	.	.	0.04	.
45	22059	88.01	62	119	380.19	0.31	0.26	0.09	.	.	0.25	.	.	.	.	.	.	.	.	.
46	22364	88.02	46	120	380.24	0.45	0.17	.	.	.	.	.	.	.	.	.	.	.	.	.
47	23120	88.02	71	169	380.26	0.24	0.22	0.18	0.13	.	.	.	.	.	.	.	.	.	.	.
48	25613	89.01	10	23	390	1.36	.	.	.	.	.	.	.	.	.	.	.	.	.	.
49	22852	88.02	64	155	400	0.50	.	1.04	.	.	.	.	.	.	.	.	.	.	.	.
50	25226	89.01	5	7	400	0.37	1.90	.	.	.	.	.	.	.	.	.	.	.	.	.
51	25515	89.01	5	7	400	.	.	0.47	.	.	.	.	.	.	.	.	.	.	.	.
52	26900	89.02	55	119	400	.	2.57	.	.	.	.	.	.	.	.	.	.	.	.	.
53	27226	89.02	57	134	400	.	1.34	.	.	.	.	.	.	.	.	.	.	.	.	.
54	27900	89.07	26	91	400	.	0.53	.	.	.	0.08	.	.	.	.	.	.	.	.	.
55	28542	89.01	45	120	400	3.72	0.36	.	.	.	.	.	.	.	.	.	.	.	.	.
56	29251	89.07	26	136	400	0.66	.	.	.	.	.	.	.	.	.	.	.	.	.	.
57	29261	89.07	26	127	400	1.01	2.03	0.64	0.25	.	.	.	.	.	.	.	.	.	.	.
58	29423	89.07	26	93	400	1.83	.	.	.	.	.	.	.	.	.	.	.	.	.	.
59	30141	89.01	53	156	400	.	7.24	.	.	.	.	.	.	.	.	.	.	.	.	.
60	30503	89.07	36	178	400	.	1.64	.	.	.	.	.	.	.	.	.	.	.	.	.
61	31045	89.07	39	209	400	0.10	.	.	.	.	.	.	.	.	.	.	.	.	.	.

Table E2a cont'd.

#	YH	Op	Locus	Lot	Stratum	Quercus	Pinus	Juniperus	Conifer	Fraxinus	Populus/ Salix	Rhamnus	Morus	Ulmus	Pyrus/ Crataegus	Prunus	Unk. 3	Unk. 4 Tamarix?	Unk. 1	Indet.
62	31259	89.07	39	214	400	.	1.01	0.47	.	.	.	.	.	.	0.07	.	.	.	.	0.07
63	31260	89.07	41	213	400	0.39	0.44	.	.	.	.	.	.	.	.	.	.	.	.	.
64	31290	89.07	46	221	400	1.62	0.88	2.18	.	.	.	.	.	.	.	.	.	.	.	.
65	31547	89.07	53	242	400	.	1.85	.	.	.	.	.	.	.	.	.	.	.	.	.
66	31960	89.07	54	244	400	3.97	0.87	1.78	.	1.56	.	.	.	.	.	.	.	.	.	.
67	22406	88.01	71	139	410.07	0.69	.	0.39	.	.	.	.	.	.	.	.	.	.	.	.
68	22895	88.02	65	166	410.15	0.64	0.05	0.13	.	.	.	.	.	0.61	.	.	.	.	0.16	.
69	26966	89.01	15	44	410.17	4.80	.	.	.	.	.	.	.	.	.	.	.	.	.	.
70	25544	89.01	7	14	410.18	1.20	.	.	.	.	.	.	.	.	.	.	.	.	.	.
71	25661	89.01	7	14	410.18	.	.	5.82	.	.	.	.	.	.	.	.	.	.	.	.
72	25679	89.01	7	16	410.18	.	.	.	0.02	.	.	.	.	.	.	.	.	.	.	0.62
73	30522	89.07	38	177	415	.	2.98	.	.	.	.	.	.	.	0.11	.	.	.	.	.
74	30824	89.07	39	191	415	0.14	.	.	.	.	.	.	.	.	.	.	.	.	.	.
75	30846	89.07	38	200	415	0.64	0.93	.	.	.	.	.	.	.	.	.	.	.	.	.
76	31003	89.07	39	199	415	.	0.17	0.59	0.01	.	.	.	.	.	.	.	.	.	.	0.16
77	31040	89.07	43	210	415	.	1.41	.	.	.	.	.	.	.	.	.	.	.	.	.
78	31305	89.07	47	225	415	1.78	1.49	0.57	.	.	.	.	.	.	.	.	.	.	.	.
79	31330	89.07	48	232	415	3.30	0.89	0.96	.	.	.	.	.	.	1.24	.	.	0.36	.	.
80	31340	89.07	39	230	415	0.16	0.47	1.31	.	.	.	.	.	.	.	.	.	.	.	.
81	31342	89.07	39	227	415	0.02	0.05	.	.	.	.	.	.	.	.	.	.	.	.	.
82	31995	89.07	57	251	415	3.38	.	.	.	.	.	.	.	.	.	.	.	.	.	.
83	31998	89.07	58	258	415	3.37	0.33	0.64	0.95	.	.	.	.	.	.	.	.	.	.	.
84	29737	89.07	31	160	415.03	0.32	0.45	.	.	.	.	.	.	.	.	.	.	.	.	.
85	31546	89.07	51	240	415.04	13.2	0.85	1.10	.	.	.	.	.	3.67	.	.	.	.	.	.
86	31966	89.07	55	245	415.05	.	9.71	5.47	.	.	.	.	.	.	.	.	.	.	.	.
87	31339	89.07	49	233	415.08	0.65	3.10	.	.	.	.	.	.	.	.	.	.	.	.	.
88	31523	89.07	50	237	415.10	.	.	1.02	.	.	.	0.31	.	.	.	.	.	.	.	.
89	22440	88.01	75	149	420	.	.	.	.	.	.	3.10	3.99	.	5.44	.	.	.	.	.
90	23138	88.02	73	173	420	2.08	0.06	0.17	.	.	.	.	.	.	.	.	.	.	.	.
91	28934	89.01	51	129	420.01	1.40	.	1.47	.	.	.	.	.	.	.	.	.	.	.	.
92	23043	88.01	93	187	420.02	2.74	.	0.41	.	.	.	.	.	.	0.71	.	.	.	.	.
93	25660	89.01	8	15	420.03	.	.	.	.	.	.	.	.	.	.	.	0.43	.	.	0.15

Table E2a cont'd.

	YH	Op	Locus	Lot	Stratum	Quercus	Pinus	Juniperus	Conifer	Fraxinus	Populus/Salix	Rhamnus	Morus	Ulmus	Pyrus/Crataegus	Prunus	Unk. 3	Unk. 4 Tamarix?	Unk. 1	Indet.
94	25664	89.01	38	15	420.03	6.10														
95	22507	88.02	55	129	430			2.23												
96	22894	88.02	69	165	430	0.70	1.68										2.58			
97	22987	88.02	67	162	430		0.56													
98	23118	88.02	58	168	430	3.24		1.50												
99	23305	88.02	69	176	430	0.84									2.58					
100	23311	88.02	64	175	430	2.43	0.41	0.83	0.43											0.17
101	23505	88.02	55	180	430	1.50			0.25					0.60						
102	27510	89.01	25	75	430		1.95													
103	28429	89.02	64	177	430	0.24	0.07			0.38	1.70									
104	26781	89.01	23	62	430.02		1.05													
105	27624	89.01	23	62	430.02	0.77														
106	25522	89.01	5	10	430.03		0.15	0.59												
107	25524	89.01	5	10	430.03	4.99														
108	26624	89.02	53	96	430.12	2.17	2.01													
109	25588	89.02	28	36	430.15		7.70								0.18					
110	25282	89.02	10	14	430.17	12.01														
111	26185	89.02	48	70	435	2.47	0.31													
112	26802	89.02	6	118	435		0.86													
113	27432	89.02	60	146	435.05		2.43													
114	25247	89.01	3	6	450.01	1.32	1.67													
115	23459	88.01	97	204	460	0.94	0.51	1.68	0.08											
116	26646	89.02	30	98	470	0.26		1.72												
117	27820	89.01	33	98	470		2.61													
118	28769	89.02	11	182	470		1.30													
119	29529	89.01	52	133	495.06	2.13														
120	31938	89.12	33	46	500	1.20														
121	30694	89.02	78	243	510.05	5.44														
122	31438	89.02	88	287	515	1.92		0.25												
123	32620	89.12	25	56	540.03	10.82														
124	32634	89.12	38	60	550.01	2.97														
125	32636	89.12	39	61	550.02		5.30													

152

APPENDIX E

Table E2a cont'd.

	YH	Op	Locus	Lot	Stratum	Quercus	Pinus	Juniperus	Conifer	Fraxinus	Populus/Salix	Rhamnus	Morus	Ulmus	Pyrus/Crataegus	Prunus	Unk. 3	Unk. 4 Tamarix?	Unk. 1	Indet.
126	32683	89.12	42	71	550.04	19.62	4.76	.	.	.	.	.	.	.	1.14	.	.	.	.	.
127	32801	89.02	87	330	570.11	0.31	.	.	.	.	.	.	.	.	.	.	.	.	.	.
128	20083	88.04	10	16	640	1.01	0.39	.	.	.	.	.	.	.	.	.	.	.	.	.
129	20098	88.06	2	1	640	1.74	1.03	.	.	.	.	.	.	.	.	.	.	.	0.17	.
130	25123	89.08	2	3	640	.	.	0.32	.	.	.	.	.	.	.	.	.	.	.	.
131	20233	88.06	10	12	660	.	0.39	.	.	.	.	.	.	.	.	.	.	.	.	.
132	20342	88.05	10	10	660	.	0.98	1.36	.	.	.	.	.	.	.	.	.	.	.	.
133	25067	89.11	4	9	660	.	0.94	1.84	.	.	.	.	.	.	.	.	.	.	.	.
134	25069	89.11	4	9	660	.	5.23	.	.	.	.	.	.	.	.	.	.	.	.	.
135	25087	89.11	4	16	660	.	0.60	.	.	.	.	.	.	.	.	.	.	.	.	.
136	25111	89.08	1	2	660	.	1.23	.	.	.	.	.	.	.	.	0.31	.	.	.	.
137	25137	89.08	3	7	660	.	1.22	0.65	.	.	.	.	.	.	.	.	.	.	.	.
138	25171	89.09	2	6	660	.	2.06	.	.	.	.	.	.	.	.	.	.	.	.	.
139	25858	89.08	3	12	660	.	0.33	0.75	0.27	.	.	.	.	.	.	.	.	.	.	.
140	25865	89.08	3	14	660	.	0.29	.	.	.	.	.	.	.	.	.	.	.	.	.
141	25129	89.08	2	6	670	0.17	2.04	0.38	.	.	.	.	.	.	.	.	.	.	.	.
142	25178	89.09	2	8	670	0.41	2.46	.	.	.	.	.	.	.	.	.	.	.	.	.
143	25879	89.08	5	17	670	.	5.00	.	0.48	.	.	.	.	.	.	.	.	.	.	.
144	20577	88.05	17	19	700	.	.	1.15	.	.	.	.	.	.	.	.	.	.	.	.
145	21280	88.05	26	30	700	.	1.03	.	.	.	.	.	.	.	.	.	.	.	.	.
146	27189	89.11	15	45	700	.	.	0.78	.	.	.	.	.	.	.	.	.	.	.	.
147	28118	89.09	21	73	700	0.27	.	.	.	.	.	.	.	.	.	.	.	.	.	.
148	28725	89.11	21	67	700	.	.	.	.	.	.	.	.	.	.	.	.	.	.	0.23
149	28745	89.11	21	73	700	1.21	.	.	.	.	.	.	.	.	.	.	.	.	.	.
150	28967	89.08	10	91	700	.	.	0.66	.	.	.	.	.	.	.	.	.	.	.	.
151	29236	89.11	20	83	700	.	0.86	0.99	.	.	.	.	.	.	.	.	.	.	.	.
152	29339	89.09	23	103	700	.	.	7.49	.	.	.	.	.	.	.	.	.	.	.	.
153	31118	89.10	11	21	700	.	0.90	.	.	.	.	.	.	.	.	.	.	.	.	.
154	31777	89.10	3	37	700	0.59	.	2.67	.	.	.	.	.	.	.	.	.	.	.	.
155	31897	89.10	25	54	700	0.65	.	40.64	.	.	.	.	.	.	.	.	.	.	.	.
156	29815	89.08	10	100	700.01	0.77	.	.	.	.	.	.	.	.	.	.	.	.	.	.
157	20865	88.04	18	33	705	.	1.94	.	.	.	.	.	.	.	.	.	.	.	.	.

Table E2a cont'd.

	YH	Op	Locus	Lot	Stratum	Quercus	Pinus	Juniperus	Conifer	Fraxinus	Populus/ Salix	Rhamnus	Morus	Ulmus	Pyrus/ Crataegus	Prunus	Unk. 3	Unk. 4 Tamarix?	Unk. 1	Indet.
158	25909	89.09	5	19	705	3.37	2.35	.	.	.	.	.	.	.	.	.	.	.	.	.
159	28252	89.11	16	47	705	.	.	.	1.97	.	.	.	.	.	.	.	.	.	.	.
160	28459	89.14	31	60	705	.	2.76	.	.	.	.	.	.	.	.	.	.	.	.	.
161	26701	89.09	12	39	705.01	.	.	.	0.18	.	.	.	.	.	.	.	.	.	.	0.55
162	26703	89.09	12	39	705.01	.	1.53	.	.	.	.	.	.	.	.	.	.	.	.	0.26
163	26719	89.09	12	42	705.01	.	5.92	.	.	.	.	.	.	.	.	.	.	.	.	.
164	26720	89.09	12	42	705.01	0.95	0.78	.	.	.	.	.	.	.	.	.	.	.	.	0.23
165	28733	89.11	14	70	705.01	0.70	.	.	.	.	.	.	.	.	.	.	.	.	.	.
166	30580	89.10	4	11	705.05	1.33	.	.	0.33	.	.	.	.	.	.	.	.	.	.	.
167	27039	89.08	9.4	44	705.10	.	0.82	.	.	.	.	.	.	.	.	.	.	.	.	.
168	28982	89.08	10	95	705.12	1.00	.	.	.	.	.	.	.	.	.	.	.	.	.	.
169	31465	89.08	30	149	705.12	3.94	4.20	2.86	.	.	.	.	.	.	.	.	.	.	.	.
170	30495	89.11	31	119	705.23	3.81	.	.	.	.	.	.	.	.	.	.	.	.	.	.
171	27474	89.09	16	63	720	.	1.33	.	.	.	.	.	.	.	.	.	.	.	.	.
172	27481	89.09	16	62	720	12.84	.	.	.	.	.	.	.	.	.	.	.	.	.	.
173	29064	89.09	21	89	720	.	2.42	.	.	.	.	.	.	.	.	.	.	.	.	.
174	29334	89.09	23	101	720	.	.	4.91	.	.	.	.	.	.	.	.	.	.	.	.
175	29348	89.09	21	102	720	.	6.51	.	.	.	.	.	.	.	.	.	.	.	.	.
176	33321	89.10	25	76	720	1.26	.	17.66	.	.	.	.	.	.	.	.	.	.	.	.
177	22756	88.03	45	81	730.02	1.00	0.40	.	.	.	.	.	.	.	.	.	.	.	.	.
178	23242	88.05	39	73	730.04	.	2.26	2.2	.	.	.	.	.	.	.	.	.	.	.	.
179	21279	88.03	23	45	735	.	0.95	.	.	.	.	.	.	.	.	.	.	.	.	.
180	31796	89.10	16	41	740	.	0.96	.	.	.	.	.	.	.	.	.	.	.	.	.
181	31876	89.10	16	51	740	.	1.23	.	.	.	.	.	.	.	.	.	.	.	.	.
182	32487	89.10	29	67	745	0.13	.	2.45	.	0.42	.	.	.	.	.	.	.	.	.	.
183	28710	89.11	19	58	750.02	.	.	0.27	.	.	.	.	.	.	.	.	.	.	.	.
184	28734	89.11	20	71	750.02	.	.	0.97	.	.	.	.	.	.	.	.	.	.	.	.
185	29216	89.11	21	77	755.02	1.04	.	0.89	0.23	.	.	.	.	.	.	.	.	.	.	.
186	30460	89.11	14	110	755.03	.	1.33	.	.	.	.	.	.	.	.	.	.	.	.	.
187	30395	89.11	25	105	755.05	1.22	.	.	.	.	.	.	.	.	.	.	.	.	.	.
188	27033	89.08	9	42	760	.	0.60	.	.	.	.	.	.	.	.	.	.	.	.	.
189	27979	89.08	9	53	760	.	.	1.59	.	.	.	.	.	.	.	.	.	.	.	.

Table E2a cont'd.

	YH	Op	Locus	Lot	Stratum	*Quercus*	*Pinus*	*Juniperus*	Conifer	*Fraxinus*	*Populus/ Salix*	*Rhamnus*	*Morus*	*Ulmus*	*Pyrus/ Crataegus*	*Prunus*	Unk. 3	Unk. 4 *Tamarix?*	Unk. 1	Indet.
190	28952	89.08	9	84	760	.	.	0.46	.	.	.	.	.	.	.	.	.	.	.	.
191	28154	89.14	19	44	770	1.66	.	1.76	.	.	.	.	.	.	.	.	.	.	.	.
192	21482	88.06	27	39	795	1.09	1.19	.	.	.	.	.	.	.	.	.	.	.	.	.
193	22704	88.06	30	46	795.02	.	0.78	0.07	.	.	.	.	.	.	.	.	.	.	.	.
194	21545	88.03	23	50	798	.	.	0.50	.	.	.	.	.	.	.	.	.	.	.	.
195	29202	89.11	21	73	798	.	1.07	.	.	.	.	.	.	.	.	.	.	.	.	.
196	22162	88.03	37	62	800	.	3.42	.	.	.	.	.	.	.	.	.	.	.	.	.
197	32736	89.11	44	184	840	.	.	5.31	.	.	.	.	.	.	.	.	.	.	.	.
198	23151	88.03	40	91	850	.	.	0.78	.	.	.	.	.	.	.	.	.	.	.	.
199	23178	88.03	40	96	850	0.35	.	8.38	.	.	.	.	.	.	.	.	.	.	.	.
200	23199	88.03	40	98	850	.	1.21	6.44	.	.	.	.	.	0.68	.	.	.	.	.	.
201	23213	88.03	40	100	850	.	3.08	13.13	.	.	.	.	.	.	.	.	.	.	.	.
202	23220	88.03	40	101	850	.	.	1.05	.	.	.	.	.	.	.	.	.	.	.	.
203	33156	89.11	46	189	850.01	.	.	7.51	.	.	.	.	.	.	.	.	.	.	.	.
204	29978	89.14	39	85	870.01	.	.	4.74	.	.	.	.	.	.	.	.	.	.	.	.
205	33143	89.14	60	167	900	.	8.82	.	.	.	.	.	.	.	.	.	.	.	.	.
206	33252	89.14	60	168	900	.	3.77	2.88	.	.	.	.	.	.	.	.	.	.	.	.
207	28836	89.14	10	72	970	.	.	0.22	.	.	.	.	.	.	.	.	.	.	.	.
208	29371	89.14	10	75	970	3.63	0.21	.	.	.	.	.	.	.	.	.	.	.	.	.
209	33263	89.14	61	170	1030	.	3.22	.	.	.	.	.	.	.	.	.	.	.	.	.

Table E2b. Charcoal from Occupation Debris (count)

	YH	Op	Locus	Lot	Stratum	No. id'ed	Quercus	Pinus	Juniperus	Conifer	Fraxinus	Populus/ Salix	Rhamnus	Morus	Ulmus	Pyrus/ Crataegus	Prunus	Unk. 3	Unk. 4 Tamarix?	Unk. 1	Indet.
1	25745	89.07	9	21	100	6	2	2	1	.	.	1	.	.	.	.	.	.	.	.	.
2	26092	89.02	42	58	100	2	.	2	.	.	.	.	.	.	.	.	.	.	.	.	.
3	27094	89.07	10	53	110.01	4	4	.	.	.	.	.	.	.	.	.	.	.	.	.	.
4	27397	89.07	10	68	110.01	3	.	1	.	.	.	.	.	.	2	.	.	.	.	.	.
5	20510	88.02	17	23	110.02	7	2	5	.	.	.	.	.	.	.	.	.	.	.	.	.
6	21089	88.02	17	23	110.02	6	2	4	.	.	.	.	.	.	.	.	.	.	.	.	.
7	26262	89.02	42	77	110.04	10	.	7	.	.	.	.	.	.	1	.	.	.	.	.	2
8	26478	89.02	42	88	110.04	15	5	2	.	.	.	1	.	.	4	2	.	.	.	.	1
9	26506	89.07	8	23	110.07	10	1	9	.	.	.	.	.	.	.	.	.	.	.	.	.
10	20500	88.01	11	21	120	2	.	1	.	1	.	.	.	.	.	.	.	.	.	.	.
11	20781	88.01	12	37	140	1	.	1	.	.	.	.	.	.	.	.	.	.	.	.	.
12	20946	88.01	24	47	150	1	.	1	.	.	.	.	.	.	.	.	.	.	.	.	.
13	20833	88.02	21	36	300	15	3	11	.	.	.	.	.	.	.	.	.	.	.	.	1
14	21086	88.02	21	45	300	10	.	10	.	.	.	.	.	.	.	.	.	.	.	.	.
15	22696	88.01	90	179	315	1	.	1	.	.	.	.	.	.	.	.	.	.	.	.	.
16	26114	89.01	13	42	315	1	.	1	.	.	.	.	.	.	.	.	.	.	.	.	.
17	22077	88.01	28	123	345.01	10	10	.	.	.	.	.	.	.	.	.	.	.	.	.	.
18	23580	88.02	27	60	355	1	.	1	.	.	.	.	.	.	.	.	.	.	.	.	.
19	26540	89.07	13	31	360	10	7	.	.	.	.	3	.	.	.	.	.	.	.	.	.
20	28066	89.07	24	94	360.03	3	3	.	.	.	.	.	.	.	.	.	.	.	.	.	.
21	25712	89.07	7	17	360.05	14	6	1	1	.	.	6	.	.	.	.	.	.	.	.	.
22	26534	89.07	8	29	360.05	10	1	.	.	.	.	9	.	.	.	.	.	.	.	.	.
23	25731	89.07	8	20	360.06	9	1	3	.	.	.	3	.	.	.	1	.	.	.	.	1
24	26204	89.07	13	33	360.06	7	1	5	.	.	.	.	.	.	1	.	.	.	.	.	.
25	27055	89.07	13	47	360.06	10	10	.	.	.	.	.	.	.	.	.	.	.	.	.	.
26	26225	89.07	13	38	360.09	3	3	.	.	.	.	.	.	.	.	.	.	.	.	.	.
27	26230	89.07	15	40	360.10	6	2	.	.	.	.	4	.	.	.	.	.	.	.	.	.
28	25748	89.07	8	22	360.13	9	1	5	1	.	.	.	.	.	.	.	.	.	2	.	.
29	26502	89.07	8	22	360.13	3	1	1	.	.	.	.	.	.	.	.	.	.	.	.	1
30	26525	89.07	8	26	360.13	2	.	1	.	.	.	1	.	.	.	.	.	.	.	.	.
31	26526	89.07	8	26	360.13	5	1	2	.	.	.	.	.	.	2	.	.	.	.	.	.
32	26209	89.07	12	34	365	4	1	.	.	1	.	.	.	.	.	.	.	.	.	2	.

Table E2b cont'd.

	YH	Op	Locus	Lot	Stratum	No. id'ed	Quercus	Pinus	Juniperus	Conifer	Fraxinus	Populus/ Salix	Rhamnus	Morus	Ulmus	Pyrus/ Crataegus	Prunus	Unk. 3	Unk. 4 Tamarix?	Unk. 1	Indet.
33	26529	89.07	12	28	365	10	.	.	.	.	.	.	.	.	.	10	.	.	.	.	.
34	27070	89.07	20	50	370	10	2	8	.	.	.	.	.	.	.	.	.	.	.	.	.
35	27372	89.07	20	57	370	10	7	3	.	.	.	.	.	.	.	.	.	.	.	.	.
36	28306	89.07	29	102	370.02	10	9	.	.	.	.	1	.	.	.	.	.	.	.	.	.
37	23397	88.07	37	56	370.03	1	.	1	.	.	.	.	.	.	.	.	.	.	.	.	.
38	28657	89.07	30	110	370.05	4	2	2	.	.	.	.	.	.	.	.	.	.	.	.	.
39	31276	89.07	45	219	370.13	10	8	2	.	.	.	.	.	.	.	.	.	.	.	.	.
40	23603	88.07	29	41	375.02	1	1	.	.	.	.	.	.	.	.	.	.	.	.	.	.
41	22021	88.02	42	114	380.01	5	5	.	.	.	.	.	.	.	.	.	.	.	.	.	.
42	22358	88.02	47	121	380.06	6	.	4	.	.	.	.	.	.	.	.	.	.	.	.	2
43	21971	88.01	38	73	380.15	10	10	.	.	.	.	.	.	.	.	.	.	.	.	.	.
44	21586	88.01	50	93	380.17	5	.	4	.	.	.	.	.	.	.	.	.	.	.	1	.
45	22059	88.01	62	119	380.19	8	1	2	1	.	.	4	.	.	.	.	.	.	.	.	.
46	22364	88.02	46	120	380.24	2	1	1	.	.	.	.	.	.	.	.	.	.	.	.	.
47	23120	88.02	71	169	380.26	6	1	2	1	2	.	.	.	.	.	.	.	.	.	.	.
48	25613	89.01	10	23	390	1	1	.	.	.	.	.	.	.	.	.	.	.	.	.	.
49	22852	88.02	64	155	400	3	1	.	2	.	.	.	.	.	.	.	.	.	.	.	.
50	25226	89.01	5	7	400	4	1	3	.	.	.	.	.	.	.	.	.	.	.	.	.
51	25515	89.01	5	7	400	1	.	.	1	.	.	.	.	.	.	.	.	.	.	.	.
52	26900	89.02	55	119	400	1	.	1	.	.	.	.	.	.	.	.	.	.	.	.	.
53	27226	89.02	57	134	400	10	.	10	.	.	.	.	.	.	.	.	.	.	.	.	.
54	27900	89.07	26	91	400	5	.	4	.	.	.	1	.	.	.	.	.	.	.	.	.
55	28542	89.01	45	120	400	5	4	1	.	.	.	.	.	.	.	.	.	.	.	.	.
56	29251	89.07	26	136	400	1	1	.	.	.	.	.	.	.	.	.	.	.	.	.	.
57	29261	89.07	26	127	400	10	2	6	1	1	.	.	.	.	.	.	.	.	.	.	.
58	29423	89.07	26	93	400	10	10	.	.	.	.	.	.	.	.	.	.	.	.	.	.
59	30141	89.01	53	156	400	1	.	1	.	.	.	.	.	.	.	.	.	.	.	.	.
60	30503	89.07	36	178	400	1	.	1	.	.	.	.	.	.	.	.	.	.	.	.	.
61	31045	89.07	39	209	400	1	1	.	.	.	.	.	.	.	.	.	.	.	.	.	.
62	31259	89.07	39	214	400	7	.	2	2	.	.	.	.	.	.	2	.	.	.	.	1
63	31260	89.07	41	213	400	4	3	1	.	.	.	.	.	.	.	.	.	.	.	.	.
64	31290	89.07	46	221	400	7	1	2	4	.	.	.	.	.	.	.	.	.	.	.	.

Table E2b cont'd.

	YH	Op	Locus	Lot	Stratum	No. id'ed	Quercus	Pinus	Juniperus	Conifer	Fraxinus	Populus/ Salix	Rhamnus	Morus	Ulmus	Pyrus/ Crataegus	Prunus	Unk. 3	Unk. 4 Tamarix?	Unk. 1	Indet.
65	31547	89.07	53	242	400	10	.	10	.	.	.	.	.	.	.	.	.	.	.	.	.
66	31960	89.07	54	244	400	15	6	4	4	.	1	.	.	.	.	.	.	.	.	.	.
67	22406	88.01	71	139	410.07	4	3	.	1	.	.	.	.	.	.	.	.	.	.	.	.
68	22895	88.02	65	166	410.15	10	3	1	1	.	.	.	.	.	4	.	.	.	.	1	.
69	26966	89.01	15	44	410.17	5	5	.	.	.	.	.	.	.	.	.	.	.	.	.	.
70	25544	89.01	7	14	410.18	1	1	.	.	.	.	.	.	.	.	.	.	.	.	.	.
71	25661	89.01	7	14	410.18	10	.	.	10	.	.	.	.	.	.	.	.	.	.	.	.
72	25679	89.01	7	16	410.18	3	.	.	.	2	.	.	.	.	.	.	.	.	.	.	1
73	30522	89.07	38	177	415	10	.	9	.	.	.	.	.	.	.	1	.	.	.	.	.
74	30824	89.07	39	191	415	1	1	.	.	.	.	.	.	.	.	.	.	.	.	.	.
75	30846	89.07	38	200	415	7	2	5	.	.	.	.	.	.	.	.	.	.	.	.	.
76	31003	89.07	39	199	415	8	.	1	5	1	.	.	.	.	.	.	.	.	.	.	1
77	31040	89.07	43	210	415	4	.	4	.	.	.	.	.	.	.	.	.	.	.	.	.
78	31305	89.07	47	225	415	10	4	5	1	.	.	.	.	.	.	.	.	.	.	.	.
79	31330	89.07	48	232	415	15	8	2	2	.	.	.	.	.	.	2	.	.	1	.	.
80	31340	89.07	39	230	415	10	1	4	5	.	.	.	.	.	.	.	.	.	.	.	.
81	31342	89.07	39	227	415	5	1	4	.	.	.	.	.	.	.	.	.	.	.	.	.
82	31995	89.07	57	251	415	10	10	.	.	.	.	.	.	.	.	.	.	.	.	.	.
83	31998	89.07	58	258	415	11	2	2	3	4	.	.	.	.	.	.	.	.	.	.	.
84	29737	89.07	31	160	415.03	5	3	2	.	.	.	.	.	.	.	.	.	.	.	.	.
85	31546	89.07	51	240	415.04	15	7	2	4	.	.	.	.	.	2	.	.	.	.	.	.
86	31966	89.07	55	245	415.05	6	.	4	2	.	.	.	.	.	.	.	.	.	.	.	.
87	31339	89.07	49	233	415.08	10	1	9	.	.	.	.	.	.	.	.	.	.	.	.	.
88	31523	89.07	50	237	415.10	10	.	.	9	.	.	.	1	.	.	.	.	.	.	.	.
89	22440	88.01	75	149	420	20	.	.	.	.	.	.	5	6	.	9	.	.	.	.	.
90	23138	88.02	73	173	420	10	8	1	1	.	.	.	.	.	.	.	.	.	.	.	.
91	28934	89.01	51	129	420.01	2	1	.	1	.	.	.	.	.	.	.	.	.	.	.	.
92	23043	88.01	93	187	420.02	10	2	.	2	.	.	.	.	.	.	6	.	.	.	.	.
93	25660	89.01	8	15	420.03	3	.	.	.	.	.	.	.	.	.	.	.	1	.	.	2
94	25664	89.01	8	15	420.03	10	10	.	.	.	.	.	.	.	.	.	.	.	.	.	.
95	22507	88.02	55	129	430	1	.	.	1	.	.	.	.	.	.	.	.	.	.	.	.
96	22894	88.02	69	165	430	2	1	1	.	.	.	.	.	.	.	.	.	.	.	.	.

Table E2b cont'd.

	YH	Op	Locus	Lot	Stratum	No. id'ed	Quercus	Pinus	Juniperus	Conifer	Fraxinus	Populus/ Salix	Rhamnus	Morus	Ulmus	Pyrus/ Crataegus	Prunus	Unk. 3	Unk. 4 Tamarix?	Unk. 1	Indet.
161	26701	89.09	12	39	705.01	3	.	.	.	2	.	.	.	.	.	.	.	.	.	.	1
162	26703	89.09	12	39	705.01	8	.	7	.	.	.	.	.	.	.	.	.	.	.	.	1
163	26719	89.09	12	42	705.01	1	.	1	.	.	.	.	.	.	.	.	.	.	.	.	.
164	26720	89.09	12	42	705.01	4	1	2	.	.	.	.	.	.	.	.	.	.	.	.	1
165	28733	89.11	14	70	705.01	5	5	.	.	.	.	.	.	.	.	.	.	.	.	.	.
166	30580	89.10	4	11	705.05	2	1	.	.	1	.	.	.	.	.	.	.	.	.	.	.
167	27039	89.08	9.4	44	705.10	2	.	2	.	.	.	.	.	.	.	.	.	.	.	.	.
168	28982	89.08	10	95	705.12	1	1	.	.	.	.	.	.	.	.	.	.	.	.	.	.
169	31465	89.08	30	149	705.12	6	4	1	1	.	.	.	.	.	.	.	.	.	.	.	.
170	30495	89.11	31	119	705.23	3	3	.	.	.	.	.	.	.	.	.	.	.	.	.	.
171	27474	89.09	16	63	720	1	.	1	.	.	.	.	.	.	.	.	.	.	.	.	.
172	27481	89.09	16	62	720	10	10	.	.	.	.	.	.	.	.	.	.	.	.	.	.
173	29064	89.09	21	89	720	1	.	1	.	.	.	.	.	.	.	.	.	.	.	.	.
174	29334	89.09	23	101	720	10	.	.	10	.	.	.	.	.	.	.	.	.	.	.	.
175	29348	89.09	21	102	720	5	.	5	.	.	.	.	.	.	.	.	.	.	.	.	.
176	33321	89.10	25	76	720	10	1	.	9	.	.	.	.	.	.	.	.	.	.	.	.
177	22756	88.03	45	81	730.02	3	2	1	.	.	.	.	.	.	.	.	.	.	.	.	.
178	23242	88.05	39	73	730.04	12	.	2	10	.	.	.	.	.	.	.	.	.	.	.	.
179	21279	88.03	23	45	735	1	.	1	.	.	.	.	.	.	.	.	.	.	.	.	.
180	31796	89.10	16	41	740	4	.	4	.	.	.	.	.	.	.	.	.	.	.	.	.
181	31876	89.10	16	51	740	1	.	1	.	.	.	.	.	.	.	.	.	.	.	.	.
182	32487	89.10	29	67	745	11	1	.	8	.	2	.	.	.	.	.	.	.	.	.	.
183	28710	89.11	19	58	750.02	1	.	.	1	.	.	.	.	.	.	.	.	.	.	.	.
184	28734	89.11	20	71	750.02	1	.	.	1	.	.	.	.	.	.	.	.	.	.	.	.
185	29216	89.11	21	77	755.02	5	3	.	1	1	.	.	.	.	.	.	.	.	.	.	.
186	30460	89.11	14	110	755.03	1	.	1	.	.	.	.	.	.	.	.	.	.	.	.	.
187	30395	89.11	25	105	755.05	10	10	.	.	.	.	.	.	.	.	.	.	.	.	.	.
188	27033	89.08	9	42	760	4	.	.	4	.	.	.	.	.	.	.	.	.	.	.	.
189	27979	89.08	9	53	760	3	.	.	3	.	.	.	.	.	.	.	.	.	.	.	.
190	28952	89.08	9	84	760	2	.	.	2	.	.	.	.	.	.	.	.	.	.	.	.
191	28154	89.14	19	44	770	10	4	.	6	.	.	.	.	.	.	.	.	.	.	.	.
192	21482	88.06	27	39	795	10	5	5	.	.	.	.	.	.	.	.	.	.	.	.	.

Table E2b cont'd.

	YH	Op	Locus	Lot	Stratum	No. id'ed	Quercus	Pinus	Juniperus	Conifer	Fraxinus	Populus/ Salix	Rhamnus	Morus	Ulmus	Pyrus/ Crataegus	Prunus	Unk. 3	Unk. 4 Tamarix?	Unk. 1	Indet.
193	22704	88.06	30	46	795.02	5	.	4	1	.	.	.	.	.	.	.	.	.	.	.	.
194	21545	88.03	23	50	798	5	.	.	5	.	.	.	.	.	.	.	.	.	.	.	.
195	29202	89.11	21	73	798	1	.	1	.	.	.	.	.	.	.	.	.	.	.	.	.
196	22162	88.03	37	62	800	1	.	1	.	.	.	.	.	.	.	.	.	.	.	.	.
197	32736	89.11	44	184	840	2	.	.	2	.	.	.	.	.	.	.	.	.	.	.	.
198	23151	88.03	40	91	850	1	.	.	1	.	.	.	.	.	.	.	.	.	.	.	.
199	23178	88.03	40	96	850	10	1	.	9	.	.	.	.	.	.	.	.	.	.	.	.
200	23199	88.03	40	98	850	10	.	1	7	.	.	.	.	.	2	.	.	.	.	.	.
201	23213	88.03	40	100	850	10	.	3	7	.	.	.	.	.	.	.	.	.	.	.	.
202	23220	88.03	40	101	850	2	.	.	2	.	.	.	.	.	.	.	.	.	.	.	.
203	33156	89.11	46	189	850.01	4	.	.	4	.	.	.	.	.	.	.	.	.	.	.	.
204	29978	89.14	39	85	870.01	7	.	.	7	.	.	.	.	.	.	.	.	.	.	.	.
205	33143	89.14	60	167	900	10	.	10	.	.	.	.	.	.	.	.	.	.	.	.	.
206	33252	89.14	60	168	900	10	.	6	4	.	.	.	.	.	.	.	.	.	.	.	.
207	28836	89.14	10	72	970	1	.	.	1	.	.	.	.	.	.	.	.	.	.	.	.
208	29371	89.14	10	75	970	2	1	1	.	.	.	.	.	.	.	.	.	.	.	.	.
209	33263	89.14	61	170	1030	1	.	1	.	.	.	.	.	.	.	.	.	.	.	.	.

Table E3a. Charcoal from Burned Buildings (weight, g)

	YH	Op	Locus	Lot	Stratum	Quercus	Pinus	Juniperus	Conifer	Fraxinus	Populus/ Salix	Ulmus	Alnus cf. viridis	Indet.	Other items, notes
	Hellenistic "Abandoned Village"														
210	20831	88.02	21	37	320	1.61	13.94	.	.	.	.	.	.	.	
211	20839	88.02	21	37	320	.	2.83	.	.	.	.	.	.	.	
212	21085	88.02	21	44	320	0.78	2.38	.	.	.	.	.	.	0.34	
213	21126	88.02	21	50	320	9.14	1.29	.	.	.	.	.	.	.	
214	23578	88.02	21	59	320	.	3.38	.	.	.	.	.	.	.	
215	21094	88.02	21	46	330	.	548.44	.	.	.	.	.	.	.	beam; reeds–25 ml (5.28 g)
216	21096	88.02	21	46	330	0.69	6.33	.	.	.	.	.	.	.	
217	21122	88.02	28	52	330	.	7.14	.	.	.	.	.	.	.	
218	22723	88.07	17	22	330	.	0.48	.	.	.	.	.	.	.	
219	23294	88.07	23	38	330	3.70	.	.	.	.	.	.	.	.	
220	23588	88.02	21	35	330	.	0.42	.	.	.	.	.	.	.	
221	25724	89.07	7	19	330	2.98	0.90	0.40	.	.	0.21	.	.	.	
222	20836	88.02	21	38	350	.	5.89	.	.	.	.	.	.	.	
223	21056	88.02	21	40	350	.	9.5	.	.	.	.	.	.	.	
224	21060	88.02	21	42	350	32.48	13.01	.	.	3.49	.	.	.	.	
225	21143	88.02	21	55	350	0.22	17.69	.	.	.	.	.	.	.	
226	21188	88.02	21	61	350	8.55	0.68	.	.	.	.	.	.	.	
227	21190	88.02	21	61	350	24.09	29.41	.	.	2.20	.	.	.	.	
228	21692	88.02	21	87	350	.	18.84	.	.	.	.	.	.	.	
229	21831	88.02	21	94	350	1.25	14.86	.	.	.	.	.	.	.	reeds–25 ml (3.34 g)
230	23581	88.02	24	54	350	0.50	.	.	.	.	.	.	.	.	
231	23590	88.02	21	42	350	.	2.20	.	.	.	.	.	.	.	
232	21814	88.02	26	93	350.07	1.72	.	.	.	.	.	.	.	.	
	Early Phrygian Destruction Level														
233	31594	89.01	97	186	610	.	60.29	.	.	.	.	.	.	.	
234	32130	89.01	97	191	610	.	37.11	.	.	.	.	.	.	.	
235	32152	89.01	97	186	610	.	185.06	.	.	.	.	.	.	.	2 bags of pine

Table E3a cont'd.

	YH	Op	Locus	Lot	Stratum	Quercus	Pinus	Juniperus	Conifer	Fraxinus	Populus/Salix	Ulmus	Alnus cf. viridis	Indet.	Other items, notes
236	32188	89.0˚	97	189	610	.	95.55	.	.	.	.	.	.	.	2 bags of pine
237	32102	89.0˚	100	192	620	.	1100	.	.	.	.	.	.	.	4 bags of pine; log
238	32132	89.0˚	100	192	620	.	239.18	.	.	.	.	.	.	.	
239	32166	89.0˚	100	188	620	.	23.23	.	.	.	.	.	.	.	
240	32182	89.0˚	100	190	620	.	75.71	.	.	.	.	.	.	.	2 bags of pine
241	32186	89.0˚	100	190	620	.	223.03	.	.	.	.	.	.	.	1 bag of pine
242	32956	89.0	100	194	620	.	793.59	.	.	.	.	.	.	.	1 bag of pine chunks
243	32966	89.0	100	198	620	.	335.60	.	.	.	.	.	.	.	1 bag of pine; "planks"
244	32972	89.0	100	198	620	.	127.83	.	.	.	.	.	.	.	1 bag of pine
245	33073	89.0	100	198	620	.	338.58	.	.	.	.	.	.	.	1 bag of pine
246	33074	89.0	100	198	620	.	50.48	.	.	.	.	.	.	.	
247	33092	89.0	100	202	620	.	355.93	.	.	.	.	.	.	.	
248	33214	89.0	100	203	620	.	109.59	.	.	.	.	.	.	.	
249	33225	89.0▮	100	203	620	.	29.79	.	.	.	.	.	.	.	
250	33531	89.0▮	100	192	620	.	112.77	.	.	.	.	.	.	.	beam, found in YH32102
251	33532	89.0▮	100	192	620	.	82.85	.	.	.	.	.	.	.	beam, found in YH32132
252	33565	89.0▮	100	205	620	.	125.13	.	.	.	.	.	.	.	
253	33584	89.0▮	100	205	620	.	23.56	.	.	.	.	.	.	.	
254	33630	89.0▮	100	209	620	.	10.41	.	.	.	.	.	.	.	
255	33645	89.0▮	100	211	620	.	69.89	.	.	.	.	.	.	.	
256	33661	89.0▮	100	211	620	27.51	.	.	.	.	.	.	.	.	including ≈ 50 rings
257	33689	89.0▮	100	211	620	51.20	35.82	.	.	.	.	.	.	.	
258	33754	89.0▮	100	216	620	.	201.89	.	.	.	.	.	.	.	
259	33755	89.0▮	100	216	620	.	155.16	.	.	.	.	.	.	.	pine log
260	33756	89.0▮	100	216	620	.	157.03	.	.	.	.	.	.	.	pine log
	Early Iron Age Burnt Reed House														
261	28137	89.0Э	23	78	725	.	.	4.47	.	.	.	.	.	.	
262	28147	89.0Э	23	80	725	.	.	6.29	.	.	.	.	.	.	

Table E3a cont'd.

	YH	Op	Locus	Lot	Stratum	*Quercus*	*Pinus*	*Juniperus*	Conifer	*Fraxinus*	*Populus/ Salix*	*Ulmus*	*Alnus* cf. *viridis*	Indet.	Other items, notes
263	28565	89.09	23	83	725	.	0.30	18.58	.	.	.	.	.	.	
264	28584	89.09	23	83	725	.	.	71.45	.	.	.	.	.	.	
265	28594	89.09	23	88	725	.	.	9.22	.	.	.	.	.	.	
266	29085	89.09	23	93	725	.	2.32	4.53	.	.	.	.	.	.	
267	29095	89.09	23	95	725	.	5.59	.	.	.	.	.	.	.	
268	29486	89.09	23	106	725	.	9.79	.	.	.	.	.	.	.	
269	29497	89.09	23	109	725	.	0.20	5.73	.	.	.	.	.	.	
270	29904	89.09	23	110	725	.	.	3.43	.	.	.	.	.	.	
271	29915	89.09	23	111	725	3.83	.	.	.	.	.	.	.	.	
272	29920	89.09	23	111	725	.	5.72	2.68	.	.	.	.	.	.	
273	29921	89.09	23	111	725	.	.	3.19	.	.	.	.	.	.	
274	30419	89.09	23	116	725	.	.	.	.	.	.	.	13.40	.	mystery wood; "planks?"
275	32466	89.10	25	57	725	4.96	16.36	78.27	.	.	.	.	.	.	
276	33332	89.10	25	80	725	.	10.64	3.98	.	.	.	1.44	.	.	
277	33336	89.10	25	80	725	11.79	13.48	13.74	1.29	.	14.18	.	.	.	
278	33416	89.10	25	93	725	.	.	6.53	.	.	7.85	.	.	.	
279	33442	89.10	25	105	725	.	1.96	.	.	.	.	.	.	.	
280	29906	89.09	23	112	725.05	.	.	7.07	.	.	.	.	.	.	
281	29916	89.09	23	112	725.05	.	8.92	1.12	.	.	.	.	.	.	
282	29326	89.09	23	99	725.06	.	.	.	.	.	.	.	.	.	reeds/grass stem

Table E3b. Charcoal from Burned Buildings (count)

	YH	Op	Locus	Lot	Stratum	Quercus	Pinus	Juniperus	Conifer	Fraxinus	Populus/ Salix	Ulmus	Alnus cf. viridis	Indet.
	Hellenistic "Abandoned Village"					.	.	.	.	.	.	.	.	.
210	20831	88.02	21	37	320	2	13	.	.	.	.	.	.	.
211	20839	88.02	21	37	320	.	8	.	.	.	.	.	.	.
212	21085	88.02	21	44	320	2	6	.	.	.	.	.	.	1
213	21126	88.02	21	50	320	8	2	.	.	.	.	.	.	.
214	23578	88.02	21	59	320	.	1	.	.	.	.	.	.	.
215	21094	88.02	21	46	330	.	8	.	.	.	.	.	.	.
216	21096	88.02	21	46	330	1	9	.	.	.	.	.	.	.
217	21122	88.02	28	52	330	.	10	.	.	.	.	.	.	.
218	22723	88.07	17	22	330	.	1	.	.	.	.	.	.	.
219	23294	88.07	23	38	330	10	.	.	.	.	.	.	.	.
220	23588	88.02	21	35	330	.	1	.	.	.	.	.	.	.
221	25724	89.07	7	19	330	5	1	1	.	.	3	.	.	.
222	20836	88.02	21	38	350	.	10	.	.	.	.	.	.	.
223	21056	88.02	21	40	350	.	10	.	.	.	.	.	.	.
224	21060	88.02	21	42	350	12	7	.	.	1	.	.	.	.
225	21143	88.02	21	55	350	1	9	.	.	.	.	.	.	.
226	21188	88.02	21	61	350	7	3	.	.	.	.	.	.	.
227	21190	88.02	21	61	350	8	11	.	.	1	.	.	.	.
228	21692	88.02	21	87	350	.	10	.	.	.	.	.	.	.
229	21831	88.02	21	94	350	2	19	.	.	.	.	.	.	.
230	23581	88.02	24	54	350	1	.	.	.	.	.	.	.	.
231	23590	88.02	21	42	350	.	1	.	.	.	.	.	.	.
232	21814	88.02	26	93	350.07	1	.	.	.	.	.	.	.	.
	Early Phrygian Destruction Level					.	.	.	.	.	.	.	.	.
233	31594	89.01	97	186	610	.	5	.	.	.	.	.	.	.
234	32130	89.01	97	191	610	.	10	.	.	.	.	.	.	.
235	32152	89.01	97	186	610	.	7	.	.	.	.	.	.	.

Table E3b cont'd.

	YH	Op	Locus	Lot	Stratum	Quercus	Pinus	Juniperus	Conifer	Fraxinus	Populus/ Salix	Ulmus	Alnus cf. viridis	Indet.
236	32188	89.01	97	189	610	.	6	.	.	.	.	.	.	.
237	32102	89.01	100	192	620	.	1	.	.	.	.	.	.	.
238	32132	89.01	100	192	620	.	19	.	.	.	.	.	.	.
239	32166	89.01	100	188	620	.	1	.	.	.	.	.	.	.
240	32182	89.01	100	190	620	.	5	.	.	.	.	.	.	.
241	32186	89.01	100	190	620	.	1	.	.	.	.	.	.	.
242	32956	89.01	100	194	620	.	1	.	.	.	.	.	.	.
243	32966	89.01	100	198	620	.	1	.	.	.	.	.	.	.
244	32972	89.01	100	198	620	.	1	.	.	.	.	.	.	.
245	33073	89.01	100	198	620	.	1	.	.	.	.	.	.	.
246	33074	89.01	100	198	620	.	10	.	.	.	.	.	.	.
247	33092	89.01	100	202	620	.	10	.	.	.	.	.	.	.
248	33214	89.01	100	203	620	.	6	.	.	.	.	.	.	.
249	33225	89.01	100	203	620	.	10	.	.	.	.	.	.	.
250	33531	89.01	100	192	620	.	1	.	.	.	.	.	.	.
251	33532	89.01	100	192	620	.	2	.	.	.	.	.	.	.
252	33565	89.01	100	205	620	.	1	.	.	.	.	.	.	.
253	33584	89.01	100	205	620	.	10	.	.	.	.	.	.	.
254	33630	89.01	100	209	620	.	2	.	.	.	.	.	.	.
255	33645	89.01	100	211	620	.	1	.	.	.	.	.	.	.
256	33661	89.01	100	211	620	5	.	.	.	.	.	.	.	.
257	33689	89.01	100	211	620	7	13	.	.	.	.	.	.	.
258	33754	89.01	100	216	620	.	1	.	.	.	.	.	.	.
259	33755	89.01	100	216	620	.	1	.	.	.	.	.	.	.
260	33756	89.01	100	216	620	.	1	.	.	.	.	.	.	.
Early Iron Age Burnt Reed House						.	.							.
261	28137	89.09	23	78	725	.	.	10	.	.	.	.	.	.
262	28147	89.09	23	80	725	.	.	1	.	.	.	.	.	.

Tab e E3b cont'd.

	YH	Op	Locus	Lot	Stratum	Quercus	Pinus	Juniperus	Conifer	Fraxinus	Populus/ Salix	Ulmus	Alnus cf. viridis	Indet.
263	28565	89.09	23	83	725	.	1	9	.	.	.	.	.	.
264	28584	89.09	23	83	725	.	.	2	.	.	.	.	.	.
265	28594	89.09	23	88	725	.	.	10	.	.	.	.	.	.
266	29085	89.09	23	93	725	.	4	6	.	.	.	.	.	.
267	29095	89.09	23	95	725	.	6	.	.	.	.	.	.	.
268	29486	89.09	23	106	725	.	1	.	.	.	.	.	.	.
269	29497	89.09	23	109	725	.	2	8	.	.	.	.	.	.
270	29904	89.09	23	110	725	.	.	1	.	.	.	.	.	.
271	29915	89.09	23	111	725	10	.	.	.	.	.	.	.	.
272	29920	89.09	23	111	725	.	8	2	.	.	.	.	.	.
273	29921	89.09	23	111	725	.	.	5	.	.	.	.	.	.
274	30419	89.09	23	116	725	.	.	.	.	.	.	.	5	.
275	32466	89.10	25	57	725	3	4	13	.	.	.	.	.	.
276	33332	89.10	25	80	725	.	6	2	.	.	.	2	.	.
277	33336	89.10	25	80	725	3	4	6	.	.	4	.	.	.
278	33416	89.10	25	93	725	.	.	4	.	.	6	.	.	.
279	33442	89.10	25	105	725	.	1	.	.	.	.	.	.	.
280	29906	89.09	23	112	725.05	.	.	10	.	.	.	.	.	.
281	29916	89.09	23	112	725.05	.	6	4	.	.	.	.	.	.
282	29326	89.09	23	99	725.06	.	.	.	.	.	.	.	.	.

Appendix F

Flotation Samples

The tables in Appendix F include the inventory of flotation samples analyzed and their contents. In the digital files accompanying this volume (YH F1–3), samples are numbered 1 to 225 in Column A or Row A in rough order by period and locus number. Occupation debris samples are listed as follows: YHSS 1:Columns 1–15; YHSS 3: Columns 16–19, excluding 24 and 39; YHSS 4: Columns 50–102; YHSS 5: Columns 103–117; YHSS 6: Columns 118–125, excluding 118, 119, 125; YHSS 7: Columns 126–191; YHSS 8/9: Columns 192–223; YHSS 10: Columns 224–225. Following those samples are the ones from the floor deposits of the three burned buildings: 226–238 (Early Phrygian Terrace Building 2A, YHSS 620); 239–240 (Hellenistic "Abandoned Village," YHSS 350); 241–252 (Early Iron Age "BRH"=Burnt Reed House, YHSS 725); 261–687 (other flotation samples taken in 1988 and 1989).

Table F1 Inventory of Samples

List of samples analyzed for this report, brief context information, summary statistics and miscellaneous contents. In the accompanying digital files, Column A: row/column number; Columns B–F: basic provenience information; Column G: stratigraphic designation; Column H: Mary Voigt's description of context; Column I: Naomi F. Miller's interpretation of context description; Column J: context simplified for sorting by category; Columns K–R: summary statistics; Column S: date analyzed; Columns T–Z: non-plant items found in the samples

Table F2 Data from Flotation Samples

Each column records one sample. Below the provenience information are data on amounts of economic plants (weight or count), seed counts of wild and weedy plants, counts of plant parts, counts of uncharred mineralized (i.e., possibly ancient) and uncharred modern seeds.

Table F3 Contents of Heavy Fractions

The samples are numbered in the same row/column order as in Appendices F1 and F2 and list charred plant remains removed from the heavy fraction without the aid of a microscope. (See digital file.)

Table F1. Inventory of Analyzed Samples with Provenience Descriptions

	YH	Op	Locus	Lot	YHSS no.	Period*	Context Description (M. M. Voigt, October 8, 1996)	Context Summary Description Interpretation	Context Category
1	20295	88.02	17	22	110.02	Med	cobble pit	pit	pit
2	26472	89.02	42	88	110.04	Med	pit	pit	pit
3	26647	89.02	42	108	110.06	Med	pit	pit	pit
4	26871	89.02	42	110	110.08	Med	pit	pit	pit
5	20605	88.01	11	21	120	Med	latest pits in Op 1 & plow zone	pit	pit
6	20608	88.01	11	25	120	Med	latest pits in Op 1 & plow zone	pit(s)	pit
7	21153	88.02	29	57	130	Med	pyrotechnic feature	pyrotechnic feature, indet.	pyro.
8	20950	88.01	24	47	150	Med	wall collapse (pithouse)	collapse	coll.
9	22343	88.01	68	160	150	Med	surface/floor (pithouse)	surface/floor	surface
10	21728	88.01	43	106	150.03	Med	coiled oven	pyrotechnic feature, oven	pyro.
11	22074	88.01	43	126	150.03	Med	coiled oven	pyrotechnic feature, oven	pyro.
12	22075	88.01	43	126	150.03	Med	coiled oven	pyrotechnic feature, oven	pyro.
13	20630	88.01	11	28	160.02	Med	south wall, pithouse; wall, floor dep.?	collapse/floor dep.osit?	coll.
14	20480	88.01	11	19	180	Med	complex of pits beneath YHSS 160	pit(s)	pit
15	20796	88.01	22	39	180	Med	complex of pits beneath YHSS 160	pit(s)	pit
16	22000	88.01	47	84	300	Hell	mixed in excavation	mixed in excavation	mixed
17	22410	88.01	72	141	315	Hell	pit in Hellenistic robber trench	pit	pit
18	22442	88.01	74	147	315	Hell	trash, robber pit?	trash, rcbber pit?	pit
19	22608	88.01	84	163	315	Hell	robber trench?	robber trench?	mixed
20	20524	88.02	16	27	320	Hell	wall collapse (pink deposit)	collapse	coll.
21	20825	88.02	21	36	320	Hell	collapse/floor deposit	collapse/floor deposit	coll.
22	20771	88.01	20	36	340	Hell	floor deposit	floor deposit	surface
23	20624	88.01	13	27	345	Hell	exterior trash, collapse	trash, collapse	debris
24	20785	88.01	17	34	345	Hell	exterior courtyard?	surface?, exterior	debris
25	21369	88.01	28	54	345.01	Hell?	pit for pithos	pit	pit
26	21926	88.02	28	101	350.05	Hell	central hearth/pillar base	central hearth/pillar base	debris
27	23634	88.07	16	47	360	Hell	floor deposit	floor deposit	surface
28	23649	88.07	16	54	360	Hell	mixed in excavation	mixed in excavation	mixed
29	23633	88.07	31	50	360.01	Hell	hearth	pyrotechnic feature, hearth	pyro.

Table F1 cont'd.

	YH	Op	Locus	Lot	YHSS no.	Period*	Context Description (M. M. Voigt, October 8, 1996)	Context Summary Description Interpretation	Context Category
145	27981	89.08	9	55	705.16	EIA	pit complex	pit(s)	pit
146	27277	89.09	18	54	705.18	EIA	hearth	pyrotechnic feature, hearth	pyro
147	30494	89.11	31	119	705.23	EIA	pit, bell-shaped	pit	pit
148	26715	89.09	13	41	705.25	EIA	hearth, casual	pyrotechnic feature, casual hearth	pyro
149	28130	89.09	21	75	720	EIA	collapse, waterlaid, in BRH	collapse	coll
150	28176	89.14	25	49	730	EIA	pits, mixed in excavation	pits	pit
151	30349	89.14	27	97	730.01	EIA	pit	pit	pit
152	28845	89.14	36	73	730.02	EIA	pit ("party pit")	pit	pit
153	20987	88.05	22	26	730.02	EIA	pit ("party pit")	pit	pit
154	21000	88.05	22	26	730.02	EIA	pit ("party pit")	pit	pit
155	22281	88.05	37	63	730.02	EIA	pit ("party pit")	pit	pit
156	22496	88.03	45	81	730.02	EIA	pit ("party pit")	pit	pit
157	22192	88.05	35	55	730.03	EIA	pit, w/ silt lenses	pit	pit
158	28491	89.14	25	65	730.04	EIA	pit ("deer pot")	pit	pit
159	23237	88.05	39	73	730.04	EIA	pit ("deer pot")	pit	pit
160	23246	88.05	40	74	730.04	EIA	contents, pot YH32993 in "deer pot pit"	pot contents	jar
161	23571	88.05	40	82	730.04	EIA	pit ("deer pot")	pit	pit
162	22466	88.03	42	74	730.06	EIA	pit	pit	pit
163	21301	88.05	24	27	730.07	EIA	pit	pit	pit
164	21526	88.03	17	48	735.01	EIA	domed oven	pyrotechnic feature, oven	pyro
165	29473	89.09	26	107	740.01	EIA	pit	pit	pit
166	32486	89.10	29	67	745	EIA	SEB structure; floor deposit?	floor deposit?	debris
167	33319	89.10	35	74	745.09	EIA	feature, brick-lined	pit	pit
168	32550	89.11	40	172	755.04	EIA	pit/hearth	pit/hearth	pit
169	21888	88.03	15	59	755.06	EIA	plaster feature; exterior	feature, plaster	debris
170	30355	89.11	19	99	755.15	EIA	hearth	pyrotechnic feature, hearth	pyro
171	28951	89.08	9.10	88	760.04	EIA	floor deposit, burned	floor deposit	surf
172	27021	89.08	9.08	39	760.07	EIA	hearth/oven	pyrotechnic feature, hearth/oven	pyro
173	28629	89.08	9.08	81	760.07	EIA	hearth/oven	pyrotechnic feature, hearth/oven	pyro

Table F1 cont'd.

	YH	Op	Locus	Lot	YHSS no.	Period*	Context Description (M. M. Voigt, October 8, 1996)	Context Summary Description Interpretation	Context Category
174	27348	89.14	19	44	770	EIA	collapse, CKD structure	collapse	coll
175	29999	89.14	26	88	775	EIA	CKD structure; floor deposit?	floor deposit?	debris
176	30631	89.14	38	105	775.06	EIA	bin, CKD structure	bin	debris
177	28188	89.14	18	54	775.08	EIA	oven, CKD structure	pyrotechnic feature, oven	pyro
178	30229	89.08	16	117	780.01	EIA	pit	pit	pit
179	28989	89.08	10	96	785	EIA	surface, exterior; burned debris	surface, exterior	surf
180	29822	89.08	10	104	785	EIA	lensed trash	ash lens?	debris
181	21024	88.06	18	27	790	EIA	contents, pot YH33603l from collapse of WFL structure	pot contents	jar
182	21453	88.06	22	33	790	EIA	collapse, WFL structure	collapse	coll
183	21463	88.06	23	34	790	EIA	collapse, WFL structure	collapse	coll
184	21046	88.06	19	32	790.03	EIA	oven	pyrotechnic feature, oven	pyro
185	21047	88.06	19	32	790.03	EIA	oven	pyrotechnic feature, oven	pyro
186	21484	88.06	19	40	790.03	EIA	oven fabric	ven fabr c	debris
187	21489	88.06	27	42	795.02	EIA	floor deposit, white	floor deposit	surf
188	21498	88.06	27	44	795.02	EIA	floor deposit, white	floor deposit	surf
189	22708	88.06	30	46	795.02	EIA	floor deposit, white	floor deposit	surf
190	28165	89.14	24	48	795.08	EIA	pit	pit	pit
191	21542	88.03	23	50	798	EIA	mixed	mixed	mixed
192	22178	88.03	38	63	800	LBA	occupation debris	occupat on debris	debris
193	22270	88.05	37	61	800	LBA	bldg collapse/floor deposit	collapse/floor deposit	coll
194	22482	88.03	44	77	800	LBA	building collapse	collapse	coll
195	31053	89.14	44	110	840	LBA	wall collapse, CBH	collapse	coll
196	22276	88.05	37	62	840	LBA	wall collapse, CBH	collapse	coll
197	22491	88.03	40	79	840	LBA	wall collapse, CBH	collapse	coll
198	22780	88.03	46	86	840.01	LBA	pit	pit	pit
199	30623	89.14	43	103	840.02	LBA	posthole	posthole	debris
200	33165	89.11	45	191	845	LBA	occupation debris	occupat on debris	debris
201	22775	88.05	37	66	850	LBA	floor deposit	floor deposit?	surf

Table F1 cont'd.

	YH	Op	Locus	Lot	YHSS no.	Period*	Context Description (M. M. Voigt, October 8, 1996)	Context Summary Description Interpretation	Context Category
202	22799	88.03	40	91	850	LBA	floor deposit	floor deposit?	surf
203	23208	88.03	40	99	850	LBA	floor deposit	floor deposit?	surf
204	23221	88.03	40	101	850	LBA	floor deposit	floor deposit?	surf
205	33187	89.11	46	197	850.01	LBA	hearth	pyrotechnic feature, hearth	pyro
206	23200	88.05	37	70	850.06	LBA	surface/floor (pithouse)	floor deposit?	surf
207	29965	89.14	39	85	870.01	LBA	pit	pit	pit
208	32447	89.14	55	154	870.02	LBA	pit	pit	pit
209	31603	89.11	32	135	870.03	LBA	pit	pit	pit
210	31836	89.14	52	132	870.04	LBA	pit	pit	pit
211	31837	89.14	52	132	870.04	LBA	pit	pit	pit
212	29943	89.09	24	115	900	LBA	surfaces, exterior	surfaces, exterior	surf
213	31089	89.14	17	117	900	LBA	mixed in excavation	mixed	mixed
214	32407	89.14	60	147	950	LBA	Collapse, red building-exterior surfaces	collapse/exterior surfaces?	coll
215	32414	89.14	60	148	950	LBA	Collapse, red building-exterior surfaces	collapse/exterior surfaces?	coll
216	23168	88.05	38	68	950	LBA	Collapse, red building-exterior surfaces	collapse/exterior surfaces?	coll
217	26342	89.14	10	21	970	LBA	trash, green lensed	trash	debris
218	26682	89.14	10	29	970	LBA	trash, green lensed	trash	debris
219	28838	89.14	10	72	970	LBA	trash, green lensed	trash	debris
220	33122	89.14	60	165	970	LBA	trash, green lensed	trash	debris
221	33125	89.14	60	164	970	LBA	trash, green lensed	trash	debris
222	23236	88.05	38	72	970	LBA	trash, green lensed	trash	debris
223	23238	88.05	38	72	970	LBA	trash, green lensed	trash	debris
224	33280	89.14	61	173	1000	MBA	mixed in excavation	mixed in excavation	mixed
225	33270	89.14	61	171	1030	MBA	erosion surface	surface	surf
226	32144	89.01	100	197	620	E Phryg	floor deposit	floor deposit	surf
227	33234	89.01	100	203	620	E Phryg	floor deposit, fr pot YH 33218 [categorized as 'carbon']	floor deposit	jar
228	33230	89.01	100	203	620	E Phryg	floor deposit (barley)	floor deposit	surf
229	33243	89.01	100	205	620	E Phryg	floor deposit, fr pot YH 33233 (lentil) [sent for C-14 dating]	floor deposit	jar

Table F1 cont'd.

	YH	Op	Locus	Lot	YHSS no.	Period*	Context Description (M. M. Voigt, October 8, 1996)	Context Summary Description Interpretation	Context Category
230	33246	89.01	100	205	620	E Phryg	floor deposit (not floated)	floor deposit	surf
231	33554	89.01	100	205	620	E Phryg	floor deposit (barley in pot)	floor deposit	jar
232	33555	89.01	100	205	620	E Phryg	floor deposit, fr pot YH 33231(flax) [sent for C-14 dating]	floor deposit	jar
233	33573	89.01	100	205	620	E Phryg	floor deposit (barley)	floor deposit	surf
234	33575	89.01	100	205	620	E Phryg	floor deposit (lentil)	floor deposit	jar
235	33580	89.01	100	205	620	E Phryg	Floor deposit, fr pot YH 33561 (wheat) [categorized as 'carbon']	floor deposit	surf
236	33587	89.01	100	205	620	E Phryg	floor deposit (barley)	floor deposit	surf
237	33590	89.01	100	205	620	E Phryg	floor deposit, fr pot YH 33589 (barley) [categorized as 'carbon']	floor deposit	jar
238	33613	89.01	100	208	620	E Phryg	floor deposit (barley)	floor deposit	surf
239	21068	88.02	21	42	350	Hell	floor deposit	floor deposit	surf
240	21183	88.02	21	61	350	Hell	floor deposit ("primary floor)	floor deposit	surf
241	29483	89.09	23	106	725	EIA	burned building collapse, floor deposit	collapse/floor deposit	surf
242	29924	89.09	23	111	725	EIA	floor deposit	floor deposit	surf
243	30416	89.09	23	116	725	EIA	floor deposit [sent for C14]	floor deposit	surf
244	33335	89.10	25	80	725	EIA	floor deposit-bitter vetch	floor deposit	surf
245	33368	89.10	25	87	725	EIA	floor deposit-barley	floor deposit	surf
246	33379	89.10	25	87	725	EIA	floor deposit (wheat) [sent for C14]	Floor deposit-wheat-C14?	surf
247	33382	89.10	25	87	725	EIA	floor deposit-wheat	floor deposit	surf
248	33402	89.10	25	87	725	EIA	floor deposit-wheat	floor deposit	surf
249	33414	89.10	38	94	725.04	EIA	oven-wheat	pyrotechnic feature, oven	pyro
250	33422	89.10	38	94	725.04	EIA	oven-wheat	pyrotechnic feature, oven	pyro
251	29923	89.09	23	112	725.05	EIA	burned building collapse	collapse	coll
252	33394	89.10	37	90	725.08	EIA	pit inside BRH-wheat-C14?	pit	pit

*Med (Medieval, YHSS 1), Hell (Hellenistic, YHSS 3), L Phryg (Late Phrygian, YHSS 4), M Phryg (Middle Phrygian, YHSS 5), E Phryg (Early Phrygian, YHSS 6), EIA (Early Iron Age, YHSS 7), LBA (Late Bronze Age, YHSS 8/9)

Table F2. Flotation sample contents. Note for summary tables 5.13, 5.14: YHSS 1: Columns 1–15; YHSS 3: Columns 16–19, excluding 24 and 39; YHSS 4: Columns 50–102; YHSS 5: Columns 103–117; YHSS 6: Columns 118–125, excluding 118, 119, 125; YHSS 7: Columns 126–191; YHSS 8/9: Columns 192–223. An * indicates minor economic taxa listed on p. 263; a † indicates additional wild and weedy components listed on pp. 263–264

Column no.	1	2†	3	4	5*†	6	7	8	9†	10†	11†	12*	13	14*†	15
YH no.	20295	26472	26647	26871	20605	20608	21153	20950	22343	21728	22074	22075	20630	20480	20796
Context type	pit	pit	pit	pit	pit	pit	pyro	coll	surf	pyro	pyro	pyro	coll	pit	pit
Operation	2	2	2	2	1	1	2	1	1	1	1	1	1	88.01	1
Locus	17	42	42	42	11	11	29	24	68	43	43	43	11	11	22
Lot	22	88	108	110	21	25	57	47	160	106	126	126	28	19	39
Phase	110.02	110.04	110.06	110.08	120	120	130	150	150	150.03	150.03	150.03	160.02	180	180
Soil volume (liters)	4	7.5	3	3.5	3	4	5	8	4	10	16	16	4	3	6.5
Charcoal (>2 mm, g)	0.45	19.09	0.91	0.43	2.04	1.33	0.31	0.69	2.06	5.67	6.71	9.77	2.09	0.64	4.56
Seed (>2 mm, g)	0.04	0.07	0.01	0.02	0.05	0.04	0.01	0.02	0.68	0.57	0.35	0.76	0.08	0.05	0.04
Other (>2 mm, g)	0.01	0.75	-	-	0.04	+	-	-	0.03	0.03	0.01	0.06	-	0.07	-
Wild/weedy (#)	13	120	3	-	59	-	1	16	214	911	446	566	13	158	13
ECONOMIC PLANTS															
Hordeum vulgare (g)	0.02	0.02	0.01	-	0.01	0.03	-	-	0.16	0.18	0.04	0.16	0.03	0.01	0.01
H. vulgare var. nudum (g)	-	-	-	-	-	-	-	-	-	-	-	-	-	-	-
Triticum aestivum/durum (g)	-	0.02	-	-	0.02	0.01	0.03	0.01	0.09	0.08	0.08	0.16	0.02	0.01	0.01
Triticum monococcum (g)	-	-	-	-	-	-	-	-	+	-	-	-	-	-	-
Triticum dicoccum (g)	-	-	-	-	-	-	-	-	0.01	-	-	-	-	0.01	-
Triticum sp. (g)	-	-	-	-	-	-	-	+	0.09	-	0.04	-	-	0.01	-
Cereal, indet. (g)	-	-	-	-	0.01	-	-	-	0.28	0.12	0.04	0.06	0.03	-	0.01
Secale cereale (g)	-	-	-	0.02	-	-	-	-	-	-	+	-	-	-	-
Setaria italica (#)	-	-	-	-	-	-	-	-	-	30	58	-	-	3	-
Oryza (g)	-	-	-	-	-	-	-	-	-	+	0.03	0.05	-	-	-
Vicia ervilia (g)	0.02	-	-	-	-	-	-	-	-	0.1	0.02	0.06	-	-	-
Lens (g)	-	-	-	-	+	-	-	-	-	-	0.02	0.06	-	-	-
Pulse (g)	+	-	-	-	-	-	-	-	0.01	-	-	0.08	-	-	0.01
cf. Pistacia/nutshell (g)	-	-	-	-	-	-	-	-	-	-	-	-	-	-	-
cf. Quercus (g)	-	-	-	-	-	-	-	-	-	-	-	-	-	-	-
Ficus carica (#)	-	-	-	-	-	-	-	-	-	-	-	-	-	-	2
Vitis vinifera (#, g)	-	-	-	-	-	-	-	-	-	-	-	2	-	-	-
WILD AND WEEDY (counts)															
Bupleurum	-	-	-	-	-	-	-	-	-	-	-	-	-	-	-
cf. Daucus	-	-	-	-	-	-	-	-	-	-	-	-	-	-	-
Torilis leptophylla?	-	-	-	-	-	-	-	-	-	-	-	-	-	-	-
YH-Apiaceae 2	-	-	-	-	-	-	-	-	-	-	-	-	-	-	-
YH-Apiaceae 4/8	-	1	-	-	-	-	-	-	-	-	1	-	-	-	-
YH-Apiaceae 6	-	-	-	-	-	-	-	-	-	-	-	-	-	-	-
YH-Apiaceae 7	-	-	-	-	-	-	-	-	-	-	-	-	-	-	-
YH-Apiaceae 10	-	-	-	-	-	-	-	-	-	-	-	-	-	-	-
YH-unknown 31 (Apiaceae)	-	-	-	-	-	-	-	-	-	-	-	-	-	-	-
APIACEAE	-	-	-	-	-	-	-	-	-	-	2	-	-	-	-
Anthemis/Matricaria	-	-	-	-	-	-	-	-	-	-	-	1	-	-	-
Artemisia	-	-	-	-	-	-	-	-	-	2	-	-	-	-	-
Carthamus	-	1	-	-	-	-	-	-	-	-	-	-	-	-	-
Centaurea cyanus-type	-	-	-	-	-	-	-	-	-	-	-	-	-	-	-

Table F2, cont'd.: Columns 1–15

Column no.	1	2†	3	4	5*†	6	7	8	9†	10†	11†	12*	13	14*†	15
YH no.	20295	26472	26647	26871	20605	20608	21153	20950	22343	21728	22074	22075	20630	20480	20796
Centaurea	-	-	-	-	-	-	-	-	-	3	2	1	-	-	-
Onopordum	-	-	-	-	-	-	-	-	1	-	-	-	-	-	-
YH-Asteraceae 1	-	-	-	-	-	-	-	-	-	-	-	-	-	-	-
YH-Asteraceae 2	-	-	-	-	-	-	-	-	1	-	2	-	-	-	-
YH-Asteraceae 5	-	-	-	-	-	-	-	-	-	1	-	-	-	-	-
YH-Asteraceae 9	-	-	-	-	1	-	-	-	-	-	-	-	-	-	-
ASTERACEAE	-	-	-	-	-	-	-	-	-	6	-	1	-	-	-
Arnebia/Lithospermum	-	-	-	-	-	-	-	-	-	-	-	2	-	-	-
Heliotropium	-	-	-	-	-	-	-	-	1	2	1	-	-	-	-
cf. Alyssum	-	-	-	-	-	-	-	-	-	-	-	-	-	-	-
cf. Camelina rumelica	-	-	-	-	-	-	-	-	-	-	-	-	-	-	-
cf. Camelina sativa	-	-	-	-	-	-	-	-	-	-	-	-	-	-	-
Conringia	-	-	-	-	-	-	-	-	-	-	-	-	-	-	-
cf. Lepidium	-	-	-	-	-	-	-	-	1	-	-	-	-	-	-
Sisymbrium altissimum-type	-	-	-	-	-	-	-	-	-	-	-	-	-	-	-
Thlaspi	-	-	-	-	-	-	-	-	-	-	-	-	-	-	-
YH-Brassicaceae 2	-	-	-	-	-	-	-	-	6	-	2	-	-	-	-
YH-Brassicaceae 3/5	-	-	-	-	-	-	-	-	1	-	-	-	-	-	-
YH-Brassicaceae 7	-	-	-	-	-	-	-	-	2	18	1	5	-	-	-
YH-Brassicaceae 10	-	-	-	-	-	-	-	-	1	-	-	2	-	-	-
YH-Brassicaceae 11	-	-	-	-	-	-	-	-	-	2	3	-	-	1	-
YH-Brassicaceae 12	-	-	-	-	1	-	-	-	-	-	2	-	-	-	-
BRASSICACEAE	-	-	-	-	-	-	-	-	3	5	2	5	-	-	1
Bufonia	-	-	-	-	-	-	-	-	1	-	-	-	-	-	-
Gypsophila	1	1	-	-	2	-	-	-	1	4	4	2	-	-	-
Silene	-	-	-	-	1	-	-	-	2	-	1	-	1	-	-
Vaccaria (YH-unknown 6)	-	-	-	-	-	-	-	-	-	-	-	-	-	-	-
CARYOPHYLLACEAE	-	-	-	-	-	-	-	-	1	-	-	-	-	1	-
YH-Caryophyllaceae 1	-	-	-	-	-	-	-	-	-	-	-	1	-	-	-
Atriplex	-	-	-	-	-	-	-	-	-	-	-	1	-	1	-
Chenopodium	-	-	-	-	-	-	1	-	2	19	15	9	-	-	-
Salsola kali-type	-	13	-	-	4	-	-	1	-	-	41	3	-	-	-
Salsola soda-type	-	-	-	-	-	-	-	-	-	-	-	-	-	-	-
Salsola	-	-	-	-	-	-	-	2	-	-	-	-	2	1	-
Suaeda	-	2	-	-	-	-	-	1	-	-	5	-	-	-	-
YH-Chenopodiaceae 2	-	4	1	-	-	-	-	-	-	2	1	-	-	-	-
CHENOPODIACEAE	-	4	1	-	2	-	-	-	-	38	-	-	-	-	-
Helianthemum	-	1	-	-	-	-	-	-	-	-	-	-	-	-	-
Carex	1	3	-	-	1	-	-	-	-	39	20	55	2	1	1
Carex 3	-	1	-	-	-	-	-	-	2	-	1	-	-	-	-
Eleocharis (YH-Cyperaceae a)	3	2	-	-	-	-	-	-	-	14	4	15	-	-	1
YH-Cyperaceae 1	-	10	-	-	-	-	-	8	2	102	111	227	-	80	-
YH-Cyperaceae 3	-	-	-	-	-	-	-	-	-	-	-	-	-	-	-
YH-Cyperaceae 4	-	-	-	-	-	-	-	-	-	-	-	-	-	-	-
YH-Cyperaceae 5	1	-	-	-	1	-	-	-	1	-	-	-	-	-	-
YH-Cyperaceae 7	-	-	-	-	-	-	-	-	-	-	1	-	-	3	-
YH-Cyperaceae 8	2	1	-	-	2	-	-	-	-	-	6	5	-	25	-
CYPERACEAE	-	3	-	-	3	-	-	-	-	53	30	26	-	-	-

Table F2, cont'd.: Columns 1–15

Column no.	1	2†	3	4	5*†	6	7	8	9†	10†	11†	12*	13	14*†	15
YH no.	20295	26472	26647	26871	20605	20608	21153	20950	22343	21728	22074	22075	20630	20480	20796
Cephalaria	-	1	-	-	-	-	-	-	-	-	7	4	-	-	-
Scabiosa	-	-	-	-	-	-	-	-	-	-	-	-	-	-	-
Euphorbia	-	-	-	-	-	-	-	-	-	-	-	-	-	-	-
Alhagi	-	-	-	-	3	-	-	-	-	-	-	-	1	-	-
Astragalus	-	1	-	-	2	-	-	-	-	-	1	5	-	-	-
Medicago	-	-	-	-	-	-	-	-	2	3	-	6	-	-	-
Onobrychis	-	-	-	-	1	-	-	-	-	-	-	-	-	-	-
Trifolium/Melilotus	-	4	-	-	-	-	-	1	-	5	-	3	-	-	-
Trigonella	1	9	-	-	5	-	-	-	10	24	37	44	-	-	1
Trigonella cf. astroites	-	-	-	-	-	-	-	-	2	5	5	6	-	2	-
YH-Fabaceae 2	-	-	-	-	-	-	-	-	-	-	-	-	-	-	-
FABACEAE	-	19	1	-	8	-	-	-	8	8	18	17	-	-	2
cf. Erodium	-	-	-	-	-	-	-	-	-	-	-	-	-	-	-
Teucrium	-	-	-	-	-	-	-	-	-	-	-	1	-	-	-
Ziziphora	-	-	-	-	-	-	-	-	-	4	2	1	-	-	-
YH-Lamiaceae 2	-	-	-	-	-	-	-	-	-	-	-	-	-	-	-
YH-Lamiac 3 (Nepeta?)	-	-	-	-	-	-	-	-	-	-	-	-	-	-	-
YH-Lamiaceae 5	-	-	-	-	-	-	-	-	-	-	-	-	-	-	-
LAMIACEAE	-	-	-	-	-	-	-	-	1	-	-	1	-	-	-
LILIACEAE	-	-	-	-	-	-	-	-	-	-	-	-	-	-	-
cf. Malva	-	-	-	-	-	-	-	-	-	2	1	-	-	-	-
Glaucium	-	-	-	-	-	-	-	-	-	-	-	-	-	-	-
Papaver	-	-	-	-	-	-	-	-	6	-	1	-	-	-	-
Fumaria	-	-	-	-	-	-	-	-	-	-	1	1	-	-	-
Plantago	-	-	-	-	-	-	-	-	2	-	-	-	-	-	-
Aegilops	-	-	-	-	-	-	-	-	-	-	-	-	-	-	-
Avena	-	-	-	-	-	-	-	-	1	-	-	1	-	-	-
Bromus japonicus-type	-	-	-	-	-	-	-	-	-	3	2	-	1	-	-
Bromus tectorum-type	-	-	-	-	-	-	-	-	5	1	1	-	-	-	-
Bromus	-	-	-	-	1	-	-	-	-	5	-	3	-	-	-
Eremopyrum	-	-	-	-	-	-	-	-	-	-	1	2	-	-	-
Hordeum cf. murinum	1	1	-	-	-	-	-	-	1	86	6	9	-	-	-
Hordeum spontaneum?	-	-	-	-	-	-	-	-	-	-	-	-	-	-	1
Hordeum misc.	-	-	-	-	-	-	-	-	-	-	-	1	-	-	-
cf. Lolium	-	-	-	-	-	-	-	-	-	-	-	-	-	-	-
cf. Phalaris	-	-	-	-	-	-	-	-	1	-	-	-	-	-	-
cf. Poa bulbosa	-	-	-	-	-	-	-	-	-	-	-	-	1	-	-
Setaria	-	-	-	-	-	-	-	-	-	-	1	-	-	-	-
Stipa	-	-	-	-	-	-	-	-	-	-	-	2	-	-	-
Taeniatherum	-	-	-	-	-	-	-	-	1	1	1	-	-	-	-
"Triticoid"	-	-	-	-	-	-	-	-	-	-	-	-	-	-	-
Triticum boeoticum	-	-	-	-	-	-	-	-	-	-	-	-	-	-	-
YH-Poaceae 1	-	-	-	-	-	-	-	1	-	2	-	-	-	1	1
YH-Poaceae 2	-	-	-	-	-	-	-	-	-	2	-	-	-	-	2
YH-Poaceae 3	-	-	-	-	2	-	-	-	-	13	-	2	-	-	-
YH-Poaceae 4	-	1	-	-	-	-	-	-	-	-	4	1	-	-	-
YH-Poaceae 5	-	-	-	-	1	-	-	-	89	-	-	-	-	-	-
YH-Poaceae 8	-	1	-	-	1	-	-	-	-	-	19	34	-	-	-

Table F2, cont'd.: Columns 1–15

Column no.	1	2†	3	4	5*†	6	7	8	9†	10†	11†	12*	13	14*†	15
YH no.	20295	26472	26647	26871	20605	20608	21153	20950	22343	21728	22074	22075	20630	20480	20796
YH-Poaceae 10/15	-	-	-	-	-	-	-	-	-	4	2	1	-	-	-
YH-Poaceae 11	-	-	-	-	-	-	-	-	-	-	-	-	-	-	-
YH-Poaceae 13	1	1	-	-	-	-	-	-	-	16	-	1	-	-	-
YH-Poaceae 14	-	-	-	-	-	-	-	-	-	1	-	-	-	-	-
YH-Poaceae 16	-	-	-	-	-	-	-	-	-	-	-	-	-	-	-
YH-Poaceae 17/18	-	-	-	-	-	-	-	-	-	-	-	-	-	-	-
YH-Poaceae 20 Aeg 1a	-	-	-	-	-	-	-	-	-	-	-	-	-	-	-
YH-Poaceae 21 Aeg 1b	-	-	-	-	-	-	-	-	-	-	-	-	-	-	-
POACEAE	1	22	-	-	12	-	-	1	40	41	42	42	1	3	2
Polygonum	-	-	-	-	-	-	-	-	-	-	-	-	-	-	-
Polygonum (was YH-Cyperaceae 2)	-	2	-	-	-	-	-	-	1	7	4	-	1	-	-
Polygonum (was YH-Cyperaceae 6)	-	-	-	-	-	-	-	-	-	-	-	-	-	36	-
Rumex	-	6	-	-	-	-	-	-	1	-	-	1	1	-	-
Portulaca	-	-	-	-	-	-	-	-	-	-	-	-	-	-	-
Androsace	-	-	-	-	-	-	-	-	-	-	-	1	-	-	-
YH-Primulaceae 1	-	-	-	-	-	-	-	-	-	-	-	-	-	-	-
YH-Primulaceae 2	-	-	-	-	-	-	-	-	-	-	-	-	-	-	-
Aconitum	-	-	-	-	-	-	-	-	-	-	-	-	-	-	-
Adonis	-	-	-	-	-	-	-	-	-	4	1	1	-	-	-
Ceratocephalus (YH-pp 2)	-	-	-	-	-	-	-	-	-	-	-	-	-	-	-
Ranunculus (YH-unknown 4)	-	-	-	-	3	-	-	-	-	3	1	-	-	-	-
Ranunculus arvensis	-	-	-	-	-	-	-	-	-	-	-	-	-	-	-
Reseda	-	1	-	-	-	-	-	1	-	-	-	-	-	-	-
cf. Asperula (YH-Rubiaceae 2)	-	-	-	-	-	-	-	-	-	-	2	-	-	-	-
Galium	-	1	-	-	1	-	-	-	1	4	6	3	2	-	1
YH-Rubiaceae 1	-	2	-	-	-	-	-	-	-	1	7	3	-	-	-
YH-Scrophulariaceae 1	-	-	-	-	-	-	-	-	-	-	-	-	-	-	-
Hyoscyamus	1	-	-	-	-	-	-	-	3	-	1	-	-	-	-
SOLANACEAE	-	-	-	-	-	-	-	-	-	-	-	-	-	-	-
Thymelaea	-	-	-	-	-	-	-	-	-	-	-	-	-	-	-
Valerianella coronata	-	-	-	-	-	-	-	-	1	-	-	-	-	-	-
Valerianella vesicaria	-	-	-	-	-	-	-	-	-	-	-	-	-	-	-
Valerianella	-	-	-	-	1	-	-	-	-	-	-	1	-	-	-
Peganum harmala	-	-	-	-	-	-	-	-	-	-	5	4	-	-	-
Zygophyllum	-	-	-	-	-	-	-	-	-	-	-	-	-	-	-
YH-unknown 9	-	-	-	-	-	-	-	-	-	359	9	-	-	2	-
YH-unknown 21	-	-	-	-	-	-	-	-	-	-	-	4	-	1	-
YH-unknown 23	-	1	-	-	-	-	-	-	1	-	-	-	-	-	-
YH-unknown 27	-	-	-	-	-	-	-	-	1	-	-	-	-	-	-
YH-unknown 29	-	-	-	-	-	-	-	-	3	-	-	-	-	-	-
YH-unknown 30	-	-	-	-	-	-	-	-	-	-	-	-	-	-	-
YH-unknown 35	-	-	-	-	-	-	-	-	-	-	-	-	-	-	-
Unknown	5	31	-	-	7	-	4	-	14	121	33	38	5	28	-
PLANT PARTS (pp)															
Hordeum internode (2-row)	-	1	-	-	1	-	-	-	-	17	21	23	-	1	-
Hordeum internode (6-row)	-	-	-	-	1	-	-	-	-	-	8	2	-	-	-
Hordeum internode (compact)	-	-	-	-	-	-	-	-	1	3	5	3	-	-	-

Table F2, cont'd.: Columns 1–15

Column no.	1	2†	3	4	5*†	6	7	8	9†	10†	11†	12*	13	14*†	15
YH no.	20295	26472	26647	26871	20605	20608	21153	20950	22343	21728	22074	22075	20630	20480	20796
Triticum aestivum int.	-	-	-	-	-	-	-	-	-	-	-	-	-	-	-
T. aestivum/durum int.	-	-	-	-	-	-	-	-	9	7	3	2	-	-	-
T. aestivum/durum int. 1	-	-	-	-	-	-	-	-	-	-	-	-	-	-	-
T. aestivum/durum int. 2	-	-	-	-	1	-	-	-	-	-	-	-	-	-	-
T. durum internode	-	-	-	-	-	-	-	-	-	-	-	-	-	-	-
T. monococcum spikelet fork	-	-	-	-	-	-	-	-	-	-	-	-	-	-	-
T. dicoccum spikelet fork	-	-	-	-	-	-	-	-	-	-	-	-	-	-	-
T. monococcum/dicoccum spikelet fork	-	-	-	-	-	-	-	-	-	-	-	-	-	-	-
Triticum rachis base	-	-	-	-	-	-	-	-	-	-	-	-	-	-	-
Triticum, rachis fragment, square cross-section	-	-	-	-	-	-	1	-	4	-	-	1	-	-	-
Cereal culm node	2	2	-	-	4	1	-	-	-	2	11	20	-	3	1
Cereal culm root base?	-	-	-	-	-	-	-	-	-	-	-	-	-	4	-
cf. Oryza glume (charred)	-	-	-	-	-	-	-	-	-	-	+	+	-	-	-
Oryza glume (silicified phytolith)	+	1	-	-	-	-	-	-	-	+	-	-	-	-	-
Vitis peduncle	-	-	-	-	-	-	-	-	1	1	-	4	-	-	-
Onopordum achene collar	-	-	-	-	-	-	-	-	-	-	-	-	-	-	-
Asteraceae head, misc.	-	-	-	-	-	-	-	-	-	-	-	-	-	-	-
Euclidium silique	-	-	-	-	-	-	-	-	-	-	-	-	-	-	-
Capparis thorn	-	-	-	-	-	-	-	-	-	5	-	1	-	-	-
Atriplex bract (YH-pp 12)	-	-	-	-	-	-	-	-	-	4	-	-	-	-	-
Salsola inflor (YH-pp 7, YH-unk 8)	-	-	-	-	-	-	-	-	-	-	-	-	-	-	-
Cyperaceae stem fragment	-	1	-	-	-	-	-	-	-	3	-	-	-	-	-
Alhagi pod section	-	-	-	-	-	1	-	-	2	-	-	-	-	-	-
Alhagi leaf?	-	-	-	-	-	-	-	-	-	-	-	-	-	-	-
Aegilops glume base	-	-	-	-	-	-	-	-	-	-	-	-	-	-	-
Hordeum internode (wild)	-	-	-	-	-	-	-	-	-	-	-	-	-	-	-
Taeniatherum int. (YH-pp 8)	-	-	-	-	1	-	-	-	-	-	-	-	-	-	-
Poaceae, mystery 1 rachis frag.	-	-	-	-	-	-	-	-	-	1	-	-	-	-	-
YH-plant part 9	-	-	-	-	-	-	-	-	-	-	-	-	-	-	-
UNCHARRED-MINERALIZED															
Arnebia linearis	-	-	-	-	-	-	-	-	-	-	-	-	-	-	-
Arnebia/Lithospermum	-	-	-	-	1	1	-	-	2	-	-	2	3	-	3
Lithospermum tenuifolium	-	-	-	-	-	-	-	-	-	-	-	-	-	-	-
Lithospermum arvense	-	-	-	-	-	-	-	-	-	-	-	-	-	-	-
Lithospermum, misc.	-	-	-	-	-	-	-	-	-	-	-	-	-	-	-
Moltkia	-	-	-	-	-	-	-	-	-	-	1	-	-	-	-
BORAGINACEAE	-	-	-	-	-	-	-	-	-	-	-	-	-	-	-
Eleocharis (Cyperaceae-a)	-	-	-	-	-	-	-	-	2	5	1	7	1	-	-
Carex	-	-	-	-	-	-	-	-	-	-	-	-	-	-	-
Ficus	-	-	-	-	-	-	-	-	-	-	1	-	-	-	4
Glaucium	-	-	-	-	-	-	-	-	-	-	-	-	-	-	5
Papaver (white)	-	-	-	-	3	-	-	-	-	-	-	-	-	-	3

Table F2, cont'd.: Columns 16–30

Column no.	16	17	18	19	20	21†	22	23	24	25†	26	27†	28	29†	30†
YH no.	22000	22410	22442	22608	20524	20825	20771	20624	20785	21369	21926	23634	23649	23633	23637
Context type	mixed	pit	pit	mixed	coll	coll	surf	debris	debris	pit	debris	surf	mixed	pyro	pyro
Operation	1	1	1	1	2	2	1	1	1	1	2	7	7	7	7
Locus	47	72	74	85	16	21	20	13	17	28	28	16	16	31	31
Lot	84	141	147	163	17	36	36	27	34	54	101	47	54	50	50
Phase	300	315	315	315	320	320	340	345	345	345.01	350.05	360	360	360.01	360.01
Soil volume (liters)	13	14	15	15	4	7	10	4	6	11	3	10	13	18	18
Charcoal (>2 mm, g)	1.38	9.90	2.72	2.56	0.77	1.24	3.97	0.67	0.67	3.78	1.54	5.28	2.24	2.64	6.68
Seed (>2 mm, g)	0.06	0.31	0.23	0.29	0.04	0.12	0.03	0.12	+	0.06	0.04	0.77	0.15	1.83	1.03
Other (>2 mm, g)	-	0.03	0.02	-	-	0.48	-	-	-	-	-	0.02	-	0.10	0.03
Wild/weedy (#)	10	60	37	20	4	12	25	15	2	26	32	110	24	328	571
ECONOMIC PLANTS															
Hordeum vulgare (g)	-	0.07	0.07	0.15	-	0.07	0.02	0.04	-	0.01	-	0.42	0.04	0.64	0.54
H. vulgare var. nudum (g)	-	-	-	-	-	-	-	-	-	-	-	-	-	-	0.04
Triticum aestivum/durum (g)	0.03	0.10	0.06	0.09	0.03	-	0.01	0.03	-	0.01	0.03	0.17	0.06	0.60	0.21
Triticum monococcum (g)	-	-	-	-	-	-	-	-	-	-	-	-	-	-	-
Triticum dicoccum (g)	-	0.01	-	-	-	-	-	-	-	-	-	-	-	-	0.01
Triticum sp. (g)	-	-	-	0.02	0.01	0.01	-	-	-	-	-	-	-	-	0.12
Cereal, indet. (g)	0.01	0.14	0.12	0.05	-	0.05	0.02	0.02	+	0.03	+	0.36	0.08	0.65	0.18
Secale cereale (g)	-	-	-	-	-	-	-	-	-	0.01	-	-	0.01	-	-
Setaria italica (#)	-	-	-	-	-	-	1	-	-	2	-	-	1	19	-
Oryza (g)	-	-	-	-	-	-	-	-	-	-	-	-	-	-	-
Vicia ervilia (g)	-	-	-	-	-	-	-	-	-	-	-	-	-	-	0.02
Lens (g)	-	-	-	-	-	-	-	-	-	-	-	-	-	0.03	0.02
Pulse (g)	-	-	-	-	-	-	-	0.02	-	0.01	-	+	-	-	0.01
cf. Pistacia/nutshell (g)	-	-	-	-	-	-	-	-	-	-	-	-	-	-	-
cf. Quercus (g)	-	-	-	-	-	-	-	-	-	-	-	-	-	-	-
Ficus carica (#)	-	-	-	-	-	-	-	-	-	-	-	-	-	-	-
Vitis vinifera (# whole, total g)	-	-	1 (.03)	-	-	-	-	-	-	-	-	-	-	-	1 (.02)
WILD AND WEEDY (counts)															
Bupleurum	-	-	-	-	-	-	-	-	-	-	-	-	-	-	-
cf. Daucus	-	-	-	-	-	-	-	-	-	-	-	-	-	-	-
Torilis leptophylla?	-	-	-	-	-	-	-	-	-	-	-	-	-	-	-
YH-Apiaceae 2	-	-	-	-	-	-	-	-	-	-	-	-	-	-	-
YH-Apiaceae 4/8	-	-	-	-	-	-	-	-	-	-	-	-	-	-	-
YH-Apiaceae 6	-	-	-	-	-	-	-	-	-	-	-	-	-	-	-
YH-Apiaceae 7	-	-	-	-	-	-	-	-	-	-	-	-	-	-	-
YH-Apiaceae 10	-	-	-	-	-	-	-	-	-	-	-	-	-	-	-
YH-unknown 31 (Apiaceae)	-	-	-	-	-	-	-	-	-	-	-	-	-	-	-
APIACEAE	-	-	-	-	-	-	-	-	-	-	-	-	-	-	-
Anthemis/Matricaria	-	-	-	-	-	-	-	-	-	-	-	-	-	-	-
Artemisia	-	-	-	-	-	-	-	-	-	-	-	-	-	-	-
Carthamus	-	-	-	-	-	-	-	-	-	-	-	-	-	-	-
Centaurea cyanus-type	-	-	-	-	-	-	-	-	-	-	-	-	-	-	-
Centaurea	-	-	-	-	-	-	-	-	-	1	-	-	1	1	-
Onopordum	-	-	-	-	-	-	-	-	-	-	-	-	-	-	-
YH-Asteraceae 1	-	1	-	-	-	-	-	-	-	-	-	-	-	-	-

Table F2, cont'd.: Columns 16–30

Column no.	16	17	18	19	20	21†	22	23	24	25†	26	27†	28	29†	30†
YH no.	22000	22410	22442	22608	20524	20825	20771	20624	20785	21369	21926	23634	23649	23633	23637
YH-Asteraceae 2	-	-	-	-	1	-	-	-	-	-	-	-	-	-	-
YH-Asteraceae 5	-	-	-	-	-	-	-	-	-	-	-	-	-	-	-
YH-Asteraceae 9	-	-	-	-	-	-	-	-	-	-	-	-	-	-	-
ASTERACEAE	-	-	-	-	-	-	-	-	-	-	-	3	-	1	3
Arnebia/Lithospermum	-	-	-	-	-	1	-	-	-	-	-	-	-	-	-
Heliotropium	-	1	-	-	-	-	-	-	-	-	-	-	-	4	11
cf. Alyssum	-	-	-	-	-	-	-	-	-	-	-	-	-	-	-
cf. Camelina rumelica	-	-	-	-	-	-	-	-	-	-	-	-	-	-	-
cf. Camelina sativa	-	-	-	-	-	-	-	-	-	-	-	-	-	-	-
Conringia	-	-	-	-	-	-	-	-	-	-	-	-	-	-	-
cf. Lepidium	-	-	-	-	-	-	-	-	-	-	-	-	-	-	-
Sisymbrium altissimum-type	-	-	-	-	-	-	-	-	-	-	-	1	-	2	-
Thlaspi	-	-	-	-	-	-	-	-	-	-	-	-	-	-	-
YH-Brassicaceae 2	-	-	-	1	-	-	-	-	-	-	-	-	-	5	1
YH-Brassicaceae 3/5	-	-	1	-	-	-	-	-	-	-	-	-	-	1	-
YH-Brassicaceae 7	-	-	-	-	-	-	-	-	-	-	-	-	-	-	-
YH-Brassicaceae 10	-	-	-	-	-	-	-	-	-	1	-	2	-	-	1
YH-Brassicaceae 11	-	-	-	-	-	-	1	-	-	-	-	2	1	22	19
YH-Brassicaceae 12	-	-	-	-	-	-	-	-	-	-	1	3	-	-	1
BRASSICACEAE	-	2	-	-	-	1	-	3	-	1	-	10	-	16	3
Bufonia	-	1	-	-	-	-	-	-	-	-	-	-	-	-	-
Gypsophila	-	-	-	-	-	-	-	-	-	-	-	-	-	1	-
Silene	-	-	-	1	-	-	-	-	-	-	-	-	1	-	1
Vaccaria (YH-unknown 6)	-	-	-	-	-	-	-	-	-	-	-	-	-	2	-
CARYOPHYLLACEAE	-	-	-	-	-	-	-	-	-	1	-	-	-	-	-
YH-Caryophyllaceae 1	-	-	-	-	-	-	-	-	-	-	-	-	-	-	-
Atriplex	-	-	-	-	-	-	-	-	-	-	-	-	-	1	-
Chenopodium	-	3	-	-	-	1	1	-	-	-	-	9	3	10	48
Salsola kali-type	-	-	-	-	-	-	-	-	-	-	-	-	-	-	1
Salsola soda-type	-	-	-	-	-	-	-	-	-	-	-	-	-	3	-
Salsola	-	2	1	1	-	-	-	-	-	-	-	-	-	-	-
Suaeda	1	-	-	-	-	2	-	2	-	-	-	4	-	2	-
YH-Chenopodiaceae 2	-	-	-	-	-	-	-	-	-	-	-	1	-	-	90
CHENOPODIACEAE	-	-	-	2	-	-	-	3	-	1	-	-	-	28	-
Helianthemum	-	-	-	-	-	-	-	-	-	-	-	-	-	-	-
Carex	1	-	5	-	1	1	-	-	-	1	3	-	2	7	6
Carex 3	-	-	-	-	-	-	-	-	-	-	-	-	-	-	-
Eleocharis (YH-Cyperaceae a)	-	-	-	-	-	-	-	-	-	-	-	2	-	1	-
YH-Cyperaceae 1	1	3	1	2	-	1	2	-	-	1	2	-	-	2	2
YH-Cyperaceae 3	-	-	-	-	-	-	-	-	-	-	-	-	-	-	-
YH-Cyperaceae 4	-	-	-	-	-	-	-	-	-	-	-	-	-	-	-
YH-Cyperaceae 5	-	-	-	-	-	-	-	-	-	-	-	2	-	-	3
YH-Cyperaceae 7	-	-	-	-	-	-	-	-	-	-	-	-	-	-	-
YH-Cyperaceae 8	-	-	-	-	-	-	-	-	-	-	-	-	-	-	-
CYPERACEAE	-	2	1	-	-	-	-	-	-	-	-	8	4	-	5	19
Cephalaria	-	-	-	-	-	-	-	-	-	-	-	-	-	-	-
Scabiosa	-	-	-	-	-	-	-	-	-	-	-	-	-	-	-
Euphorbia	-	-	-	-	-	-	-	-	-	-	-	1	-	2	-

Table F2, cont'd.: Columns 16–30

Column no.	16	17	18	19	20	21†	22	23	24	25†	26	27†	28	29†	30†
YH no.	22000	22410	22442	22608	20524	20825	20771	20624	20785	21369	21926	23634	23649	23633	23637
Alhagi	-	2	2	-	-	-	-	-	-	1	-	14	-	55	86
Astragalus	-	1	-	-	-	-	-	-	-	-	-	1	-	1	-
Medicago	-	-	-	-	1	1	-	-	-	-	1	-	-	-	-
Onobrychis	-	-	-	-	-	-	-	-	-	-	-	-	-	-	-
Trifolium/Melilotus	1	1	-	-	-	-	-	-	-	-	-	-	-	-	1
Trigonella	-	4	2	1	1	-	-	1	1	-	1	5	8	7	9
Trigonella cf. astroites	-	-	-	-	-	-	2	-	-	-	-	-	-	-	2
YH-Fabaceae 2	-	-	-	2	-	1	2	-	-	-	-	-	-	-	-
FABACEAE	-	2	2	-	-	1	3	-	-	3	-	4	-	18	15
cf. Erodium	-	-	-	-	-	-	-	-	-	-	-	-	-	-	-
Teucrium	-	-	-	-	-	-	-	-	-	-	-	-	-	-	1
Ziziphora	-	-	2	1	-	-	-	-	-	-	-	1	-	-	3
YH-Lamiaceae 2	-	-	-	-	-	-	-	-	-	-	-	-	-	-	-
YH-Lamiac 3 (Nepeta?)	-	-	-	-	-	-	-	-	-	-	-	-	-	-	1
YH-Lamiaceae 5	-	-	-	-	-	-	-	-	-	-	-	-	-	-	-
LAMIACEAE	-	-	-	-	-	-	-	-	-	-	-	2	-	-	-
LILIACEAE	-	-	1	-	-	-	-	-	-	-	-	-	-	-	-
cf. Malva	-	1	-	-	-	1	-	-	-	-	-	-	-	-	1
Glaucium	-	-	-	-	-	-	1	-	-	-	-	-	-	7	1
Papaver	-	1	-	-	-	-	-	-	-	-	1	1	-	1	3
Fumaria	-	-	-	1	-	-	-	-	-	-	-	-	-	1	1
Plantago	-	-	-	-	-	-	-	-	-	-	-	-	-	-	-
Aegilops	-	-	-	-	-	-	-	-	-	-	-	-	-	-	-
Avena	-	-	1	-	-	-	-	-	-	-	-	-	-	-	-
Bromus japonicus-type	-	-	-	-	-	-	-	-	-	-	-	-	-	-	8
Bromus tectorum-type	-	-	-	-	-	-	-	-	-	-	-	-	-	-	4
Bromus	-	-	-	-	-	-	1	-	-	-	-	-	-	4	2
Eremopyrum	-	-	1	-	-	1	1	-	-	2	-	1	1	-	-
Hordeum cf. murinum	-	-	-	-	-	-	-	-	-	1	1	-	-	-	-
Hordeum spontaneum?	-	-	-	-	-	-	-	-	-	-	-	2	-	-	-
Hordeum misc.	-	1	-	-	-	-	-	-	-	-	-	-	-	3	-
cf. Lolium	-	-	-	-	-	-	-	-	-	-	-	-	-	-	-
cf. Phalaris	-	-	-	-	-	-	-	-	-	-	-	-	-	-	-
cf. Poa bulbosa	-	-	-	-	-	-	-	-	-	-	-	-	-	-	-
Setaria	-	-	-	-	-	-	-	-	-	-	-	-	-	3	4
Stipa	-	-	-	-	-	-	-	-	-	-	-	1	-	2	-
Taeniatherum	-	-	-	-	-	-	-	-	-	-	-	-	-	-	-
"Triticoid"	-	1	-	-	-	-	-	-	-	-	1	-	-	5	-
Triticum boeoticum	-	-	-	1	-	-	-	-	-	-	-	-	-	-	-
YH-Poaceae 1	-	5	-	-	-	-	-	-	-	-	1	-	2	1	1
YH-Poaceae 2	-	1	-	-	-	-	-	-	-	-	-	-	-	-	-
YH-Poaceae 3	-	2	3	-	-	-	3	-	-	1	-	-	-	1	-
YH-Poaceae 4	-	-	-	-	-	-	-	-	-	-	-	1	-	-	1
YH-Poaceae 5	-	1	-	-	-	-	-	-	-	-	-	-	-	-	-
YH-Poaceae 8	-	-	1	-	-	-	-	-	-	-	2	3	-	4	10
YH-Poaceae 10/15	-	-	-	-	-	-	-	-	-	-	-	-	-	-	1
YH-Poaceae 11	-	-	-	-	-	-	-	-	-	-	-	-	-	-	-
YH-Poaceae 13	-	-	-	-	-	-	-	-	-	-	-	-	1	-	-

Table F2, cont'd.: Columns 16–30

Column no.	16	17	18	19	20	21†	22	23	24	25†	26	27†	28	29†	30†
YH no.	22000	22410	22442	22608	20524	20825	20771	20624	20785	21369	21926	23634	23649	23633	23637
YH-Poaceae 14	-	-	-	-	-	-	-	-	-	-	-	-	-	-	-
YH-Poaceae 16	-	-	-	-	-	-	-	-	-	-	-	-	-	-	-
YH-Poaceae 17/18	-	-	-	-	-	-	-	-	-	-	-	-	-	-	-
YH-Poaceae 20 Aeg 1a	-	-	-	-	-	-	-	-	-	-	-	-	-	-	-
YH-Poaceae 21 Aeg 1b	-	-	-	-	-	-	-	-	-	-	-	-	-	-	-
POACEAE	6	18	10	6	-	-	3	6	-	7	4	23	2	43	20
Polygonum	-	-	-	-	-	-	-	-	-	-	-	-	-	-	-
Polygonum (was YH-Cyperaceae 2)	-	-	-	-	-	-	-	-	-	-	-	1	-	2	-
Polygonum (was YH-Cyperaceae 6)	-	-	-	-	-	-	-	-	-	-	-	-	-	4	3
Rumex	-	-	-	-	-	-	1	-	-	-	-	-	-	-	-
Portulaca	-	-	-	-	-	-	-	-	-	-	-	-	-	2	-
Androsace	-	-	-	-	-	-	-	-	-	-	-	-	-	-	1
YH-Primulaceae 1	-	-	-	-	-	-	-	-	-	-	-	-	-	-	4
YH-Primulaceae 2	-	-	-	-	-	-	-	-	-	-	-	-	-	-	-
Aconitum	-	-	-	-	-	-	-	-	-	-	-	-	-	-	-
Adonis	-	1	2	-	-	-	-	-	-	-	-	-	1	-	-
Ceratocephalus (YH-pp 2)	-	-	-	-	-	-	-	-	-	-	-	-	-	-	-
Ranunculus (YH-unknown 4)	-	-	-	-	-	-	-	-	-	-	-	-	-	-	-
Ranunculus arvensis	-	-	-	-	-	-	-	-	-	-	-	-	-	-	-
Reseda	-	-	-	-	-	-	-	-	-	-	-	-	-	-	-
cf. Asperula (YH-Rubiaceae 2)	-	-	-	-	-	-	-	-	-	-	-	-	-	-	-
Galium	-	3	1	1	-	-	-	-	-	1	3	-	-	2	12
YH-Rubiaceae 1	-	-	-	-	-	-	-	-	-	-	-	4	-	22	161
YH-Scrophulariaceae 1	-	-	-	-	-	-	-	-	-	-	-	-	-	10	1
Hyoscyamus	-	-	-	-	-	-	-	-	-	-	-	-	-	6	-
SOLANACEAE	-	-	-	-	-	-	2	-	-	-	-	-	-	-	1
Thymelaea	-	-	-	-	-	-	-	-	-	-	-	1	-	2	-
Valerianella coronata	-	-	-	-	-	-	-	-	-	-	-	-	-	-	-
Valerianella vesicaria	-	-	-	-	-	-	-	-	-	-	-	-	-	-	-
Valerianella	-	-	-	-	-	-	-	-	-	-	-	-	-	-	-
Peganum harmala	-	-	-	-	-	-	-	-	-	1	-	-	-	5	2
Zygophyllum	-	-	-	-	-	-	1	-	-	2	-	-	-	-	-
YH-unknown 9	-	-	-	-	-	-	-	-	-	-	-	-	-	-	-
YH-unknown 21	-	-	-	-	-	-	-	-	1	-	2	-	-	-	-
YH-unknown 23	-	-	-	-	-	-	1	-	-	-	1	1	1	1	2
YH-unknown 27	-	-	-	-	-	-	-	-	-	-	-	-	-	-	-
YH-unknown 29	-	-	-	-	-	-	-	-	-	-	-	-	-	-	-
YH-unknown 30	-	-	-	-	-	-	-	-	-	-	-	-	-	-	-
YH-unknown 35	-	-	-	-	-	-	-	-	-	-	-	-	-	-	-
Unknown	1	10	2	3	-	-	9	-	-	-	13	50	6	104	64
PLANT PARTS (pp)															
Hordeum internode (2-row)	-	3	1	-	-	-	1	-	-	-	-	19	-	36	10
Hordeum internode (6-row)	-	-	1	-	-	-	-	-	-	1	-	20	-	7	3
Hordeum internode (compact)	-	2	-	-	-	-	-	-	1	-	-	6	-	4	4
Triticum aestivum int.	-	-	-	-	-	-	-	-	-	1	-	-	-	-	-
T. aestivum/durum int.	-	5	7	4	-	-	-	-	1	-	-	7	3	19	39
T. aestivum/durum int. 1	-	-	-	-	-	-	-	-	-	-	-	-	-	-	-

Table F2, cont'd.: Columns 16–30

Column no.	16	17	18	19	20	21†	22	23	24	25†	26	27†	28	29†	30†
YH no.	22000	22410	22442	22608	20524	20825	20771	20624	20785	21369	21926	23634	23649	23633	23637
T. aestivum/durum int. 2	-	-	-	-	-	-	-	-	-	-	-	-	-	-	-
T. durum internode	-	-	-	-	-	-	-	-	-	-	-	-	-	-	-
T. monococcum spikelet fork	-	-	-	-	-	-	-	-	-	-	-	-	-	-	-
T. dicoccum spikelet fork	-	-	-	-	-	-	-	-	-	-	-	-	-	-	-
T. monococcum/dicoccum spikelet fork	-	-	-	-	-	-	-	-	-	-	-	-	-	-	-
Triticum rachis base	-	-	-	-	-	-	-	-	-	-	-	1	-	-	-
Triticum, rachis fragment, square cross-section	-	-	-	-	-	-	-	-	-	-	-	-	-	-	1
Cereal culm node	-	6	4	1	-	-	-	-	-	-	-	7	-	-	6
Cereal culm root base?	-	-	-	-	-	-	-	-	-	-	-	-	-	5	-
cf. Oryza glume (charred)	-	-	-	-	-	-	-	-	-	-	-	-	-	-	-
Oryza glume (silicified phytolith)	-	-	-	-	-	-	-	-	-	-	-	-	-	-	-
Vitis peduncle	-	-	-	-	-	-	-	-	-	-	-	-	-	-	-
Onopordum achene collar	-	-	-	-	-	-	-	-	-	-	-	-	-	-	-
Asteraceae head, misc.	-	-	-	-	-	-	-	-	-	-	-	-	-	-	-
Euclidium silique	-	-	-	-	-	-	-	-	-	-	-	-	-	-	-
Capparis thorn	-	-	-	-	-	-	-	-	-	-	-	-	-	-	-
Atriplex bract (YH-pp 12)	-	-	-	-	-	-	-	-	-	-	-	-	-	-	1
Salsola inflor (YH-pp 7, YH-unk 8)	-	-	-	-	-	-	-	-	-	-	-	-	-	-	-
Cyperaceae stem fragment	-	-	-	-	-	-	-	-	-	-	-	-	-	-	-
Alhagi pod section	-	1	2	1	-	-	-	-	-	-	-	3	-	4	6
Alhagi leaf?	-	-	-	-	-	-	-	-	-	-	-	-	-	-	-
Aegilops glume base	-	-	-	-	-	-	-	-	-	-	-	-	-	-	-
Hordeum internode (wild)	-	-	-	-	-	-	-	-	-	-	-	-	-	-	-
Taeniatherum int. (YH-pp 8)	-	-	-	-	-	-	-	-	-	-	-	-	-	-	-
Poaceae, mystery 1 rachis frag.	-	-	-	-	-	-	-	-	-	-	-	-	-	-	-
YH-plant part 9	-	-	-	-	-	-	-	-	-	-	-	1	-	-	-
UNCHARRED-MINERALIZED															
Arnebia linearis	-	-	-	-	-	-	-	-	-	-	-	-	-	-	-
Arnebia/Lithospermum	1	-	9	8	-	-	-	-	-	1	1	-	3	22	457
Lithospermum tenuifolium	-	1	-	-	-	-	-	-	-	-	-	-	-	-	-
Lithospermum arvense	-	-	-	-	-	-	-	-	-	-	-	-	-	-	-
Lithospermum, misc.	-	-	-	-	-	-	-	-	-	-	-	-	-	-	-
Moltkia	-	-	-	-	-	-	-	-	-	-	-	-	1	-	-
BORAGINACEAE	-	-	-	-	-	-	-	-	-	-	-	-	-	-	-
Eleocharis (Cyperaceae-a)	-	3	2	-	-	-	-	-	-	-	-	-	8	-	-
Carex	-	-	10	-	-	-	-	-	-	1	-	1	-	-	-
Ficus	-	-	+	-	-	-	-	-	-	-	-	-	-	-	-
Glaucium	-	-	1	-	-	-	-	-	1	-	1	2	-	4	2
Papaver (white)	-	-	-	-	-	-	-	-	-	-	-	-	-	-	-

Table F2, cont'd.: Columns 31–45

	31†	32†	33	34	35	36†	37*†	38	39	40	41	42	43	44	45*
Column no.															
YH no.	23393	25713	26223	22734	27658	28338	31277	20924	22022	22023	21352	21937	21366	21212	21207
Context type	debris	pyro	pyro	debris	pyro	pyro	pit	mixed	pyro	debris	pyro	pyro	pyro	pyro	pyro
Operation	7	7	7	7	7	7	7	1	2	2	1	2	1	1	1
Locus	36	17	13	20	22	30	45	21	42	43	23	41	25	31	32
Lot	55	7	35	25	66	105	219	42	114	115	50	104	53	65	63
Phase	360.02	360.05	360.09	365	370.04	370.05	370.13	380	380.01	380.03	380.04	380.05	380.12	380.13	380.14
Soil volume (liters)	15	10	2.5	9	6	15	12	5	0.5	0.9	7	2	13	8	13
Charcoal (>2 mm, g)	3.41	5.35	0.59	1.64	4.23	2.20	6.83	0.73	20.61	0.19	1.42	0.82	3.28	4.08	6.32
Seed (>2 mm, g)	0.90	0.77	0.04	0.30	0.65	2.83	0.79	0.03	+	0.53	0.10	0.09	1.02	0.10	0.35
Other (>2 mm, g)	0.02	0.06	0.02	+	0.14	0.37	0.02	0.01	-	0.42	+	0.02	+	0.01	+
Wild/weedy (#)	169	360	50	28	188	2729	62	9	1	157	158	5	159	23	50
ECONOMIC PLANTS															
Hordeum vulgare (g)	0.40	0.37	0.01	0.11	0.12	0.77	0.41	0.01	+	0.24	0.01	0.01	0.94	0.02	0.17
H. vulgare var. nudum (g)	-	-	-	-	-	-	-	-	-	-	-	-	-	-	-
Triticum aestivum/durum (g)	0.32	0.05	0.02	0.05	0.27	0.83	0.13	0.01	-	-	0.03	0.03	0.07	0.03	0.07
Triticum monococcum (g)	-	-	-	-	-	0.02	-	-	-	-	-	-	0.01	-	-
Triticum dicoccum (g)	0.06	-	-	-	-	0.01	-	-	-	0.01	-	-	0.02	-	-
Triticum sp. (g)	0.05	-	-	-	-	0.34	+	+	-	0.17	0.01	0.02	-	-	-
Cereal, indet. (g)	0.14	0.16	0.03	0.11	0.19	0.72	0.26	0.01	-	0.12	0.06	0.01	0.26	0.03	0.13
Secale cereale (g)	-	0.01	-	-	-	-	-	-	-	-	-	-	-	-	+
Setaria italica (#)	11	35	-	-	-	3	-	-	-	-	-	-	-	-	3
Oryza (g)	-	-	-	-	-	-	-	-	-	-	-	-	-	-	-
Vicia ervilia (g)	0.05	+	-	-	0.02	0.09	-	-	-	-	-	0.01	-	-	-
Lens (g)	-	-	-	-	-	-	-	-	-	-	-	-	-	-	-
Pulse (g)	-	-	-	0.02	+	0.05	0.01	-	-	-	-	+	+	0.02	-
cf. Pistacia/nutshell (g)	-	-	-	-	-	-	-	-	-	-	-	-	-	-	-
cf. Quercus (g)	-	-	-	-	-	0.01	-	-	-	-	-	-	-	-	-
Ficus carica (#)	-	-	-	-	-	-	-	-	-	-	-	-	-	-	-
Vitis vinifera (# whole, total g)	-	2 (.02)	-	-	-	3 (.05)	-	-	-	-	-	-	-	-	-
WILD AND WEEDY (counts)															
Bupleurum	-	-	-	-	-	-	-	-	-	-	-	-	-	-	-
cf. Daucus	-	-	-	-	-	-	-	-	-	-	-	-	-	-	-
Torilis leptophylla?	-	-	-	-	-	-	-	-	-	-	-	-	-	-	-
YH-Apiaceae 2	-	1	-	-	-	2	-	-	-	-	-	-	1	-	-
YH-Apiaceae 4/8	-	-	-	-	-	-	-	-	-	-	-	-	-	-	-
YH-Apiaceae 6	-	-	-	-	-	-	-	-	-	-	-	-	-	-	-
YH-Apiaceae 7	-	-	-	-	-	-	-	-	-	2	-	-	-	-	-
YH-Apiaceae 10	-	-	-	-	-	3	-	-	-	-	-	-	-	-	1
YH-unknown 31 (Apiaceae)	-	-	-	-	-	-	-	-	-	-	-	-	-	-	-
APIACEAE	-	1	-	-	-	3	-	-	-	-	-	-	1	1	-
Anthemis/Matricaria	-	1	-	-	-	-	-	-	-	-	-	-	-	-	-
Artemisia	-	1	-	-	-	-	-	-	-	-	-	-	-	-	-
Carthamus	-	-	-	-	1	-	-	-	-	-	-	-	-	-	-
Centaurea cyanus-type	-	-	-	-	-	3	-	-	-	-	-	-	-	-	-
Centaurea	4	4	1	1	-	21	2	-	-	-	-	-	-	-	-
Onopordum	-	1	-	-	1	33	2	-	-	-	-	-	-	-	-
YH-Asteraceae 1	1	-	-	-	-	-	-	-	-	-	-	-	-	-	-

Table F2, cont'd.: Columns 31–45

Column no.	31†	32†	33	34	35	36†	37*†	38	39	40	41	42	43	44	45*
YH no.	23393	25713	26223	22734	27658	28338	31277	20924	22022	22023	21352	21937	21366	21212	21207
YH-Asteraceae 2	-	-	-	-	1	2	-	-	-	1	-	-	-	-	-
YH-Asteraceae 5	-	-	-	-	-	-	-	-	-	-	-	-	-	-	-
YH-Asteraceae 9	-	-	-	-	-	-	-	-	-	-	-	-	-	-	-
ASTERACEAE	-	-	-	-	-	5	-	-	-	-	-	-	1	-	-
Arnebia/Lithospermum	-	-	-	-	-	-	-	-	-	-	-	-	-	-	-
Heliotropium	-	2	1	1	17	2	1	-	-	1	-	-	-	-	1
cf. Alyssum	-	-	-	-	-	3	-	-	-	-	-	-	-	-	-
cf. Camelina rumelica	-	-	-	-	-	-	-	-	-	-	-	-	-	-	-
cf. Camelina sativa	-	-	-	-	-	-	-	-	-	-	-	-	-	-	-
Conringia	-	-	-	-	-	1	-	-	-	-	-	-	-	-	-
cf. Lepidium	-	-	-	-	1	-	-	-	-	-	-	-	-	-	-
Sisymbrium altissimum-type	-	-	-	-	-	-	-	-	-	1	-	-	-	-	-
Thlaspi	-	-	-	-	-	-	-	-	-	-	-	-	-	-	-
YH-Brassicaceae 2	-	2	-	-	1	-	-	-	-	-	-	-	-	-	-
YH-Brassicaceae 3/5	-	8	-	-	1	-	-	-	-	-	-	-	2	-	-
YH-Brassicaceae 7	-	-	-	-	1	-	-	-	-	-	-	-	-	-	-
YH-Brassicaceae 10	-	-	-	-	-	-	-	-	-	-	-	-	-	-	-
YH-Brassicaceae 11	-	1	-	-	-	3	-	-	-	-	-	-	-	3	-
YH-Brassicaceae 12	-	-	-	-	8	1	-	-	-	-	-	-	-	-	-
BRASSICACEAE	1	112	-	-	5	38	1	-	-	-	1	-	4	-	1
Bufonia	-	-	-	1	-	1	-	-	-	-	-	-	-	-	-
Gypsophila	-	-	-	-	4	1	-	-	-	1	3	-	5	1	2
Silene	-	1	-	-	2	2	1	-	-	4	1	-	2	-	-
Vaccaria (YH-unknown 6)	-	1	-	-	2	-	-	-	-	-	-	1	6	-	-
CARYOPHYLLACEAE	3	-	-	1	-	-	-	-	-	-	-	-	4	-	-
YH-Caryophyllaceae 1	-	-	-	-	-	-	-	-	-	-	-	-	-	-	-
Atriplex	-	2	-	-	-	7	-	-	-	-	-	-	-	-	-
Chenopodium	7	9	3	-	-	-	-	-	-	-	-	-	2	-	2
Salsola kali-type	-	1	2	-	-	1	-	-	-	-	-	-	-	-	1
Salsola soda-type	1	-	-	-	-	3	1	-	-	-	-	-	-	-	-
Salsola	-	-	-	-	-	-	-	-	-	-	-	-	-	-	1
Suaeda	1	1	2	-	-	2	-	-	-	-	3	-	-	1	4
YH-Chenopodiaceae 2	-	-	-	-	-	-	-	-	-	-	12	-	-	-	-
CHENOPODIACEAE	6	-	-	3	3	50	1	-	-	-	2	-	20	-	-
Helianthemum	-	-	-	-	-	-	-	-	-	-	-	-	-	-	-
Carex	2	4	2	-	17	44	10	1	1	5	39	-	13	2	3
Carex 3	-	-	-	-	-	-	-	-	-	-	-	-	-	-	-
Eleocharis (YH-Cyperaceae a)	2	-	-	-	1	-	-	-	-	-	9	-	-	-	-
YH-Cyperaceae 1	3	-	1	3	-	3	3	1	-	-	4	-	19	1	3
YH-Cyperaceae 3	-	-	-	-	-	1	-	-	-	-	-	-	-	-	-
YH-Cyperaceae 4	-	-	-	-	-	4	1	-	-	-	-	-	-	-	-
YH-Cyperaceae 5	-	-	-	-	3	-	-	-	-	-	3	-	-	-	2
YH-Cyperaceae 7	-	-	-	-	-	-	-	-	-	-	-	1	-	-	-
YH-Cyperaceae 8	-	-	-	-	-	-	-	-	-	-	-	-	-	-	-
CYPERACEAE	3	4	1	-	5	10	-	1	-	1	25	-	20	4	10
Cephalaria	-	-	-	-	-	1	-	-	-	1	-	-	-	-	-
Scabiosa	-	-	-	-	-	-	-	-	-	-	-	1	-	-	-
Euphorbia	-	1	-	1	-	1	-	-	-	-	-	-	-	-	-

Table F2, cont'd.: Columns 31–45

Column no.	31†	32†	33	34	35	36†	37*†	38	39	40	41	42	43	44	45*
YH no.	23393	25713	26223	22734	27658	28338	31277	20924	22022	22023	21352	21937	21366	21212	21207
Alhagi	26	124	2	-	1	1	4	-	-	71	1	-	-	-	-
Astragalus	-	-	-	3	-	50	-	-	-	-	-	-	-	-	-
Medicago	1	-	1	-	1	99	-	-	-	-	2	-	6	-	1
Onobrychis	-	-	-	-	-	-	-	-	-	-	-	-	-	-	-
Trifolium/Melilotus	4	-	2	-	6	75	1	-	-	1	1	-	1	-	-
Trigonella	19	9	10	-	6	1003	4	1	-	-	2	-	1	1	1
Trigonella cf. astroites	-	-	-	-	-	69	-	-	-	-	-	-	5	-	-
YH-Fabaceae 2	-	-	-	-	-	-	-	-	-	-	-	-	-	-	-
FABACEAE	3	-	6	-	3	556	6	1	-	8	-	2	9	-	2
cf. Erodium	-	-	-	-	-	-	-	-	-	-	-	-	-	-	-
Teucrium	-	-	-	-	-	1	-	-	-	1	-	-	-	-	-
Ziziphora	2	1	2	-	5	96	-	-	-	3	1	-	-	1	-
YH-Lamiaceae 2	-	-	-	-	-	-	-	-	-	-	-	-	-	-	-
YH-Lamiac 3 (Nepeta?)	-	-	-	-	2	2	-	-	-	-	-	-	2	-	-
YH-Lamiaceae 5	-	2	-	-	-	28	-	-	-	-	-	-	-	-	-
LAMIACEAE	-	-	3	-	1	-	-	-	-	-	-	-	-	-	-
LILIACEAE	-	-	-	-	-	1	-	-	-	-	-	-	-	-	-
cf. Malva	1	-	-	-	-	1	1	-	-	-	1	-	1	-	-
Glaucium	-	-	-	-	13	1	-	-	-	1	1	-	1	-	-
Papaver	4	3	-	1	3	14	-	-	-	6	-	-	-	-	-
Fumaria	-	-	-	-	-	-	-	-	-	-	-	-	-	-	-
Plantago	-	-	-	-	-	3	-	-	-	-	1	-	-	-	-
Aegilops	1	-	-	-	-	-	-	-	-	-	-	-	-	-	-
Avena	-	-	-	-	1	2	-	-	-	-	-	-	-	-	-
Bromus japonicus-type	1	1	1	-	-	13	-	-	-	-	-	-	-	-	-
Bromus tectorum-type	-	-	1	-	-	8	-	-	-	1	-	-	-	-	-
Bromus	-	-	-	-	-	27	3	-	-	-	-	-	-	-	-
Eremopyrum	2	-	-	-	3	6	-	-	-	2	-	-	1	-	-
Hordeum cf. murinum	1	1	2	2	6	4	1	-	-	-	1	-	-	1	-
Hordeum spontaneum?	-	-	-	-	-	-	-	-	-	-	-	-	-	-	-
Hordeum misc.	-	-	-	-	-	-	-	-	-	-	-	-	-	-	-
cf. Lolium	-	-	-	-	-	-	-	-	-	-	-	-	-	-	-
cf. Phalaris	1	-	-	-	-	-	1	-	-	-	-	-	2	-	-
cf. Poa bulbosa	-	-	-	-	1	35	1	-	-	6	-	-	-	-	-
Setaria	1	-	1	-	-	-	-	-	-	-	-	-	-	-	2
Stipa	2	-	-	-	3	-	1	-	-	-	-	1	-	-	-
Taeniatherum	1	-	-	-	-	1	-	-	-	-	-	-	1	-	-
"Triticoid"	1	-	-	-	-	-	-	-	-	-	1	-	-	-	-
Triticum boeoticum	-	-	-	-	-	-	-	-	-	-	-	-	2	-	-
YH-Poaceae 1	1	-	-	-	2	7	3	1	-	7	2	-	-	-	1
YH-Poaceae 2	1	-	-	-	-	5	1	-	-	-	-	-	-	2	-
YH-Poaceae 3	2	-	-	-	-	10	-	-	-	-	1	-	-	-	-
YH-Poaceae 4	-	-	-	-	-	1	-	-	-	-	-	-	-	-	-
YH-Poaceae 5	-	-	-	-	-	-	-	-	-	-	-	-	-	-	-
YH-Poaceae 8	4	-	-	-	1	11	-	-	-	1	1	-	-	-	2
YH-Poaceae 10/15	5	-	-	-	-	-	-	-	-	-	-	-	-	-	-
YH-Poaceae 11	-	-	-	-	-	-	-	-	-	-	-	-	-	-	-
YH-Poaceae 13	5	-	1	-	-	1	-	-	-	-	-	-	-	-	-

Table F2, cont'd.: Columns 31–45

Column no.	31†	32†	33	34	35	36†	37*†	38	39	40	41	42	43	44	45*
YH no.	23393	25713	26223	22734	27658	28338	31277	20924	22022	22023	21352	21937	21366	21212	21207
YH-Poaceae 14	-	-	-	-	-	4	-	-	-	-	-	-	-	-	-
YH-Poaceae 16	1	-	-	-	-	-	-	-	-	-	-	-	-	-	1
YH-Poaceae 17/18	-	-	-	-	1	7	-	-	-	-	1	-	-	-	-
YH-Poaceae 20 Aeg 1a	-	-	-	-	-	2	-	-	-	-	-	-	-	-	-
YH-Poaceae 21 Aeg 1b	-	-	-	-	-	2	-	-	-	-	-	-	-	-	-
POACEAE	18	11	4	5	32	264	4	2	-	17	29	-	21	4	3
Polygonum	-	-	-	-	2	1	-	-	-	-	-	-	-	-	-
Polygonum (was YH-Cyperaceae 2)	3	12	-	-	-	-	1	-	-	-	-	-	-	-	-
Polygonum (was YH-Cyperaceae 6)	-	-	1	-	1	25	-	-	-	-	-	-	-	-	-
Rumex	-	-	-	-	-	-	-	-	-	-	-	-	-	-	-
Portulaca	1	-	-	1	1	-	-	-	-	-	-	-	-	-	-
Androsace	-	-	-	-	2	2	-	-	-	-	1	-	1	-	-
YH-Primulaceae 1	-	-	-	-	-	-	-	-	-	-	-	-	-	-	-
YH-Primulaceae 2	-	-	-	-	-	-	-	-	-	-	-	-	-	-	-
Aconitum	-	-	-	-	-	-	-	-	-	1	-	-	-	-	-
Adonis	-	2	-	-	-	-	-	1	-	2	-	-	2	-	-
Ceratocephalus (YH-pp 2)	-	-	-	-	-	1	-	-	-	4	1	-	-	-	-
Ranunculus (YH-unknown 4)	-	-	-	-	-	-	-	-	-	-	-	-	-	-	-
Ranunculus arvensis	-	-	-	-	-	-	-	-	-	1	-	-	-	-	-
Reseda	-	12	-	-	-	-	-	-	-	-	-	-	1	-	-
cf. Asperula (YH-Rubiaceae 2)	-	-	-	-	2	-	-	-	-	-	-	-	-	-	-
Galium	7	2	-	5	-	11	3	-	-	5	-	-	-	1	-
YH-Rubiaceae 1	5	5	-	-	-	8	-	-	-	-	-	-	-	-	-
YH-Scrophulariaceae 1	-	2	-	-	2	-	-	-	-	-	-	-	-	-	1
Hyoscyamus	5	7	-	-	3	9	-	-	-	-	1	-	1	-	-
SOLANACEAE	-	-	-	-	-	-	-	-	-	-	-	-	-	-	-
Thymelaea	-	-	-	-	-	2	-	-	-	-	-	-	-	-	1
Valerianella coronata	-	2	-	-	-	3	1	-	-	-	-	-	-	-	-
Valerianella vesicaria	-	-	-	-	-	-	-	-	-	-	-	-	-	-	-
Valerianella	-	-	-	-	4	-	-	-	-	-	-	-	-	-	-
Peganum harmala	4	1	-	-	1	-	-	-	-	-	-	-	1	-	1
Zygophyllum	1	-	-	-	-	-	1	-	-	1	-	-	-	-	-
YH-unknown 9	-	-	-	-	-	-	-	-	-	-	-	-	-	-	-
YH-unknown 21	-	-	-	-	1	-	-	-	-	-	6	-	-	-	-
YH-unknown 23	-	4	-	-	-	1	-	-	-	-	-	-	-	-	3
YH-unknown 27	1	-	-	-	4	6	-	-	-	1	-	-	-	-	-
YH-unknown 29	-	-	-	-	-	-	1	-	-	-	-	-	-	-	-
YH-unknown 30	-	-	-	-	-	-	-	-	-	-	-	-	-	-	-
YH-unknown 35	-	1	-	-	-	-	-	-	-	-	-	-	-	-	-
Unknown	85	128	1	5	87	275	24	3	-	5	23	-	20	-	28
PLANT PARTS (pp)															
Hordeum internode (2-row)	2	31	1	3	8	64	5	1	-	30	-	-	-	-	-
Hordeum internode (6-row)	-	5	-	-	-	6	2	-	-	5	-	-	-	-	-
Hordeum internode (compact)	-	15	-	1	2	25	-	-	-	2	-	-	-	-	-
Triticum aestivum int.	2	-	-	-	5	256	2	-	-	97	1	3	-	-	-
T. aestivum/durum int.	7	5	-	4	37	80	14	-	-	102	-	-	-	-	-
T. aestivum/durum int. 1	-	-	-	-	-	-	-	-	-	-	-	-	-	-	-

Table F2, cont'd.: Columns 31–45

Column no.	31†	32†	33	34	35	36†	37*†	38	39	40	41	42	43	44	45*
YH no.	23393	25713	26223	22734	27658	28338	31277	20924	22022	22023	21352	21937	21366	21212	21207
T. aestivum/durum int. 2	-	-	-	-	-	-	-	-	-	-	-	-	-	-	-
T. durum internode	-	-	-	-	-	2	-	-	-	2	-	-	-	-	-
T. monococcum spikelet fork	-	-	-	-	-	-	-	-	-	4	-	-	-	-	1
T. dicoccum spikelet fork	-	-	-	-	-	-	-	-	-	-	-	-	-	-	-
T. monococcum/dicoccum spikelet fork	-	-	-	1	-	4	-	-	-	-	1	-	-	-	-
Triticum rachis base	-	-	-	-	-	-	-	-	-	2	-	-	-	-	-
Triticum, rachis fragment, square cross-section	-	-	-	-	8	12	-	-	-	4	-	-	-	-	-
Cereal culm node	7	10	1	-	14	83	8	-	-	-	-	2	1	1	2
Cereal culm root base?	-	-	-	-	-	-	1	-	-	2	-	-	-	-	-
cf. Oryza glume (charred)	-	-	-	-	-	-	-	-	-	-	-	-	-	-	-
Oryza glume (silicified phytolith)	-	-	-	-	-	-	-	-	-	-	-	-	-	-	-
Vitis peduncle	-	-	-	-	-	3	-	-	-	-	-	-	-	-	-
Onopordum achene collar	-	-	-	-	-	38	-	-	-	-	-	-	-	-	-
Asteraceae head, misc.	-	-	-	-	-	2	-	-	-	-	-	-	-	-	-
Euclidium silique	-	-	-	-	-	-	-	-	-	-	-	-	-	-	-
Capparis thorn	-	-	-	-	-	-	-	-	-	-	-	-	-	-	-
Atriplex bract (YH-pp 12)	-	1	-	-	-	4	-	-	-	-	-	-	-	-	-
Salsola inflor (YH-pp 7, YH-unk 8)	-	-	-	-	-	-	-	-	-	-	-	-	-	-	-
Cyperaceae stem fragment	-	-	-	-	-	1	-	-	-	-	-	-	-	-	-
Alhagi pod section	-	80	-	3	-	2	-	-	-	6	-	-	-	-	-
Alhagi leaf?	-	-	-	-	-	1	-	-	-	-	-	-	-	-	-
Aegilops glume base	1	-	-	-	-	1	-	-	-	-	-	1	-	-	-
Hordeum internode (wild)	-	-	-	-	-	-	-	-	-	-	-	-	-	-	-
Taeniatherum int. (YH-pp 8)	-	-	-	-	-	-	-	-	-	-	-	-	-	-	-
Poaceae, mystery 1 rachis frag.	-	-	-	-	3	1	-	-	-	-	-	-	-	-	-
YH-plant part 9	-	-	-	-	-	-	-	-	-	-	-	-	-	-	-
UNCHARRED-MINERALIZED															
Arnebia linearis	-	-	-	-	-	-	-	-	-	-	-	-	-	-	-
Arnebia/Lithospermum	-	2	2	-	12	1	47	2	1	6	-	1	2	-	-
Lithospermum tenuifolium	-	-	-	-	-	-	-	-	-	-	-	-	1	1	-
Lithospermum arvense	-	-	-	-	-	-	1	-	-	-	-	-	-	-	-
Lithospermum, misc.	-	-	-	-	-	2	-	-	-	-	-	-	-	-	-
Moltkia	1	-	-	-	-	2	-	-	-	-	-	-	-	-	-
BORAGINACEAE	2	-	-	-	-	-	-	-	-	-	-	-	-	-	-
Eleocharis (Cyperaceae-a)	1	-	-	-	1	-	-	-	-	-	-	-	-	-	-
Carex	-	-	-	-	-	1	2	1	1	-	-	-	-	-	-
Ficus	-	-	-	-	-	-	-	-	-	-	-	-	-	-	-
Glaucium	1	1	1	-	1	1	1	-	-	-	-	-	-	-	-
Papaver (white)	-	-	-	-	-	4	2	-	-	2	-	-	-	-	-

Table F2, cont'd.: Columns 46–60

Column no.	46†	47†	48†	49†	50†	51†	52*	53†	54†	55	56	57	58†	59†	60
YH no.	21245	21585	21719	22669	27718	22085	22084	22096	22109	22138	22691	22407	22337	29295	26418
Context type	pyro	pit	pyro	pit	pit	pit	pit	pit	pit	pit	pit	pit	pit	pit	pit
Operation	1	1	1	1	2	1	1	1	1	1	1	1	1	7	1
Locus	38	50	52	88	62	64	63	65	65	70	70	71	81	26	16
Lot	73	93	102	175	151	124	127	128	130	138	177	139	158	145	52
Phase	380.15	380.17	380.18	390.01	410.01	410.02	410.03	410.04	410.04	410.06	410.06	410.07	410.09	410.11	410.17
Soil volume (liters)	16	14	13	15	14	15	14	15	15.5	15	15	14	13	8	7
Charcoal (>2 mm, g)	3.54	3.54	10.06	3.63	3.91	9.33	7.78	7.37	4.03	19.68	5.45	4.43	6.17	17.83	6.94
Seed (>2 mm, g)	0.07	0.13	0.84	1.76	0.26	0.26	0.21	0.45	0.86	0.70	1.20	0.27	0.22	0.53	0.44
Other (>2 mm, g)	-	+	0.49	0.17	0.08	0.04	-	0.19	0.10	0.02	0.03	0.01	+	0.02	0.06
Wild/weedy (#)	15	141	633	144	191	82	57	133	1211	150	384	40	42	66	46
ECONOMIC PLANTS															
Hordeum vulgare (g)	+	0.06	0.21	0.34	0.13	0.08	0.09	0.09	0.21	0.13	0.47	0.09	0.11	0.22	0.16
H. vulgare var. nudum (g)	-	-	-	-	-	-	-	-	-	-	-	-	-	-	-
Triticum aestivum/durum (g)	0.04	0.02	0.03	0.64	0.03	0.11	0.07	0.08	0.22	0.33	0.13	0.08	0.09	0.10	0.04
Triticum monococcum (g)	-	-	-	-	-	-	0.02	-	0.01	-	-	-	-	-	-
Triticum dicoccum (g)	-	-	-	-	-	-	-	-	-	-	-	-	-	-	-
Triticum sp. (g)	+	-	-	0.37	-	-	-	-	0.02	-	0.03	-	-	-	0.12
Cereal, indet. (g)	0.05	0.01	0.06	0.40	0.12	0.06	0.05	0.1	0.09	0.20	0.28	0.08	0.10	0.19	0.15
Secale cereale (g)	-	0.01	0.03	-	-	-	0.01	-	-	-	-	-	-	-	-
Setaria italica (#)	-	2	-	-	-	-	148	1	-	-	-	-	-	3	-
Oryza (g)	-	-	-	-	-	-	-	-	-	-	-	-	-	-	-
Vicia ervilia (g)	-	-	0.01	0.05	-	0.01	0.02	-	+	0.03	0.02	+	-	0.01	-
Lens (g)	-	0.01	0.40	+	-	-	-	-	-	0.01	0.04	-	-	-	+
Pulse (g)	0.01	0.02	0.06	-	-	-	0.02	-	-	0.02	+	-	-	-	-
cf. Pistacia/nutshell (g)	-	-	-	-	-	-	-	-	-	-	-	-	-	-	-
cf. Quercus (g)	-	-	+	-	-	-	-	-	-	-	-	-	-	-	-
Ficus carica (#)	-	-	-	-	-	-	-	-	-	-	-	-	-	-	-
Vitis vinifera (# whole, total g)	-	-	1	-	-	-	-	-	-	-	-	-	-	-	-
WILD AND WEEDY (counts)															
Bupleurum	-	-	-	-	-	-	-	-	-	-	-	-	-	-	-
cf. Daucus	-	-	-	-	-	-	-	-	-	-	-	-	-	-	-
Torilis leptophylla?	-	-	-	-	-	-	-	-	1	-	-	-	-	-	-
YH-Apiaceae 2	-	-	-	-	-	-	-	-	1	-	-	-	-	-	-
YH-Apiaceae 4/8	1	-	-	-	-	-	-	-	1	-	-	-	-	-	-
YH-Apiaceae 6	-	-	-	-	-	-	-	-	-	-	-	-	-	-	-
YH-Apiaceae 7	-	-	-	-	-	-	-	-	-	-	-	-	-	-	-
YH-Apiaceae 10	-	-	-	-	1	-	-	1	-	-	-	-	-	-	-
YH-unknown 31 (Apiaceae)	-	-	-	3	-	-	-	-	-	-	-	-	-	-	-
APIACEAE	-	-	-	-	3	-	-	-	1	-	-	-	-	-	1
Anthemis/Matricaria	-	-	-	-	-	-	-	-	-	-	-	-	-	-	-
Artemisia	-	-	-	-	-	-	-	-	-	-	-	-	-	-	-
Carthamus	-	-	-	-	-	-	-	-	-	-	-	-	-	-	-
Centaurea cyanus-type	-	-	-	-	-	1	-	-	-	-	-	-	-	-	-
Centaurea	-	2	-	6	14	1	2	3	2	2	-	-	2	-	-
Onopordum	-	-	-	-	16	-	-	1	-	1	1	-	-	-	-
YH-Asteraceae 1	-	-	-	-	-	3	-	-	1	-	-	-	-	-	-

Table F2, cont'd.: Columns 61–75

Column no.	61†	62*	63†	64	65	66	67	68†	69†	70	71	72	73*†	74	75
YH no.	25652	30039	28856	29567	29573	31528	22441	23009	23053	25653	25844	23150	23307	22411	25514
Context type	pit	surf	pyro	pyro	pyro	pit	pyro	pit	pit	pit	pit	pit	pit	fill	pit
Operation	1	7	7	7	7	7	1	1	1	1	2	2	2	1	1
Locus	7	31	31	31	31	51	75	91	93	8	43	64	69	73	5
Lot	14	172	120	151	151	240	149	181	188	15	50	175	176	142	10
Phase	410.18	415	415.01	415.02	415.02	415.04	420	420.01	420.02	420.03	420.06	430	430	430.02	430.03
Soil volume (liters)	2	2	15	1	1	11	1.5	14	12	2	14	14	2.3	15	7
Charcoal (>2 mm, g)	2.04	1.37	2.26	8.57	0.60	7.09	15.03	5.64	4.12	1.90	4.85	6.48	0.82	0.72	7.33
Seed (>2 mm, g)	0.45	0.04	1.54	0.08	0.67	0.41	0.02	0.97	0.22	0.09	0.16	0.32	3.11	0.07	0.23
Other (>2 mm, g)	0.19	-	0.43	-	0.71	+	+	0.10	+	+	0.01	-	2.3	-	-
Wild/weedy (#)	88	10	727	14	181	28	6	542	26	20	31	36	1260	10	11
ECONOMIC PLANTS															
Hordeum vulgare (g)	0.09	+	0.63	0.04	0.14	0.15	-	0.27	0.04	0.02	0.05	0.13	1.38	0.02	0.02
H. vulgare var. nudum (g)	-	-	-	-	-	-	-	-	-	-	-	-	-	-	-
Triticum aestivum/durum (g)	0.17	-	0.43	0.04	0.44	0.10	0.01	0.41	0.11	-	0.06	-	1.00	0.04	0.02
Triticum monococcum (g)	-	-	+	-	-	-	-	-	-	-	-	-	-	0.01	-
Triticum dicoccum (g)	-	-	-	-	0.02	-	-	-	-	-	-	-	-	-	-
Triticum sp. (g)	0.02	+	0.23	-	-	0.01	-	0.05	-	-	-	0.06	0.26	-	-
Cereal, indet. (g)	0.16	+	0.28	0.03	0.12	0.22	0.01	0.19	0.04	0.04	0.08	0.10	0.21	0.01	0.04
Secale cereale (g)	+	-	-	-	-	-	-	-	-	-	-	+	-	-	-
Setaria italica (#)	-	-	-	-	-	-	-	-	-	-	-	2	-	-	-
Oryza (g)	-	-	-	-	-	-	-	-	-	-	-	-	-	-	-
Vicia ervilia (g)	-	-	+	-	0.01	-	-	0.03	-	0.01	0.01	-	0.01	-	-
Lens (g)	-	-	0.01	-	-	-	-	-	-	-	-	-	-	-	-
Pulse (g)	-	-	0.04	-	-	-	+	0.01	-	-	+	-	-	-	-
cf. Pistacia/nutshell (g)	-	-	-	+	-	-	-	-	-	-	-	-	-	-	-
cf. Quercus (g)	-	-	-	-	-	-	-	-	-	-	-	-	-	-	-
Ficus carica (#)	-	-	1	-	-	-	-	-	-	-	-	-	-	-	-
Vitis vinifera (# whole, total g)	-	+	4 (+)	-	-	+	-	-	-	-	-	-	-	-	-
WILD AND WEEDY (counts)															
Bupleurum	-	-	-	-	-	-	-	-	-	-	-	-	-	-	-
cf. Daucus	-	-	-	-	-	-	-	-	-	-	-	-	-	-	-
Torilis leptophylla?	-	-	-	-	-	-	-	-	-	-	-	-	-	-	-
YH-Apiaceae 2	-	-	-	-	-	2	-	-	-	-	-	-	-	-	-
YH-Apiaceae 4/8	-	-	1	-	-	-	-	-	-	-	-	-	-	-	-
YH-Apiaceae 6	-	-	-	-	-	-	-	-	-	-	-	-	-	-	-
YH-Apiaceae 7	-	-	-	-	-	-	-	-	-	-	-	-	-	-	-
YH-Apiaceae 10	2	-	-	-	1	-	-	-	-	-	-	-	-	-	-
YH-unknown 31 (Apiaceae)	-	-	2	-	-	-	-	-	-	-	-	-	-	-	-
APIACEAE	1	-	-	-	-	-	-	-	-	-	-	-	-	-	2
Anthemis/Matricaria	-	-	-	-	-	-	-	-	-	-	-	-	-	-	-
Artemisia	-	-	-	2	14	-	-	-	-	-	-	-	-	-	-
Carthamus	-	-	-	-	-	-	-	-	-	-	-	-	-	-	-
Centaurea cyanus-type	-	-	-	-	-	-	-	-	-	-	-	1	-	-	-
Centaurea	-	-	5	-	-	-	-	7	-	-	-	1	2	-	-
Onopordum	-	-	4	-	1	-	-	1	-	-	-	1	-	-	-
YH-Asteraceae 1	-	-	-	-	-	-	-	1	-	-	-	-	93	-	-

Table F2, cont'd.: Columns 61–75

Column no.	61†	62*	63†	64	65	66	67	68†	69†	70	71	72	73*†	74	75
YH no.	25652	30039	28856	29567	29573	31528	22441	23009	23053	25653	25844	23150	23307	22411	25514
YH-Asteraceae 2	–	–	–	–	1	–	–	1	–	–	–	–	52	–	–
YH-Asteraceae 5	–	–	–	–	–	–	–	–	–	–	–	–	–	–	–
YH-Asteraceae 9	–	–	–	–	1	–	–	–	–	–	–	–	–	–	–
ASTERACEAE	2	–	1	–	1	–	–	18	–	–	–	–	2	–	–
Arnebia/Lithospermum	–	–	–	–	–	–	–	–	–	–	–	–	–	–	–
Heliotropium	–	–	29	1	–	–	–	–	1	–	–	–	–	–	–
cf. Alyssum	–	–	–	–	–	–	–	–	–	–	–	–	1	–	–
cf. Camelina rumelica	–	–	–	–	–	–	–	–	–	–	–	–	–	–	–
cf. Camelina sativa	–	–	–	–	–	–	–	1	–	–	–	–	–	–	–
Conringia	–	–	–	–	–	–	–	–	–	–	–	–	–	–	–
cf. Lepidium	–	–	–	–	–	–	–	–	–	–	–	–	–	–	–
Sisymbrium altissimum-type	–	–	3	–	–	–	–	–	–	–	–	–	–	–	–
Thlaspi	–	–	–	–	–	–	–	–	–	–	–	–	–	–	–
YH-Brassicaceae 2	–	–	–	–	–	–	–	–	–	–	–	–	7	–	–
YH-Brassicaceae 3/5	–	–	1	–	–	–	–	–	–	–	–	–	–	–	–
YH-Brassicaceae 7	–	–	–	–	–	–	1	–	–	–	–	1	–	–	–
YH-Brassicaceae 10	–	–	–	–	–	–	–	–	–	–	–	–	–	–	–
YH-Brassicaceae 11	–	–	5	–	–	–	–	2	–	–	–	–	10	–	–
YH-Brassicaceae 12	–	–	4	–	–	–	–	–	–	–	–	–	–	–	–
BRASSICACEAE	6	–	64	1	3	–	–	7	–	–	–	1	87	–	–
Bufonia	–	–	1	–	–	–	–	–	–	–	–	–	–	–	–
Gypsophila	3	–	9	–	1	–	–	–	–	–	–	1	4	–	–
Silene	–	–	7	–	–	–	–	3	–	1	–	–	1	–	–
Vaccaria (YH-unknown 6)	–	–	6	–	–	1	–	–	–	–	1	–	1	–	–
CARYOPHYLLACEAE	–	–	–	1	–	–	–	–	–	–	–	–	4	–	–
YH-Caryophyllaceae 1	–	–	–	–	–	–	–	–	–	–	–	–	–	–	–
Atriplex	–	–	–	–	–	–	–	–	–	–	–	–	–	–	–
Chenopodium	–	–	–	–	–	–	–	–	–	–	–	–	–	–	–
Salsola kali-type	–	–	–	–	–	–	–	–	–	–	–	–	–	–	–
Salsola soda-type	–	–	3	–	–	–	–	–	–	2	–	–	–	–	–
Salsola	–	–	–	–	–	–	–	–	1	–	–	2	–	–	–
Suaeda	–	–	4	–	–	1	–	1	–	–	1	–	–	–	–
YH-Chenopodiaceae 2	–	–	–	–	–	–	–	–	–	–	–	–	–	–	1
CHENOPODIACEAE	–	–	17	–	–	–	–	14	–	1	–	–	–	–	–
Helianthemum	–	–	–	–	–	–	–	–	–	–	–	–	–	–	–
Carex	1	4	56	–	1	2	–	97	3	6	4	6	1	2	–
Carex 3	–	–	–	–	–	–	–	–	1	–	–	–	–	–	–
Eleocharis (YH-Cyperaceae a)	–	–	2	–	–	–	–	3	1	–	–	–	–	–	–
YH-Cyperaceae 1	–	–	5	–	–	–	–	4	–	1	2	–	–	–	1
YH-Cyperaceae 3	–	–	–	–	–	–	–	–	–	–	–	–	–	–	–
YH-Cyperaceae 4	–	–	–	–	–	–	–	1	–	–	–	–	–	–	–
YH-Cyperaceae 5	1	–	–	–	1	2	–	–	3	–	–	–	–	–	–
YH-Cyperaceae 7	–	–	–	–	–	–	–	–	–	–	–	–	–	–	–
YH-Cyperaceae 8	–	–	–	–	–	–	–	–	–	–	–	–	–	–	–
CYPERACEAE	10	–	21	–	–	4	–	9	–	–	2	–	–	–	–
Cephalaria	–	–	1	–	–	–	–	–	–	–	–	–	–	–	–
Scabiosa	–	–	–	–	–	–	–	–	–	–	–	–	–	–	–
Euphorbia	–	–	–	–	–	–	–	–	–	–	–	–	1	1	–

Table F2, cont'd.: Columns 61–75

Column no.	61†	62*	63†	64	65	66	67	68†	69†	70	71	72	73*†	74	75
YH no.	25652	30039	28856	29567	29573	31528	22441	23009	23053	25653	25844	23150	23307	22411	25514
Alhagi	1	-	4	1	-	1	-	-	3	1	1	4	4	-	-
Astragalus	-	-	1	-	1	-	-	1	-	-	-	1	-	-	-
Medicago	1	-	12	-	3	-	-	1	-	-	-	2	-	-	-
Onobrychis	-	-	-	-	-	-	-	-	-	-	-	-	-	-	-
Trifolium/Melilotus	1	-	20	-	-	-	-	1	-	-	-	1	-	-	1
Trigonella	4	-	25	1	16	3	-	75	-	1	5	2	391	2	1
Trigonella cf. astroites	-	-	4	-	2	-	-	1	-	-	-	-	50	1	-
YH-Fabaceae 2	-	-	-	-	-	-	1	-	-	-	-	-	-	-	-
FABACEAE	-	1	23	-	-	3	-	9	-	-	2	-	2	-	-
cf. Erodium	-	-	-	-	-	-	-	-	-	-	-	-	-	-	-
Teucrium	-	-	1	-	-	-	-	-	-	-	-	1	-	-	-
Ziziphora	6	-	6	4	40	-	-	5	-	-	1	1	23	-	-
YH-Lamiaceae 2	-	-	-	-	-	-	-	-	-	-	-	-	-	-	-
YH-Lamiac 3 (Nepeta?)	-	-	-	-	-	-	-	1	-	-	-	-	1	-	-
YH-Lamiaceae 5	-	-	-	-	-	-	-	-	-	-	-	-	2	-	-
LAMIACEAE	-	-	2	-	-	-	-	3	-	-	-	-	-	-	-
LILIACEAE	-	-	-	-	-	-	-	-	-	-	1	-	-	-	-
cf. Malva	-	-	-	-	-	-	-	2	-	-	-	-	-	-	-
Glaucium	1	-	11	1	-	-	-	-	-	-	-	-	2	-	-
Papaver	2	-	13	-	30	-	2	2	-	-	-	-	5	-	-
Fumaria	-	-	-	-	-	-	-	-	-	-	-	-	-	-	-
Plantago	-	-	2	-	-	-	-	1	-	-	-	-	-	-	-
Aegilops	-	-	-	-	-	-	-	-	-	-	-	-	-	-	-
Avena	-	-	3	-	-	-	-	-	-	-	-	-	1	-	-
Bromus japonicus-type	-	-	1	1	-	-	-	6	-	-	-	-	26	-	-
Bromus tectorum-type	2	-	3	-	-	-	-	6	-	-	-	-	39	-	-
Bromus	-	-	5	-	-	-	-	-	-	1	1	-	1	-	-
Eremopyrum	6	-	7	-	16	1	-	24	1	-	-	-	2	-	-
Hordeum cf. murinum	-	-	18	-	-	1	-	15	-	-	1	-	-	-	-
Hordeum spontaneum?	-	-	1	-	-	-	-	-	-	-	-	-	-	-	-
Hordeum misc.	-	-	-	-	1	-	-	-	-	-	-	-	-	-	-
cf. Lolium	-	-	-	-	-	-	-	-	-	-	-	-	-	-	-
cf. Phalaris	-	-	-	-	-	-	-	-	-	-	-	-	-	-	-
cf. Poa bulbosa	-	-	4	-	-	-	-	3	-	-	-	-	-	-	-
Setaria	-	-	-	-	-	-	-	-	1	-	-	-	-	-	-
Stipa	-	-	4	-	3	-	-	2	1	-	-	-	2	-	-
Taeniatherum	-	-	-	-	-	-	-	-	-	-	-	-	7	-	-
"Triticoid"	-	-	-	-	-	1	-	-	-	-	-	-	1	-	-
Triticum boeoticum	-	-	-	-	-	-	-	-	-	-	-	-	-	-	-
YH-Poaceae 1	-	-	5	-	-	-	-	-	-	-	3	-	371	-	-
YH-Poaceae 2	-	-	-	-	-	-	2	-	-	-	-	-	-	-	-
YH-Poaceae 3	2	-	1	-	-	-	-	-	-	-	1	-	-	-	-
YH-Poaceae 4	-	-	-	-	-	-	-	-	-	-	-	-	-	-	-
YH-Poaceae 5	-	-	-	-	-	-	-	9	-	-	-	1	-	-	-
YH-Poaceae 8	-	-	3	-	-	-	-	2	-	-	1	-	-	-	-
YH-Poaceae 10/15	-	-	3	-	-	-	-	-	-	-	-	-	1	-	-
YH-Poaceae 11	-	-	-	-	-	-	-	-	-	-	-	-	-	-	-
YH-Poaceae 13	-	-	2	-	-	-	-	2	-	-	-	-	-	-	-

Table F2, cont'd.: Columns 61–75

Column no.	61†	62*	63†	64	65	66	67	68†	69†	70	71	72	73*†	74	75
YH no.	25652	30039	28856	29567	29573	31528	22441	23009	23053	25653	25844	23150	23307	22411	25514
YH-Poaceae 14	-	-	-	-	-	-	-	-	-	-	-	-	-	-	-
YH-Poaceae 16	-	-	-	-	-	-	-	-	-	-	-	-	-	-	-
YH-Poaceae 17/18	4	-	2	-	-	-	-	2	-	-	-	-	2	-	-
YH-Poaceae 20 Aeg 1a	-	-	-	-	-	-	-	-	-	-	-	-	-	-	-
YH-Poaceae 21 Aeg 1b	-	-	-	-	-	-	-	-	-	-	-	-	-	-	-
POACEAE	9	3	210	-	17	1	-	169	5	-	4	4	19	3	4
Polygonum	-	-	-	-	-	-	-	-	-	-	-	-	-	-	-
Polygonum (was YH-Cyperaceae 2)	-	-	2	-	-	1	-	-	-	1	-	-	1	1	-
Polygonum (was YH-Cyperaceae 6)	1	1	-	-	-	-	-	1	1	-	-	-	-	-	-
Rumex	-	-	-	-	-								-	-	
Portulaca	-	-	-	-	-	-	-	-	1	-	-	-	-	-	-
Androsace	-	-	2	-	2	-	-	1	-	-	-	-	-	-	-
YH-Primulaceae 1	-	-	-	-	-	-	-	-	-	-	-	-	-	-	-
YH-Primulaceae 2	-	-	-	-	-	-	-	-	-	-	-	-	1	-	-
Aconitum	1	-	-	-	-	-	-	-	-	-	-	-	-	-	-
Adonis	3	-	1	-	1	1	-	-	-	1	-	-	1	-	-
Ceratocephalus (YH-pp 2)	2	-	-	-	4	-	-	-	-	-	-	-	-	-	-
Ranunculus (YH-unknown 4)	1	-	-	-	-	-	-	-	-	-	-	-	1	-	-
Ranunculus arvensis	-	-	-	-	-	-	-	-	-	-	-	-	-	-	-
Reseda	-	-	-	-	-	-	-	-	-	-	-	-	-	-	-
cf. Asperula (YH-Rubiaceae 2)	-	-	-	-	-	-	-	-	-	-	-	-	1	-	-
Galium	9	1	31	-	7	3	-	9	-	3	-	5	2	-	1
YH-Rubiaceae 1	2	-	-	-	1	-	-	-	-	-	-	-	-	-	-
YH-Scrophulariaceae 1	-	-	1	-	-	-	-	-	-	-	-	-	-	-	-
Hyoscyamus	-	-	1	-	-	-	-	-	-	-	-	-	1	-	-
SOLANACEAE	-	-	4	-	-	-	-	-	-	-	-	-	-	-	-
Thymelaea	-	-	-	-	-	-	-	-	-	-	-	-	-	-	-
Valerianella coronata	3	-	4	-	-	-	-	17	-	-	-	-	34	-	-
Valerianella vesicaria	1	-	2	-	-	-	-	2	-	-	-	-	-	-	-
Valerianella	-	-	-	1	3	-	-	-	-	-	-	-	-	-	-
Peganum harmala	-	-	1	-	-	-	-	-	-	-	-	1	-	-	-
Zygophyllum	-	-	9	-	1	-	-	-	3	-	-	-	-	-	-
YH-unknown 9	-	-	-	-	-	-	-	-	-	-	-	-	-	-	-
YH-unknown 21	-	-	-	-	-	-	-	-	-	-	-	-	-	-	-
YH-unknown 23	-	-	1	-	-	-	-	-	-	-	-	-	-	-	-
YH-unknown 27	-	-	21	-	8	-	-	-	-	-	-	-	-	-	-
YH-unknown 29	-	-	-	-	-	-	-	-	-	-	-	-	-	-	-
YH-unknown 30	-	-	-	-	-	-	-	-	-	-	-	-	-	-	-
YH-unknown 35	-	-	-	-	-	-	-	-	-	-	-	-	-	-	-
Unknown	9	1	293	6	46	4	-	81	5	-	5	6	65	-	-
PLANT PARTS (pp)															
Hordeum internode (2-row)	41	-	54	-	29	6	-	3	-	-	-	-	295	-	-
Hordeum internode (6-row)	10	-	2	1	-	2	-	3	-	-	-	-	12	-	-
Hordeum internode (compact)	22	-	2	-	3	2	-	3	-	-	-	2	79	-	-
Triticum aestivum int.	26	-	-	1	37	-	-	7	-	-	1	-	-	-	-
T. aestivum/durum int.	406	-	441	5	328	7	2	41	1	3	8	2	283	-	-
T. aestivum/durum int. 1	-	-	-	-	-	-	-	-	-	-	-	-	-	-	-

Table F2, cont'd.: Columns 61–75

Column no.	61†	62*	63†	64	65	66	67	68†	69†	70	71	72	73*†	74	75
YH no.	25652	30039	28856	29567	29573	31528	22441	23009	23053	25653	25844	23150	23307	22411	25514
T. aestivum/durum int. 2	-	-	-	-	-	-	-	-	-	-	-	-	-	-	-
T. durum internode	10	-	4	-	4	-	-	-	-	-	-	-	-	-	-
T. monococcum spikelet fork	-	-	-	-	-	-	-	-	-	-	-	-	-	-	-
T. dicoccum spikelet fork	-	-	-	-	-	-	-	-	-	-	-	-	-	-	-
T. monococcum/dicoccum spikelet fork	-	-	-	-	99	-	-	-	-	-	-	-	-	2	-
Triticum rachis base	-	-	14	-	-	1	-	-	-	-	-	-	-	-	-
Triticum, rachis fragment, square cross-section	27	-	17	-	9	-	-	-	-	-	-	-	48	-	-
Cereal culm node	68	-	69	3	-	4	-	27	2	2	3	2	-	-	-
Cereal culm root base?	-	-	-	-	-	-	-	-	1	-	1	-	-	-	-
cf. Oryza glume (charred)	-	-	-	-	-	-	-	-	-	-	-	-	-	-	-
Oryza glume (silicified phytolith)	-	-	-	-	-	-	-	-	-	-	-	-	-	-	-
Vitis peduncle	-	-	-	-	-	-	-	-	-	-	-	-	-	-	-
Onopordum achene collar	-	-	-	-	-	-	-	-	-	1	-	-	-	-	-
Asteraceae head, misc.	-	-	-	-	-	-	-	-	-	-	-	-	-	-	-
Euclidium silique	-	-	-	-	-	-	-	-	1	-	-	-	-	-	-
Capparis thorn	-	-	-	-	-	-	-	-	-	-	-	-	-	-	-
Atriplex bract (YH-pp 12)	-	-	-	-	-	-	-	-	-	-	-	-	-	-	-
Salsola inflor (YH-pp 7, YH-unk 8)	-	-	-	-	-	-	-	-	-	-	-	-	-	-	-
Cyperaceae stem fragment	-	-	-	-	-	-	-	-	-	-	-	-	-	-	-
Alhagi pod section	-	-	-	+	-	-	-	-	-	-	-	-	2	-	-
Alhagi leaf?	-	-	1	-	-	-	-	-	-	-	-	-	-	-	-
Aegilops glume base	-	-	-	-	-	-	-	-	-	-	-	-	-	-	-
Hordeum internode (wild)	-	-	-	-	-	-	-	-	-	-	-	-	-	-	-
Taeniatherum int. (YH-pp 8)	-	-	-	-	-	-	-	-	-	-	-	-	1	-	-
Poaceae, mystery 1 rachis frag.	2	-	-	-	1	-	-	-	-	-	-	-	9	-	-
YH-plant part 9	-	-	-	-	-	-	-	-	-	-	-	-	-	-	-
UNCHARRED-MINERALIZED															
Arnebia linearis	-	-	-	-	-	-	-	1	-	many	-	-	-	-	-
Arnebia/Lithospermum	-	-	-	1	1	-	-	4	-	-	3	4	-	1	-
Lithospermum tenuifolium	-	-	-	-	-	-	-	6	-	-	-	-	-	-	-
Lithospermum arvense	-	-	-	-	-	-	-	-	-	-	-	-	-	-	-
Lithospermum, misc.	2	-	13	-	-	-	-	30	2	-	-	-	75	-	-
Moltkia	-	-	-	-	-	-	-	-	-	1	1	-	-	2	-
BORAGINACEAE	-	-	-	-	-	-	-	-	-	-	-	-	-	-	-
Eleocharis (Cyperaceae-a)	-	3	-	-	-	-	-	6	5	-	4	3	-	-	-
Carex	-	-	-	-	-	-	-	3	1	1	6	6	-	-	-
Ficus	-	-	-	-	-	-	-	-	-	-	1	-	-	-	1
Glaucium	-	-	4	-	-	-	-	3	2	-	-	-	-	-	-
Papaver (white)	-	-	1	-	2	-	-	-	-	-	-	-	-	-	-

Table F2, cont'd.: Columns 76–90

Column no.	76	77*†	78	79	80	81*	82†	83	84†	85	86	87	88	89†	90†
YH no.	23495	26789	26613	26618	26628	27239	26899	28764	26870	26482	27428	28421	25843	25248	28774
Context type	pit	pit	pit	pit	pit	pit	pit	pit	pit	pyro	pit	pit	pit	coll	coll
Operation	1	1	2	2	2	2	2	2	2	2	2	2	2	1	2
Locus	100	19	30	53	53	58	56	11	54	44	60	66	44	3.2	59
Lot	214	64	97	96	101	137	120	184	109	90	146	176	54	8	186
Phase	430.04	430.08	430.10	430.12	430.12	430.14	430.19	435.02	435.03	435.04	435.05	435.06	435.07	450.01	450.02
Soil volume (liters)	15	15	15	2.5	1	15	11.5	7	15	4.5	15	15	15	10	11
Charcoal (>2 mm, g)	2.74	15.57	5.45	7.01	2.56	10.09	5.32	5.95	12.08	1.54	8.35	5.05	5.18	10.17	7.03
Seed (>2 mm, g)	0.13	0.46	0.11	0.44	0.15	0.52	6.09	0.34	0.35	0.88	0.37	0.25	0.35	0.48	0.31
Other (>2 mm, g)	0.02	0.06	+	0.01	-	-	2.20	0.09	0.07	0.27	+	0.01	0.01	-	0.02
Wild/weedy (#)	26	102	15	17	8	12	595	28	71	257	18	29	50	338	35
ECONOMIC PLANTS															
Hordeum vulgare (g)	0.02	0.15	0.03	0.06	0.04	0.18	2.25	0.20	0.18	0.01	0.06	0.08	0.11	0.17	0.17
H. vulgare var. nudum (g)	-	-	-	-	-	-	-	-	-	-	-	-	-	-	-
Triticum aestivum/durum (g)	0.04	0.04	0.03	0.09	0.02	0.14	1.75	0.02	0.11	0.53	0.04	0.05	0.06	0.13	0.06
Triticum monococcum (g)	-	+	-	0.01	0.01	-	-	0.01	-	-	0.01	-	-	-	-
Triticum dicoccum (g)	0.01	-	-	-	-	-	-	-	-	-	0.01	-	-	-	-
Triticum sp. (g)	0.03	0.05	0.01	0.06	0.02	0.05	0.72	+	-	-	0.07	0.04	0.04	+	+
Cereal, indet. (g)	0.05	0.18	0.06	0.22	0.07	0.24	2.09	0.17	0.13	0.44	0.15	0.11	0.17	0.14	0.10
Secale cereale (g)	-	-	-	-	-	-	0.01	-	-	-	-	-	-	-	-
Setaria italica (#)	-	3	-	-	-	-	-	-	-	-	1	-	-	1	-
Oryza (g)	-	-	-	-	-	-	-	-	-	-	-	-	-	-	-
Vicia ervilia (g)	-	-	-	-	-	0.02	0.05	-	0.01	-	0.02	-	0.01	0.01	-
Lens (g)	-	-	+	-	-	0.01	0.03	-	-	-	-	+	-	-	-
Pulse (g)	-	-	-	0.03	0.02	-	-	-	-	0.01	-	-	0.01	-	+
cf. Pistacia/nutshell (g)	-	-	-	+	-	-	+	-	-	-	-	-	-	0.01	-
cf. Quercus (g)	-	-	-	-	-	-	-	-	-	-	-	-	-	-	-
Ficus carica (#)	-	-	-	-	-	-	-	-	-	-	-	-	-	-	-
Vitis vinifera (# whole, total g)	-	-	-	-	-	-	-	-	-	-	-	-	-	-	-
WILD AND WEEDY (counts)															
Bupleurum	-	-	-	-	-	-	-	-	-	-	-	-	-	-	-
cf. Daucus	-	-	-	-	-	-	-	-	-	-	-	-	-	-	-
Torilis leptophylla?	-	-	-	-	-	-	-	-	-	-	-	-	-	-	-
YH-Apiaceae 2	-	-	-	-	-	-	-	-	-	-	-	-	-	-	-
YH-Apiaceae 4/8	-	-	-	-	-	-	1	-	-	-	-	-	-	-	-
YH-Apiaceae 6	-	-	-	-	-	-	-	-	-	-	-	-	-	-	-
YH-Apiaceae 7	-	-	-	-	-	-	-	-	-	-	-	-	-	-	-
YH-Apiaceae 10	-	-	-	-	-	-	1	-	-	-	2	-	-	-	-
YH-unknown 31 (Apiaceae)	-	-	-	-	-	-	-	-	-	-	-	-	-	-	-
APIACEAE	-	-	-	-	-	-	2	-	-	-	-	-	-	-	-
Anthemis/Matricaria	-	-	-	-	-	-	-	-	-	-	-	-	-	-	-
Artemisia	-	-	-	-	1	-	-	-	-	-	-	-	-	-	-
Carthamus	-	1	-	-	-	-	-	-	-	-	-	-	-	-	-
Centaurea cyanus-type	-	-	-	-	-	-	-	-	-	1	-	-	-	1	-
Centaurea	1	2	-	-	-	-	7	1	-	4	-	-	2	1	-
Onopordum	-	1	-	-	-	-	7	-	-	-	-	-	-	-	-
YH-Asteraceae 1	-	-	1	-	-	-	-	-	-	-	-	-	-	-	-

Table F2, cont'd.: Columns 76–90

Column no.	76	77*†	78	79	80	81*	82†	83	84†	85	86	87	88	89†	90†
YH no.	23495	26789	26613	26618	26628	27239	26899	28764	26870	26482	27428	28421	25843	25248	28774
YH-Asteraceae 2	-	1	-	-	-	-	-	-	-	1	-	-	-	1	-
YH-Asteraceae 5	-	-	-	-	-	-	5	-	-	-	2	-	-	-	-
YH-Asteraceae 9	-	-	-	-	-	-	-	-	-	-	-	-	-	-	-
ASTERACEAE	-	-	-	-	-	-	5	1	-	-	-	-	-	1	1
Arnebia/Lithospermum	-	-	-	-	-	-	-	-	-	-	-	-	-	-	-
Heliotropium	-	2	-	-	-	-	1	-	-	-	-	-	1	1	-
cf. Alyssum	-	-	-	-	-	-	2	-	-	-	1	-	-	-	-
cf. Camelina rumelica	-	-	-	-	-	-	-	-	-	-	-	-	-	-	-
cf. Camelina sativa	-	-	-	-	-	-	-	-	-	-	-	-	-	-	-
Conringia	-	-	-	-	-	-	-	-	-	1	-	-	-	-	-
cf. Lepidium	-	-	-	-	-	-	-	-	-	-	-	-	-	-	-
Sisymbrium altissimum-type	-	-	-	-	-	-	-	-	-	-	-	-	-	-	-
Thlaspi	-	-	-	-	-	-	-	-	-	-	-	-	-	-	-
YH-Brassicaceae 2	-	3	-	-	-	-	-	-	-	2	-	-	-	5	-
YH-Brassicaceae 3/5	-	-	-	-	-	-	-	-	-	-	-	-	-	3	-
YH-Brassicaceae 7	-	-	-	-	-	-	-	-	-	2	-	-	-	-	-
YH-Brassicaceae 10	-	-	-	-	-	-	-	-	-	-	-	-	-	-	-
YH-Brassicaceae 11	-	-	-	-	-	-	-	-	1	-	-	-	-	-	-
YH-Brassicaceae 12	-	-	-	-	-	-	-	-	-	-	-	1	-	11	-
BRASSICACEAE	-	7	-	-	-	-	46	1	3	4	-	-	-	18	-
Bufonia	-	-	-	-	-	-	-	-	-	-	-	-	-	-	-
Gypsophila	-	4	-	-	-	-	4	1	-	3	-	-	1	-	-
Silene	-	-	-	-	-	-	3	-	-	-	-	-	-	-	9
Vaccaria (YH-unknown 6)	1	-	-	1	-	-	-	1	-	-	1	-	1	-	-
CARYOPHYLLACEAE	-	-	-	-	-	-	1	1	-	-	-	-	-	-	-
YH-Caryophyllaceae 1	-	-	-	-	-	-	-	-	-	-	-	-	-	-	-
Atriplex	-	-	-	-	-	-	1	-	-	-	-	-	-	-	-
Chenopodium	-	1	-	-	-	2	3	-	-	1	-	-	-	-	-
Salsola kali-type	-	-	-	-	-	-	-	-	-	-	-	-	-	-	-
Salsola soda-type	-	4	-	-	-	-	-	-	-	-	-	-	2	1	-
Salsola	-	-	-	-	-	-	-	-	-	-	-	1	-	-	1
Suaeda	-	-	-	-	-	-	2	1	1	-	1	1	-	-	1
YH-Chenopodiaceae 2	-	-	-	-	-	-	-	-	-	-	-	-	-	2	-
CHENOPODIACEAE	-	-	-	-	-	-	20	-	-	-	1	1	-	81	-
Helianthemum	-	-	-	-	-	-	1	-	-	-	-	-	-	-	-
Carex	3	-	2	-	1	1	61	-	1	5	3	-	1	18	1
Carex 3	-	-	-	-	-	-	1	-	-	-	-	-	-	-	-
Eleocharis (YH-Cyperaceae a)	-	-	1	-	-	-	1	-	-	-	-	-	-	1	-
YH-Cyperaceae 1	1	-	-	-	-	-	1	2	1	-	-	-	1	-	-
YH-Cyperaceae 3	-	-	-	-	-	-	-	-	-	-	-	-	-	1	-
YH-Cyperaceae 4	-	-	-	-	-	-	-	-	-	-	-	-	-	-	-
YH-Cyperaceae 5	-	-	-	-	-	-	6	-	-	-	-	-	-	4	-
YH-Cyperaceae 7	-	-	-	-	-	-	-	-	-	-	-	-	-	-	-
YH-Cyperaceae 8	-	-	-	-	-	-	-	-	-	-	-	-	-	-	-
CYPERACEAE	-	7	-	2	2	2	22	1	5	3	-	2	2	23	-
Cephalaria	-	-	-	1	-	-	-	-	-	2	-	-	-	-	-
Scabiosa	-	-	-	-	-	-	-	-	-	-	-	-	-	1	-
Euphorbia	-	-	-	-	-	-	-	-	-	-	-	-	-	-	-

Table F2, cont'd.: Columns 76–90

Column no.	76	77*†	78	79	80	81*	82†	83	84†	85	86	87	88	89†	90†
YH no.	23495	26789	26613	26618	26628	27239	26899	28764	26870	26482	27428	28421	25843	25248	28774
Alhagi	3	2	-	-	-	1	-	-	6	-	-	-	1	-	-
Astragalus	-	1	-	-	-	-	1	-	-	1	-	-	-	-	1
Medicago	-	1	-	-	-	-	2	-	-	1	-	-	1	-	-
Onobrychis	-	-	-	-	-	-	-	-	-	-	-	-	-	5	-
Trifolium/Melilotus	-	4	-	-	-	-	4	-	3	-	-	-	1	3	1
Trigonella	2	12	-	1	-	-	30	1	8	2	-	-	1	3	3
Trigonella cf. astroites	1	-	-	1	-	-	6	-	-	-	-	-	-	-	-
YH-Fabaceae 2	-	-	-	-	-	-	-	-	-	-	-	-	-	-	-
FABACEAE	-	5	1	-	-	1	18	-	1	-	-	10	3	13	-
cf. Erodium	-	-	-	-	-	-	-	-	-	-	-	-	-	-	1
Teucrium	-	1	-	-	-	-	3	-	-	1	-	-	-	-	1
Ziziphora	2	4	-	1	-	-	9	-	1	1	-	-	1	9	2
YH-Lamiaceae 2	-	-	-	-	-	-	4	-	-	-	-	-	-	-	-
YH-Lamiac 3 (Nepeta?)	-	-	-	-	-	-	-	-	-	-	-	-	-	-	1
YH-Lamiaceae 5	-	-	-	-	-	1	-	2	-	-	-	-	-	-	-
LAMIACEAE	-	-	-	-	-	-	2	-	1	-	-	-	-	-	-
LILIACEAE	-	-	-	-	-	-	-	1	-	-	-	-	-	-	-
cf. Malva	-	1	-	-	-	-	-	-	-	-	-	-	-	-	-
Glaucium	-	-	1	-	-	-	-	-	-	-	-	-	1	-	-
Papaver	-	-	-	-	-	-	-	-	1	-	-	-	-	-	-
Fumaria	-	-	-	-	-	-	-	-	-	-	-	-	-	-	-
Plantago	-	-	-	-	-	-	-	-	-	-	-	-	-	1	-
Aegilops	-	1	1	-	-	-	-	-	-	-	-	-	-	-	-
Avena	-	-	-	-	-	-	-	-	-	-	-	-	-	-	-
Bromus japonicus-type	-	-	-	-	-	-	8	-	1	-	-	-	-	1	-
Bromus tectorum-type	-	-	-	-	-	-	20	1	-	1	-	-	-	-	-
Bromus	-	-	-	-	-	-	-	-	-	-	-	-	-	2	-
Eremopyrum	4	-	1	-	-	-	4	-	1	119	-	1	14	-	-
Hordeum cf. murinum	1	2	2	-	-	1	5	-	-	-	-	2	-	-	2
Hordeum spontaneum?	-	-	-	-	-	-	-	-	-	-	-	-	-	-	-
Hordeum misc.	-	-	-	-	-	-	-	-	-	-	-	-	-	-	-
cf. Lolium	-	-	-	-	-	-	-	-	-	-	-	-	-	-	-
cf. Phalaris	-	-	-	-	-	-	-	-	-	-	-	-	-	-	-
cf. Poa bulbosa	-	-	-	-	-	-	-	-	-	-	-	1	-	-	-
Setaria	-	-	-	1	-	-	-	1	-	-	-	-	-	-	-
Stipa	1	-	-	-	-	-	17	-	-	20	-	-	4	-	-
Taeniatherum	-	-	-	-	-	-	-	-	-	-	-	-	-	-	-
"Triticoid"	-	2	-	-	-	-	1	-	-	3	-	-	-	1	-
Triticum boeoticum	-	-	-	-	-	-	-	-	-	-	-	-	-	-	-
YH-Poaceae 1	-	4	1	-	-	-	3	-	3	2	2	-	-	12	-
YH-Poaceae 2	1	-	-	-	-	-	-	-	-	-	-	-	-	-	-
YH-Poaceae 3	-	-	-	-	-	-	-	-	-	-	-	-	-	-	-
YH-Poaceae 4	-	-	-	-	-	-	-	-	-	-	-	-	-	2	-
YH-Poaceae 5	-	1	-	-	-	-	-	-	-	-	-	-	-	-	-
YH-Poaceae 8	-	-	-	-	-	-	27	-	-	-	-	-	-	30	-
YH-Poaceae 10/15	-	-	-	-	-	-	1	-	-	-	-	-	-	-	-
YH-Poaceae 11	-	-	-	-	-	-	13	-	-	-	-	-	-	-	-
YH-Poaceae 13	-	-	-	-	-	-	-	-	-	-	-	-	-	-	-

Table F2, cont'd.: Columns 76–90

Column no.	76	77*†	78	79	80	81*	82†	83	84†	85	86	87	88	89†	90†
YH no.	23495	26789	26613	26618	26628	27239	26899	28764	26870	26482	27428	28421	25843	25248	28774
YH-Poaceae 14	-	-	-	-	-	-	-	-	-	-	-	-	-	-	-
YH-Poaceae 16	-	-	-	-	-	-	-	-	-	-	-	-	-	-	-
YH-Poaceae 17/18	-	2	-	-	-	-	-	-	-	-	2	-	-	-	-
YH-Poaceae 20 Aeg 1a	-	1	-	-	-	-	-	-	-	-	-	-	-	-	-
YH-Poaceae 21 Aeg 1b	-	1	-	-	-	-	1	-	-	-	-	-	-	-	-
POACEAE	3	9	1	3	1	1	126	4	24	65	1	5	8	26	7
Polygonum	-	-	-	-	-	-	-	-	-	-	1	-	-	-	-
Polygonum (was YH-Cyperaceae 2)	-	-	-	5	3	1	39	-	1	-	1	1	1	1	-
Polygonum (was YH-Cyperaceae 6)	-	2	-	-	-	-	2	-	-	-	-	-	-	-	-
Rumex	-	-	-	-	-	-	-	-	-	-	-	-	-	1	-
Portulaca	-	-	-	-	-	-	-	-	-	-	-	-	-	1	-
Androsace	-	-	1	-	-	-	2	3	-	-	1	-	-	6	-
YH-Primulaceae 1	-	-	-	-	-	-	-	-	-	-	-	-	-	-	-
YH-Primulaceae 2	-	-	-	-	-	-	-	-	-	-	-	-	-	-	-
Aconitum	-	-	-	-	-	-	-	-	-	-	-	-	-	-	-
Adonis	1	-	-	-	-	1	2	1	-	-	-	-	-	29	-
Ceratocephalus (YH-pp 2)	-	-	-	-	-	-	-	-	-	-	-	-	-	-	-
Ranunculus (YH-unknown 4)	-	-	-	-	-	-	8	-	-	-	-	-	-	-	-
Ranunculus arvensis	-	1	-	-	-	-	-	-	-	-	-	-	-	-	-
Reseda	-	-	-	-	-	-	-	-	-	-	-	-	-	-	-
cf. Asperula (YH-Rubiaceae 2)	-	1	-	-	-	-	-	-	-	-	-	-	-	-	-
Galium	-	7	-	1	-	-	10	2	3	1	-	1	2	6	1
YH-Rubiaceae 1	-	-	-	-	-	-	4	-	-	-	-	-	-	-	-
YH-Scrophulariaceae 1	-	-	-	-	-	-	-	-	-	-	-	-	-	-	-
Hyoscyamus	-	1	-	-	-	-	-	-	1	-	-	1	-	1	-
SOLANACEAE	-	-	-	-	-	-	-	-	-	-	-	-	-	-	-
Thymelaea	-	-	1	-	-	-	-	-	-	-	-	1	-	-	-
Valerianella coronata	-	1	-	-	-	-	1	-	-	-	-	-	-	1	-
Valerianella vesicaria	-	-	-	-	-	-	-	1	1	-	-	-	-	-	1
Valerianella	-	-	-	-	-	-	-	1	-	-	-	-	-	-	-
Peganum harmala	-	1	-	-	-	-	-	-	-	-	-	-	-	4	-
Zygophyllum	-	-	-	-	-	-	-	-	1	-	-	-	-	-	-
YH-unknown 9	-	-	-	-	-	-	-	-	-	-	-	-	-	1	-
YH-unknown 21	-	1	-	-	-	-	1	-	-	-	-	-	-	1	-
YH-unknown 23	1	-	-	-	-	-	15	-	-	-	1	-	-	4	-
YH-unknown 27	-	-	1	-	-	-	2	-	-	10	-	-	1	-	1
YH-unknown 29	-	-	-	-	-	-	-	-	1	-	-	-	-	-	-
YH-unknown 30	-	-	-	-	-	-	-	-	-	-	-	-	-	-	-
YH-unknown 35	-	-	-	-	-	-	-	-	-	-	-	-	-	-	-
Unknown	10	18	6	2	-	8	165	12	15	22	10	6	3	66	13
PLANT PARTS (pp)															
Hordeum internode (2-row)	-	3	1	5	-	-	21	1	11	-	-	1	-	9	4
Hordeum internode (6-row)	-	-	-	-	-	-	67	-	-	-	-	-	-	1	-
Hordeum internode (compact)	-	2	-	-	-	-	14	-	8	-	-	-	-	3	-
Triticum aestivum int.	1	3	-	-	-	-	343	-	-	-	-	-	-	-	-
T. aestivum/durum int.	3	7	2	4	-	1	382	10	7	485	1	1	31	7	9
T. aestivum/durum int. 1	-	-	-	-	-	-	-	-	-	-	-	-	-	-	-

Table F2, cont'd.: Columns 76–90

Column no.	76	77*†	78	79	80	81*	82†	83	84†	85	86	87	88	89†	90†
YH no.	23495	26789	26613	26618	26628	27239	26899	28764	26870	26482	27428	28421	25843	25248	28774
T. aestivum/durum int. 2	-	-	-	-	-	-	-	-	-	-	-	-	-	-	-
T. durum internode	-	-	-	-	-	-	52	-	1	-	-	-	-	-	-
T. monococcum spikelet fork	-	-	-	5	2	-	-	10	-	-	10	-	-	-	-
T. dicoccum spikelet fork	-	-	-	-	1	-	-	1	-	-	-	-	-	-	-
T. monococcum/dicoccum spikelet fork	-	-	1	27	6	-	25	53	-	-	16	-	-	2	-
Triticum rachis base	-	-	2	-	-	-	40	-	-	11	-	-	1	-	-
Triticum, rachis fragment, square cross-section	-	-	-	-	-	-	83	1	-	11	-	-	2	-	-
Cereal culm node	2	1	1	2	-	1	342	5	7	97	1	4	10	-	2
Cereal culm root base?	-	2	-	-	-	-	-	-	-	24	-	-	-	-	-
cf. Oryza glume (charred)	-	-	-	-	-	-	-	-	-	-	-	-	-	-	-
Oryza glume (silicified phytolith)	-	-	-	-	-	-	-	-	-	-	-	-	-	-	-
Vitis peduncle	-	-	-	-	-	-	-	-	-	-	-	-	-	-	-
Onopordum achene collar	-	-	-	1	-	-	1	-	-	-	-	-	-	-	-
Asteraceae head, misc.	-	-	-	-	-	-	-	-	-	1	-	-	-	-	-
Euclidium silique	-	-	-	-	-	-	1	-	-	-	-	-	-	-	-
Capparis thorn	-	-	-	-	-	-	-	-	-	-	-	-	-	-	-
Atriplex bract (YH-pp 12)	-	-	-	-	-	-	1	-	-	-	-	-	-	-	-
Salsola inflor (YH-pp 7, YH-unk 8)	-	1	-	-	-	-	1	1	-	-	-	-	-	-	-
Cyperaceae stem fragment	-	-	-	-	-	-	-	-	-	-	-	-	-	-	-
Alhagi pod section	-	3	-	-	-	+	1	-	1	-	-	-	+	-	-
Alhagi leaf?	-	-	-	-	-	-	-	-	-	-	-	-	-	1	-
Aegilops glume base	-	-	-	-	-	-	10	-	-	6	-	-	2	-	-
Hordeum internode (wild)	-	-	-	-	-	-	-	-	-	-	-	-	-	-	-
Taeniatherum int. (YH-pp 8)	-	-	-	-	-	-	-	-	-	-	-	-	-	-	-
Poaceae, mystery 1 rachis frag.	-	-	-	-	-	-	4	-	-	1	-	-	-	-	-
YH-plant part 9	-	-	-	-	-	-	-	-	1	-	-	1	-	-	-
UNCHARRED-MINERALIZED															
Arnebia linearis	-	-	-	-	-	-	-	-	-	-	-	-	-	-	-
Arnebia/Lithospermum	-	-	1	3	-	-	6	-	20	1	-	4	1	4	1
Lithospermum tenuifolium	-	-	-	-	-	2	1	1	1	-	-	-	10	-	-
Lithospermum arvense	-	-	-	-	-	-	5	-	-	-	-	-	-	-	-
Lithospermum, misc.	-	-	-	-	-	-	3	2	-	2	-	-	-	4	-
Moltkia	-	-	-	-	-	-	1	-	-	-	-	-	-	-	1
BORAGINACEAE	-	-	-	-	-	-	-	-	2	-	-	-	-	-	-
Eleocharis (Cyperaceae-a)	6	4	-	-	-	-	20	-	9	14	-	1	3	1	-
Carex	2	-	-	-	-	-	-	-	-	-	-	1	1	-	-
Ficus	2	-	-	-	1	-	-	-	-	-	-	-	3	-	-
Glaucium	-	-	-	4	-	1	-	2	-	-	-	-	-	4	2
Papaver (white)	2	-	-	-	-	-	-	2	-	-	-	-	-	-	6

Table F2, cont'd.: Columns 91–105

Column no.	91	92†	93	94*†	95	96	97	98†	99	100*†	101*	102*†	103*	104	105
YH no.	27569	30664	23774	26944	28547	27937	28548	28522	30194	30191	31560	29540	31203	30692	32334
Context type	pyro	pit	mixed	pit	pit	pit	pit	pit	pit	debris	pit	pit	mixed	pit	pit
Operation	1	2	1	1	1	2	1	1	1	1	1	1	2	2	2
Locus	27	76	98	23	46	55	47	45	77	79	91	52	82	78	88
Lot	88	237	237	72	122	167	123	117	163	164	179	139	257	243	319
Phase	450.07	450.10	460	470.01	470.03	470.07	470.10	475.02	480.01	480.02	480.03	495.04	510.04	510.05	510.08
Soil volume (liters)	9	15	6	7	7	9	13	4.5	4	3	6	8	10	13	2.8
Charcoal (>2 mm, g)	3.32	16.06	5.48	10.38	3.75	3.79	3.96	3.24	5.75	3.29	8.42	5.67	8.34	12.62	7.54
Seed (>2 mm, g)	0.08	9.41	0.41	0.25	0.46	0.08	0.20	0.16	0.19	0.38	1.59	2.32	0.13	0.41	0.18
Other (>2 mm, g)	0.09	0.02	-	-	-	-	+	+	-	0.01	-	0.05	+	+	+
Wild/weedy (#)	19	724	29	16	26	1	14	43	30	611	70	77	17	33	8
ECONOMIC PLANTS															
Hordeum vulgare (g)	0.07	6.97	0.16	0.04	0.25	0.02	0.08	0.07	0.03	0.15	0.61	1.41	0.02	0.16	0.12
H. vulgare var. nudum (g)	-	-	-	-	-	-	-	-	-	-	-	-	-	-	-
Triticum aestivum/durum (g)	0.01	0.32	0.14	0.01	0.14	0.02	0.02	0.01	0.12	0.10	0.63	0.33	0.05	0.03	0.03
Triticum monococcum (g)	-	0.03	0.01	-	-	-	-	-	-	-	0.01	-	-	-	-
Triticum dicoccum (g)	-	-	-	-	-	-	-	-	-	-	-	0.03	+	-	-
Triticum sp. (g)	0.03	0.14	-	0.04	+	-	0.02	-	-	+	-	0.05	-	0.03	-
Cereal, indet. (g)	0.03	1.54	0.06	0.10	0.18	0.05	0.10	0.06	0.09	0.11	0.78	0.47	0.04	0.14	0.08
Secale cereale (g)	-	-	-	-	-	-	-	-	-	-	-	-	-	-	-
Setaria italica (#)	-	-	2	-	-	-	-	-	-	-	-	-	-	-	-
Oryza (g)	-	-	-	-	-	-	-	-	-	-	-	-	-	-	-
Vicia ervilia (g)	+	0.03	-	-	-	-	-	-	-	+	0.04	0.01	-	0.03	-
Lens (g)	-	-	0.01	0.01	-	+	+	-	-	-	-	0.02	0.01	-	-
Pulse (g)	-	-	-	0.03	-	-	-	-	-	-	0.04	-	-	+	-
cf. Pistacia/nutshell (g)	-	+	-	-	0.01	-	-	-	-	-	-	-	+	-	-
cf. Quercus (g)	-	-	-	-	-	-	-	-	-	-	-	-	-	-	-
Ficus carica (#)	-	-	-	-	-	-	-	-	-	-	-	-	-	-	-
Vitis vinifera (# whole, total g)	-	-	-	-	-	-	-	-	-	-	-	-	-	-	-
WILD AND WEEDY (counts)															
Bupleurum	-	1	-	-	-	-	-	-	-	-	-	-	-	-	-
cf. Daucus	-	-	-	-	-	-	-	-	-	-	-	-	-	-	-
Torilis leptophylla?	-	-	-	-	-	-	-	-	-	-	-	-	-	-	-
YH-Apiaceae 2	-	4	-	-	-	-	-	-	-	-	1	-	-	-	-
YH-Apiaceae 4/8	-	1	-	-	-	-	-	-	1	-	1	-	-	-	-
YH-Apiaceae 6	-	2	-	-	-	-	-	-	1	-	-	-	-	-	-
YH-Apiaceae 7	-	1	-	-	-	-	-	-	-	-	-	-	-	-	-
YH-Apiaceae 10	-	-	-	-	-	-	-	-	-	-	-	-	-	1	-
YH-unknown 31 (Apiaceae)	-	-	-	-	-	-	-	-	-	-	-	-	-	-	-
APIACEAE	-	9	-	-	-	-	-	1	-	1	-	-	-	1	-
Anthemis/Matricaria	-	-	-	-	-	-	-	-	-	-	-	-	-	-	-
Artemisia	-	1	-	-	-	-	-	-	-	-	-	-	-	-	-
Carthamus	-	-	-	-	-	-	-	-	-	-	-	-	-	-	-
Centaurea cyanus-type	-	-	-	-	-	-	-	-	-	-	-	-	-	-	-
Centaurea	-	2	-	1	-	-	-	-	-	-	1	1	1	-	-
Onopordum	-	-	-	-	-	-	-	-	-	-	-	-	-	-	-
YH-Asteraceae 1	-	-	-	-	-	-	-	-	-	-	-	-	-	-	-

Table F2, cont'd.: Columns 91–105

Column no.	91	92†	93	94*†	95	96	97	98†	99	100*†	101*	102*†	103*	104	105
YH no.	27569	30664	23774	26944	28547	27937	28548	28522	30194	30191	31560	29540	31203	30692	32334
YH-Asteraceae 2	-	-	-	-	-	-	-	-	-	57	-	-	-	-	-
YH-Asteraceae 5	-	-	-	-	-	-	-	-	-	-	-	-	-	-	-
YH-Asteraceae 9	-	-	-	-	-	-	-	-	-	-	-	-	-	-	-
ASTERACEAE	-	-	-	-	-	-	-	-	-	16	-	-	-	-	-
Arnebia/Lithospermum	-	-	-	-	-	-	-	-	-	1	-	-	-	-	-
Heliotropium	-	-	-	-	-	-	-	-	-	-	-	-	-	1	-
cf. Alyssum	-	-	-	-	-	-	-	-	-	-	-	-	-	-	-
cf. Camelina rumelica	-	-	-	-	-	-	-	-	-	-	-	-	-	-	-
cf. Camelina sativa	-	-	-	-	-	-	-	-	-	-	-	-	-	-	-
Conringia	-	1	-	-	-	-	-	-	-	-	-	-	-	-	-
cf. Lepidium	-	-	-	-	-	-	-	-	-	-	-	-	-	-	-
Sisymbrium altissimum-type	-	-	-	-	-	-	-	-	-	-	-	-	1	-	-
Thlaspi	-	-	-	-	-	-	-	-	-	-	-	-	-	-	-
YH-Brassicaceae 2	-	-	-	-	-	-	-	-	-	10	-	-	-	-	-
YH-Brassicaceae 3/5	1	2	-	-	-	-	-	-	-	3	-	-	-	-	-
YH-Brassicaceae 7	-	-	-	-	-	-	-	-	-	3	-	-	-	-	-
YH-Brassicaceae 10	-	1	-	-	-	-	-	-	-	-	-	-	-	-	-
YH-Brassicaceae 11	-	3	1	-	-	-	-	-	-	25	-	-	-	-	-
YH-Brassicaceae 12	-	-	1	-	-	-	-	-	-	20	-	-	-	-	-
BRASSICACEAE	1	4	-	-	-	-	-	-	1	79	-	2	-	1	1
Bufonia	-	-	-	-	-	-	-	-	-	-	-	-	-	-	-
Gypsophila	-	14	-	1	-	-	-	1	-	6	2	-	-	-	-
Silene	-	30	-	1	-	-	1	1	-	-	-	1	-	-	-
Vaccaria (YH-unknown 6)	-	1	-	-	-	-	-	-	-	-	1	-	1	-	-
CARYOPHYLLACEAE	-	-	-	-	-	-	-	-	-	-	5	-	-	-	-
YH-Caryophyllaceae 1	-	-	-	-	-	-	-	-	-	-	-	-	-	-	-
Atriplex	-	-	-	-	-	-	-	-	-	-	-	-	-	-	-
Chenopodium	-	-	-	2	-	-	-	-	-	-	-	-	-	-	-
Salsola kali-type	-	-	-	-	-	-	-	-	-	-	-	-	-	-	-
Salsola soda-type	-	-	-	-	-	-	-	-	-	1	-	2	-	-	-
Salsola	-	-	-	-	-	-	-	-	-	-	-	-	-	-	-
Suaeda	1	9	-	-	-	-	1	2	1	8	-	-	2	2	1
YH-Chenopodiaceae 2	-	-	-	-	-	-	-	-	-	-	4	-	-	-	-
CHENOPODIACEAE	-	40	-	-	-	-	-	-	-	-	13	-	-	-	-
Helianthemum	1	-	-	-	-	-	-	-	-	-	-	-	-	-	-
Carex	1	1	1	-	1	-	-	1	2	15	1	-	2	-	-
Carex 3	-	-	-	-	-	-	-	-	-	-	-	-	-	-	-
Eleocharis (YH-Cyperaceae a)	-	-	-	-	1	-	-	-	-	-	-	1	-	-	-
YH-Cyperaceae 1	1	1	2	-	-	-	-	-	-	1	1	1	3	1	1
YH-Cyperaceae 3	-	-	-	-	-	-	-	-	-	-	-	-	-	-	-
YH-Cyperaceae 4	-	-	-	-	-	-	-	-	-	-	-	-	-	-	-
YH-Cyperaceae 5	-	-	-	-	-	-	-	1	-	-	3	1	-	-	-
YH-Cyperaceae 7	-	-	1	-	-	-	-	-	-	-	-	-	-	-	-
YH-Cyperaceae 8	-	-	-	-	-	-	-	-	-	-	-	-	-	-	-
CYPERACEAE	-	8	3	-	-	-	-	-	-	-	-	-	2	4	-
Cephalaria	-	1	-	-	-	-	-	-	-	1	4	5	-	-	-
Scabiosa	-	-	-	-	-	-	-	-	-	-	-	-	-	-	-
Euphorbia	-	-	-	-	-	-	-	1	-	-	-	-	-	-	-

Table F2, cont'd.: Columns 91–105

Column no.	91	92†	93	94*†	95	96	97	98†	99	100*†	101*	102*†	103*	104	105
YH no.	27569	30664	23774	26944	28547	27937	28548	28522	30194	30191	31560	29540	31203	30692	32334
Alhagi	-	2	-	1	7	-	2	-	1	-	-	2	-	-	-
Astragalus	-	-	-	-	-	-	-	-	-	-	-	1	-	-	-
Medicago	-	2	-	-	-	-	1	-	-	1	-	-	-	1	-
Onobrychis	-	-	-	-	-	-	-	-	-	-	-	-	-	-	-
Trifolium/Melilotus	-	3	2	1	-	-	1	-	-	22	2	-	-	-	-
Trigonella	3	5	5	1	1	-	-	-	1	183	-	1	-	-	-
Trigonella cf. astroites	-	-	-	-	-	-	-	1	-	30	1	-	-	-	-
YH-Fabaceae 2	-	-	-	-	-	-	-	-	-	-	-	-	-	-	-
FABACEAE	1	26	-	-	1	-	-	1	-	16	2	1	1	1	2
cf. Erodium	-	-	-	-	-	-	-	-	-	1	-	-	-	-	-
Teucrium	-	-	-	-	1	-	-	-	-	2	-	-	-	-	-
Ziziphora	1	4	-	1	1	-	1	-	1	13	-	1	-	-	-
YH-Lamiaceae 2	-	-	-	-	-	-	-	-	-	-	-	-	-	-	-
YH-Lamiac 3 (Nepeta?)	-	-	-	-	-	-	-	-	-	-	-	-	-	-	-
YH-Lamiaceae 5	-	-	-	-	-	-	-	-	1	-	-	-	-	1	-
LAMIACEAE	-	-	-	-	-	-	-	-	-	1	-	-	-	-	-
LILIACEAE	-	-	-	-	-	-	-	-	-	-	-	-	-	-	-
cf. Malva	-	2	-	-	-	-	-	-	-	1	-	-	-	-	-
Glaucium	-	-	-	-	-	-	-	-	1	3	-	-	-	-	-
Papaver	-	2	-	-	-	-	-	-	-	1	2	-	-	-	-
Fumaria	-	-	-	-	-	-	-	-	-	-	-	-	-	-	-
Plantago	-	-	-	-	-	-	-	-	-	1	-	-	-	-	-
Aegilops	-	-	-	-	-	-	-	-	-	-	-	1	-	-	-
Avena	-	-	-	-	-	-	-	-	-	1	-	3	-	-	-
Bromus japonicus-type	-	-	-	-	-	-	-	-	-	-	-	1	-	-	-
Bromus tectorum-type	-	-	-	-	-	-	-	-	-	-	-	4	-	-	-
Bromus	-	-	-	-	-	-	-	-	1	3	-	-	-	-	-
Eremopyrum	-	1	-	1	-	-	1	-	7	-	-	-	1	-	-
Hordeum cf. murinum	-	1	1	-	-	-	-	-	-	3	-	-	-	-	-
Hordeum spontaneum?	-	-	-	-	-	-	-	-	-	-	-	-	-	-	-
Hordeum misc.	-	-	1	-	-	-	-	-	-	-	1	-	-	1	-
cf. Lolium	-	1	-	-	-	-	-	-	-	-	4	-	-	-	-
cf. Phalaris	-	1	-	-	-	-	-	-	-	4	-	-	-	-	-
cf. Poa bulbosa	-	-	-	-	-	-	-	-	1	-	-	-	-	-	-
Setaria	-	4	3	-	-	-	-	-	-	-	-	3	-	-	1
Stipa	-	3	-	-	-	-	-	-	-	2	1	-	-	-	-
Taeniatherum	-	-	-	1	-	-	-	1	-	-	-	-	-	-	-
"Triticoid"	-	1	-	-	-	-	-	-	-	-	2	-	-	-	-
Triticum boeoticum	-	-	1	-	-	-	-	-	-	-	4	-	-	-	-
YH-Poaceae 1	-	1	2	1	-	-	2	-	2	10	-	1	-	-	-
YH-Poaceae 2	-	-	-	-	-	-	-	1	-	-	-	-	-	-	-
YH-Poaceae 3	-	1	-	-	1	-	-	-	-	-	-	-	-	-	-
YH-Poaceae 4	-	-	-	-	-	-	-	-	-	-	-	-	-	-	-
YH-Poaceae 5	1	-	-	-	-	-	-	-	-	1	-	-	-	-	-
YH-Poaceae 8	-	-	-	-	-	-	-	-	1	-	-	1	-	-	-
YH-Poaceae 10/15	-	-	-	-	-	-	-	-	-	-	-	-	-	-	-
YH-Poaceae 11	-	-	-	-	-	-	-	-	-	-	-	-	-	-	-
YH-Poaceae 13	-	-	-	-	-	-	-	-	-	1	-	-	-	-	-

Table F2, cont'd.: Columns 91–105

Column no.	91	92†	93	94*†	95	96	97	98†	99	100*†	101*	102*†	103*	104	105
YH no.	27569	30664	23774	26944	28547	27937	28548	28522	30194	30191	31560	29540	31203	30692	32334
YH-Poaceae 14	-	-	-	-	-	-	-	-	-	-	-	-	-	-	-
YH-Poaceae 16	-	-	-	-	-	-	-	-	-	7	-	-	-	-	1
YH-Poaceae 17/18	-	1	-	-	-	-	-	-	-	-	-	-	-	-	1
YH-Poaceae 20 Aeg 1a	-	-	-	-	-	-	-	-	1	-	-	-	-	-	-
YH-Poaceae 21 Aeg 1b	-	-	-	-	-	-	-	-	2	-	-	-	-	-	-
POACEAE	5	37	3	2	1	-	-	5	2	22	2	9	5	13	-
Polygonum	-	-	-	1	-	-	-	-	-	1	-	-	-	-	-
Polygonum (was YH-Cyperaceae 2)	-	2	-	-	-	-	-	1	-	1	-	1	-	1	-
Polygonum (was YH-Cyperaceae 6)	-	-	-	-	-	-	-	-	-	1	-	-	-	-	-
Rumex	-	-	-	-	-	-	-	1	-	-	-	-	-	-	-
Portulaca	1	6	-	-	-	-	-	-	-	-	-	-	-	-	-
Androsace	-	-	-	-	1	-	-	-	-	7	-	-	-	-	-
YH-Primulaceae 1	-	-	-	-	-	-	-	-	-	-	-	-	-	-	-
YH-Primulaceae 2	-	-	-	-	-	-	-	-	-	-	-	-	-	-	-
Aconitum	-	-	-	-	-	-	-	-	-	-	-	-	-	-	-
Adonis	-	2	-	-	-	-	-	1	-	1	-	1	-	-	-
Ceratocephalus (YH-pp 2)	-	-	-	-	-	-	-	-	-	1	-	-	-	-	-
Ranunculus (YH-unknown 4)	-	-	-	-	-	-	-	-	-	3	1	-	-	-	-
Ranunculus arvensis	-	-	-	-	-	-	-	-	-	-	-	-	-	-	-
Reseda	-	-	-	-	-	-	-	-	-	-	-	-	-	-	-
cf. Asperula (YH-Rubiaceae 2)	-	-	-	-	1	-	-	-	-	3	3	-	-	-	-
Galium	1	474	1	-	8	1	3	23	2	8	5	22	2	4	-
YH-Rubiaceae 1	-	-	-	-	-	-	-	-	-	-	-	-	-	-	-
YH-Scrophulariaceae 1	-	-	-	-	-	-	-	-	-	-	-	-	-	-	-
Hyoscyamus	-	3	-	-	-	-	-	-	1	1	-	-	-	-	-
SOLANACEAE	-	-	-	-	-	-	-	-	-	-	-	-	-	-	-
Thymelaea	-	1	-	1	-	-	-	-	-	-	-	-	-	-	-
Valerianella coronata	-	-	1	-	1	-	-	-	-	1	-	-	-	-	-
Valerianella vesicaria	-	-	-	-	-	-	-	-	-	2	-	-	-	-	-
Valerianella	-	-	-	-	-	-	-	-	-	-	-	2	-	-	-
Peganum harmala	-	-	-	-	-	-	-	-	-	-	-	1	-	-	-
Zygophyllum	-	-	-	-	-	-	-	-	-	1	-	-	-	-	-
YH-unknown 9	-	-	-	-	-	-	-	-	-	-	-	-	-	-	-
YH-unknown 21	-	-	-	-	-	-	-	-	-	3	-	1	-	-	-
YH-unknown 23	-	1	-	-	-	-	-	-	-	-	-	-	-	-	-
YH-unknown 27	-	-	-	-	-	-	-	-	-	-	-	6	-	-	-
YH-unknown 29	-	-	-	-	-	-	-	-	-	-	-	-	-	-	-
YH-unknown 30	-	-	-	-	-	-	-	-	-	-	-	-	-	-	-
YH-unknown 35	-	-	-	-	-	-	1	-	-	-	-	-	-	-	-
Unknown	2	117	11	-	9	1	10	5	7	85	24	20	-	19	2
PLANT PARTS (pp)															
Hordeum internode (2-row)	-	19	-	-	3	-	1	-	-	-	-	3	-	-	-
Hordeum internode (6-row)	-	1	-	-	-	-	-	-	-	-	-	-	-	-	-
Hordeum internode (compact)	-	3	-	-	-	-	-	-	-	-	1	-	1	-	-
Triticum aestivum int.	-	-	-	-	-	-	-	-	-	1	-	8	-	-	-
T. aestivum/durum int.	-	4	1	-	2	-	5	2	-	20	4	26	-	-	-
T. aestivum/durum int. 1	-	-	-	-	-	-	-	-	-	-	-	-	-	-	-

Table F2, cont'd.: Columns 91–105

Column no.	91	92†	93	94*†	95	96	97	98†	99	100*†	101*	102*†	103*	104	105
YH no.	27569	30664	23774	26944	28547	27937	28548	28522	30194	30191	31560	29540	31203	30692	32334
T. aestivum/durum int. 2	-	-	-	-	-	-	-	-	-	-	-	-	-	-	-
T. durum internode	-	-	-	-	-	-	-	-	-	2	-	-	-	-	-
T. monococcum spikelet fork	-	4	4	-	-	-	-	-	-	-	-	-	-	-	-
T. dicoccum spikelet fork	-	-	-	-	-	-	-	-	-	-	-	-	-	-	-
T. monococcum/dicoccum spikelet fork	3	36	-	-	-	-	-	-	-	1	-	-	-	-	-
Triticum rachis base	-	1	2	-	-	-	-	-	-	2	-	-	-	-	-
Triticum, rachis fragment, square cross-section	-	-	-	-	-	-	1	-	-	5	1	-	-	-	-
Cereal culm node	2	8	-	-	-	-	3	-	-	6	-	5	1	-	-
Cereal culm root base?	-	-	-	-	-	-	-	-	-	-	-	-	-	-	-
cf. Oryza glume (charred)	-	-	-	-	-	-	-	-	-	-	-	-	-	-	-
Oryza glume (silicified phytolith)	-	-	-	-	-	-	-	-	-	-	-	-	-	-	-
Vitis peduncle	-	-	-	-	-	-	-	-	-	-	-	-	-	-	-
Onopordum achene collar	-	-	-	-	-	-	-	-	-	-	-	-	-	-	-
Asteraceae head, misc.	-	-	-	-	-	-	-	-	-	-	-	-	-	-	-
Euclidium silique	-	-	-	-	-	-	-	-	-	2	-	-	1	-	-
Capparis thorn	-	-	-	-	-	-	-	-	-	-	-	-	-	-	-
Atriplex bract (YH-pp 12)	-	-	-	-	-	-	-	-	-	-	-	-	-	-	-
Salsola inflor (YH-pp 7, YH-unk 8)	-	-	-	-	-	-	-	-	-	-	-	3	-	-	-
Cyperaceae stem fragment	-	-	-	-	-	-	-	-	-	-	-	-	-	-	-
Alhagi pod section	-	1	-	+	-	-	-	-	-	-	-	-	-	-	-
Alhagi leaf?	-	-	-	-	-	-	-	-	-	8	-	-	-	-	-
Aegilops glume base	-	-	-	-	-	-	-	-	-	-	-	-	-	-	-
Hordeum internode (wild)	-	-	-	-	-	-	-	-	-	-	-	-	-	-	-
Taeniatherum int. (YH-pp 8)	-	-	-	-	-	-	-	-	-	1	-	-	-	-	-
Poaceae, mystery 1 rachis frag.	-	-	-	-	-	-	-	-	-	-	-	-	-	-	-
YH-plant part 9	-	5	-	-	-	-	-	-	-	-	1	-	-	-	-
UNCHARRED-MINERALIZED															
Arnebia linearis	-	-	-	-	-	-	-	-	-	-	-	-	-	-	-
Arnebia/Lithospermum	3	7	5	-	-	-	2	1	-	-	-	6	1	1	-
Lithospermum tenuifolium	-	-	-	-	-	-	-	1	-	-	-	-	-	-	-
Lithospermum arvense	-	-	-	-	-	-	-	-	-	-	-	2	-	-	-
Lithospermum, misc.	-	-	-	-	-	-	-	-	-	-	1	-	-	-	-
Moltkia	-	1	-	-	-	-	-	-	-	-	-	-	-	-	-
BORAGINACEAE	-	-	-	-	-	-	-	-	-	-	-	-	-	-	1
Eleocharis (Cyperaceae-a)	-	-	-	-	2	-	-	-	9	-	-	3	-	-	-
Carex	-	-	-	1	1	-	-	-	-	-	-	-	-	-	-
Ficus	-	-	-	2	-	-	-	-	-	-	-	-	-	1	-
Glaucium	-	2	7	-	-	-	1	-	-	2	-	-	-	-	-
Papaver (white)	-	-	-	6	-	-	-	-	-	-	-	-	-	1	-

Table F2, cont'd.: Columns 106–120

Column no.	106	107†	108	109	110	111	112	113	114*	115	116	117	118	119	120
YH no.	31439	30990	32303	32578	29541	31931	32692	30171	32598	32818	32590	32860	20127	20303	20144
Context type	pit	pit	debris	debris	debris	pit	jar	debris	pit	pyro	debris	pit	debris	debris	debris
Operation	2	2	2	2	1	12	12	1	2	2	2	12	5	3	5
Locus	88	70	88	87	55	27	42	74	87	87	87	45	2	6	5
Lot	287	256	312	324	136	36	71	161	330	328	326	77	2	10	5
Phase	515	515.03	530	530	535	540	550.04	560	570.11	570.13	570.15	590.01	640	640	650
Soil volume (liters)	2.5	11	15	5.5	7	10	0.7	6.5	7	8.5	7	2	8	7	8
Charcoal (>2 mm, g)	17.72	28.36	23.26	5.08	3.89	5.76	18.62	3.65	19.97	4.64	7.30	4.69	0.59	0.26	0.40
Seed (>2 mm, g)	0.33	0.48	0.70	0.10	0.32	0.20	0.02	0.26	0.34	0.06	0.20	0.03	-	-	0.03
Other (>2 mm, g)	-	+	0.02	-	-	-	-	-	-	+	+	-	-	-	0
Wild/weedy (#)	4	44	46	23	25	5	3	14	13	57	64	3			9
ECONOMIC PLANTS															
Hordeum vulgare (g)	0.03	0.19	0.34	0.05	0.07	0.07	-	0.03	0.23	0.12	0.10	-	-	-	0.01
H. vulgare var. nudum (g)	-	-	-	-	-	-	-	-	-	-	-	-	-	-	-
Triticum aestivum/durum (g)	0.03	0.07	0.07	0.03	0.06	0.05	-	0.12	+	0.05	0.06	0.01	-	-	0.01
Triticum monococcum (g)	-	-	-	-	-	-	-	-	+	+	-	-	-	-	-
Triticum dicoccum (g)	-	-	-	-	-	-	-	-	-	-	-	-	-	-	-
Triticum sp. (g)	-	-	0.03	-	-	0.05	-	-	-	-	0.04	-	-	-	-
Cereal, indet. (g)	0.15	0.19	0.32	0.07	0.11	0.09	-	0.26	0.19	0.04	0.04	0.01	-	-	0.01
Secale cereale (g)	-	-	-	-	-	-	-	-	-	-	-	-	-	-	-
Setaria italica (#)	-	-	1	-	-	-	-	1	1	-	-	-	-	-	-
Oryza (g)	-	-	-	-	-	-	-	-	-	-	-	-	-	-	-
Vicia ervilia (g)	-	-	-	-	-	-	-	-	-	-	-	-	-	-	-
Lens (g)	-	-	0.03	-	-	-	-	-	-	-	0.01	-	-	-	-
Pulse (g)	-	0.02	0.04	0.01	-	-	0.02	0.02	-	-	-	-	-	-	-
cf. Pistacia/nutshell (g)	-	-	-	-	-	-	-	-	-	-	-	-	-	-	-
cf. Quercus (g)	-	-	-	-	-	-	-	-	-	-	-	-	-	-	-
Ficus carica (#)	-	-	1	-	-	-	-	-	-	1	-	-	-	-	-
Vitis vinifera (# whole, total g)	-	-	-	-	-	-	-	-	-	-	-	-	-	-	-
WILD AND WEEDY (counts)															
Bupleurum	-	-	-	-	-	-	-	-	-	-	-	-	-	-	-
cf. Daucus	-	-	-	-	-	-	-	-	-	-	-	-	-	-	-
Torilis leptophylla?	-	-	-	-	-	-	-	-	-	-	-	-	-	-	-
YH-Apiaceae 2	-	-	-	-	-	-	-	-	-	2	2	-	-	-	-
YH-Apiaceae 4/8	-	-	-	-	-	-	-	-	-	-	-	-	-	-	-
YH-Apiaceae 6	-	-	-	-	-	-	-	-	1	-	-	-	-	-	-
YH-Apiaceae 7	-	1	-	-	-	-	-	-	-	-	-	-	-	-	-
YH-Apiaceae 10	-	-	-	-	-	-	-	-	-	-	-	-	-	-	-
YH-unknown 31 (Apiaceae)	-	-	-	-	-	-	-	-	-	-	-	-	-	-	-
APIACEAE	-	-	1	1	-	-	-	-	-	-	-	-	-	-	-
Anthemis/Matricaria	-	-	-	-	-	-	-	-	-	-	-	-	-	-	-
Artemisia	-	-	-	-	-	-	-	-	-	-	-	-	-	-	-
Carthamus	-	-	-	-	-	-	-	-	-	-	-	-	-	-	-
Centaurea cyanus-type	-	-	-	-	-	-	-	-	-	-	-	-	-	-	-
Centaurea	-	-	-	-	1	-	-	3	1	1	1	-	-	-	-
Onopordum	-	-	-	-	-	-	-	-	-	-	-	-	-	-	-
YH-Asteraceae 1	-	-	1	-	-	-	-	11	-	-	-	-	-	-	-

Table F2, cont'd.: Columns 106–120

Column no.	106	107†	108	109	110	111	112	113	114*	115	116	117	118	119	120
YH no.	31439	30990	32303	32578	29541	31931	32692	30171	32598	32818	32590	32860	20127	20303	20144
YH-Asteraceae 2	-	-	-	-	-	-	-	-	-	-	-	-	-	-	-
YH-Asteraceae 5	-	-	-	-	-	-	-	-	-	-	-	-	-	-	-
YH-Asteraceae 9	-	-	-	-	-	-	-	-	-	-	-	-	-	-	-
ASTERACEAE	-	-	-	-	-	-	-	-	-	-	1	-	-	-	-
Arnebia/Lithospermum	-	-	-	-	-	-	-	-	-	-	-	-	-	-	-
Heliotropium	-	-	-	-	-	-	-	-	-	2	2	-	-	-	-
cf. Alyssum	-	-	-	-	-	-	-	-	-	-	-	-	-	-	-
cf. Camelina rumelica	-	-	-	-	-	-	-	-	-	-	-	-	-	-	-
cf. Camelina sativa	-	-	-	-	-	-	-	-	-	-	-	-	-	-	-
Conringia	-	-	-	-	-	-	-	-	-	-	-	-	-	-	-
cf. Lepidium	-	-	-	-	-	-	-	-	-	-	-	-	-	-	-
Sisymbrium altissimum-type	-	-	-	-	-	-	-	-	-	-	-	-	-	-	-
Thlaspi	-	-	-	-	-	-	-	-	-	-	-	-	-	-	-
YH-Brassicaceae 2	-	-	-	-	3	-	-	-	-	-	-	-	-	-	-
YH-Brassicaceae 3/5	-	2	-	-	-	-	-	-	-	-	-	-	-	-	-
YH-Brassicaceae 7	-	-	-	-	-	-	-	-	-	-	-	-	-	-	-
YH-Brassicaceae 10	-	-	-	-	-	-	-	-	-	-	-	-	-	-	-
YH-Brassicaceae 11	-	-	-	-	-	-	-	-	2	1	-	-	-	-	-
YH-Brassicaceae 12	-	-	-	-	-	-	-	-	-	-	-	-	-	-	-
BRASSICACEAE	-	-	-	-	-	-	-	-	-	-	2	-	-	-	-
Bufonia	-	-	-	-	-	-	-	-	-	-	-	-	-	-	-
Gypsophila	-	2	1	1	1	1	-	-	-	-	-	-	-	-	-
Silene	-	-	-	-	-	-	-	-	-	-	-	-	-	-	-
Vaccaria (YH-unknown 6)	-	1	2	-	-	-	-	-	-	-	-	-	-	-	-
CARYOPHYLLACEAE	-	-	-	-	-	-	-	-	-	-	-	-	-	-	-
YH-Caryophyllaceae 1	-	-	-	-	-	-	-	-	-	-	-	-	-	-	-
Atriplex	-	1	-	-	-	-	-	-	-	-	-	-	-	-	-
Chenopodium	-	-	-	1	2	-	-	-	-	-	4	-	-	-	-
Salsola kali-type	-	-	-	-	-	-	-	-	-	-	-	-	-	-	-
Salsola soda-type	-	-	-	-	-	-	-	-	-	-	-	-	-	-	-
Salsola	-	-	-	-	-	-	-	-	-	-	-	-	-	-	-
Suaeda	-	-	4	2	-	-	-	-	-	2	4	-	-	-	-
YH-Chenopodiaceae 2	-	-	-	-	-	-	-	-	-	1	-	-	-	-	-
CHENOPODIACEAE	-	-	1	-	-	-	-	-	-	-	-	-	-	-	-
Helianthemum	-	-	-	-	-	-	-	-	-	-	-	-	-	-	-
Carex	-	1	-	1	-	-	-	-	-	3	-	-	-	-	-
Carex 3	-	-	-	-	-	-	-	-	-	-	-	-	-	-	-
Eleocharis (YH-Cyperaceae a)	-	-	-	-	-	-	-	-	2	-	-	-	-	-	-
YH-Cyperaceae 1	1	-	1	1	2	-	-	-	-	7	3	-	-	-	-
YH-Cyperaceae 3	-	-	-	-	-	-	-	-	-	-	-	-	-	-	-
YH-Cyperaceae 4	-	-	-	-	-	-	-	-	-	-	-	-	-	-	-
YH-Cyperaceae 5	-	-	-	-	1	-	-	-	-	-	-	-	-	-	-
YH-Cyperaceae 7	-	-	-	-	-	-	-	-	-	-	-	-	-	-	-
YH-Cyperaceae 8	-	-	-	-	-	-	-	-	-	-	-	-	-	-	-
CYPERACEAE	-	2	-	1	1	-	-	-	-	3	3	-	-	-	-
Cephalaria	-	-	-	-	-	-	-	-	-	-	1	-	-	-	-
Scabiosa	-	-	-	-	-	-	-	-	-	-	-	-	-	-	-
Euphorbia	-	-	-	-	-	-	-	-	-	-	-	-	-	-	-

Table F2, cont'd.: Columns 106–120

Column no.	106	107†	108	109	110	111	112	113	114*	115	116	117	118	119	120
YH no.	31439	30990	32303	32578	29541	31931	32692	30171	32598	32818	32590	32860	20127	20303	20144
Alhagi	-	-	-	-	-	-	-	-	-	3	-	-	-	-	-
Astragalus	-	-	-	-	-	-	-	-	-	-	-	-	-	-	-
Medicago	-	-	1	-	-	-	-	-	-	-	-	-	-	-	1
Onobrychis	-	-	-	-	-	-	-	-	-	-	-	-	-	-	-
Trifolium/Melilotus	-	1	-	-	1	-	-	-	-	7	1	-	-	-	-
Trigonella	-	6	-	-	1	-	-	-	1	9	4	1	-	-	2
Trigonella cf. astroites	-	-	-	-	-	-	-	-	-	-	-	-	-	-	1
YH-Fabaceae 2	-	-	-	-	-	-	-	-	-	-	-	-	-	-	-
FABACEAE	-	3	2	1	3	-	-	-	-	3	-	-	-	-	-
cf. Erodium	-	-	-	-	-	-	-	-	-	-	-	-	-	-	-
Teucrium	-	-	-	1	-	-	-	-	-	-	-	-	-	-	-
Ziziphora	-	1	-	-	2	-	-	-	-	1	-	-	-	-	-
YH-Lamiaceae 2	-	-	-	-	-	-	-	-	-	-	-	-	-	-	-
YH-Lamiac 3 (Nepeta?)	-	-	-	-	-	-	-	-	-	-	-	-	-	-	-
YH-Lamiaceae 5	-	-	-	-	-	-	-	-	-	-	1	-	-	-	-
LAMIACEAE	-	-	-	-	-	-	-	-	-	-	-	-	-	-	-
LILIACEAE	-	-	-	-	-	-	-	-	-	-	-	-	-	-	-
cf. Malva	-	-	1	-	-	-	-	-	-	-	-	-	-	-	-
Glaucium	-	-	-	-	-	-	-	-	-	-	-	-	-	-	-
Papaver	-	-	-	-	-	-	-	-	-	-	-	-	-	-	-
Fumaria	-	-	-	-	-	-	-	-	-	-	-	-	-	-	-
Plantago	-	-	-	-	-	-	-	-	-	-	-	-	-	-	-
Aegilops	-	-	-	-	-	-	-	-	-	-	-	-	-	-	-
Avena	1	-	-	-	-	-	-	-	-	-	-	-	-	-	-
Bromus japonicus-type	-	1	-	1	-	-	-	-	-	-	1	-	-	-	-
Bromus tectorum-type	-	1	1	-	-	-	-	-	-	-	-	-	-	-	-
Bromus	-	2	-	-	-	-	-	-	-	1	-	-	-	-	-
Eremopyrum	-	-	1	-	-	-	-	-	-	-	-	-	-	-	-
Hordeum cf. murinum	-	-	-	-	-	-	-	-	-	-	-	-	-	-	-
Hordeum spontaneum?	-	-	-	-	-	-	-	-	-	-	-	-	-	-	-
Hordeum misc.	-	-	-	-	-	-	-	-	-	-	-	-	-	-	-
cf. Lolium	-	-	2	-	-	-	1	-	-	-	-	-	-	-	-
cf. Phalaris	-	-	-	-	-	-	-	-	-	-	-	-	-	-	-
cf. Poa bulbosa	-	-	-	-	-	-	-	-	-	-	-	-	-	-	-
Setaria	-	-	1	-	-	-	-	-	-	-	1	-	-	-	-
Stipa	-	-	-	-	-	-	-	-	-	-	-	-	-	-	-
Taeniatherum	-	-	-	-	-	-	-	-	-	-	-	-	-	-	-
"Triticoid"	-	-	-	-	-	-	-	-	-	-	-	-	-	-	-
Triticum boeoticum	-	-	-	-	-	1	-	-	-	-	-	-	-	-	-
YH-Poaceae 1	-	-	-	-	-	-	-	-	-	-	-	-	-	-	-
YH-Poaceae 2	-	-	-	1	-	-	-	-	-	-	-	-	-	-	-
YH-Poaceae 3	-	-	-	-	-	-	-	-	-	-	-	-	-	-	-
YH-Poaceae 4	-	-	-	-	-	-	-	-	-	-	-	-	-	-	-
YH-Poaceae 5	-	-	-	-	-	-	-	-	-	-	-	-	-	-	-
YH-Poaceae 8	-	2	-	-	1	-	-	-	-	-	1	-	-	-	1
YH-Poaceae 10/15	-	-	-	-	-	-	-	-	-	-	-	-	-	-	-
YH-Poaceae 11	-	-	-	-	-	-	-	-	-	-	-	-	-	-	-
YH-Poaceae 13	-	-	-	-	-	-	-	-	-	-	-	-	-	-	-

Table F2, cont'd.: Columns 106–120

Column no.	106	107†	108	109	110	111	112	113	114*	115	116	117	118	119	120
YH no.	31439	30990	32303	32578	29541	31931	32692	30171	32598	32818	32590	32860	20127	20303	20144
YH-Poaceae 14	-	-	-	-	1	-	-	-	-	1	-	-	-	-	-
YH-Poaceae 16	-	-	-	-	-	-	-	-	-	-	-	-	-	-	-
YH-Poaceae 17/18	-	3	4	2	-	-	-	-	-	2	2	-	-	-	-
YH-Poaceae 20 Aeg 1a	-	-	-	-	-	-	-	-	-	-	-	-	-	-	-
YH-Poaceae 21 Aeg 1b	-	-	-	-	-	-	-	-	-	-	-	-	-	-	-
POACEAE	2	3	4	5	1	1	1	-	1	3	10	-	-	-	4
Polygonum	-	-	-	-	-	-	1	-	-	-	-	1	-	-	-
Polygonum (was YH-Cyperaceae 2)	-	-	-	-	-	-	-	-	-	-	3	-	-	-	-
Polygonum (was YH-Cyperaceae 6)	-	-	1	-	-	1	-	-	-	-	-	-	-	-	-
Rumex	-	2	5	1	-	-	-	-	-	-	-	-	-	-	-
Portulaca	-	-	-	-	-	-	-	-	-	-	1	-	-	-	-
Androsace	-	-	-	-	1	-	-	-	-	-	1	-	-	-	-
YH-Primulaceae 1	-	-	-	-	-	-	-	-	-	-	-	-	-	-	-
YH-Primulaceae 2	-	-	-	-	-	-	-	-	-	-	-	-	-	-	-
Aconitum	-	-	-	-	-	-	-	-	-	-	-	-	-	-	-
Adonis	-	-	-	1	-	-	-	-	1	-	-	-	-	-	-
Ceratocephalus (YH-pp 2)	-	-	-	-	-	-	-	-	-	-	-	-	-	-	-
Ranunculus (YH-unknown 4)	-	-	-	-	-	-	-	-	-	-	-	-	-	-	-
Ranunculus arvensis	-	-	-	-	-	-	-	-	-	-	-	-	-	-	-
Reseda	-	-	-	-	-	-	-	-	-	-	-	-	-	-	-
cf. Asperula (YH-Rubiaceae 2)	-	-	-	-	-	-	-	-	-	-	-	-	-	-	-
Galium	-	7	11	1	2	1	-	-	3	3	13	-	-	-	-
YH-Rubiaceae 1	-	-	-	-	-	-	-	-	-	-	1	-	-	-	-
YH-Scrophulariaceae 1	-	-	-	-	-	-	-	-	-	-	-	-	-	-	-
Hyoscyamus	-	-	-	-	-	-	-	-	-	-	-	-	-	-	-
SOLANACEAE	-	1	-	-	-	-	-	-	-	-	-	-	-	-	-
Thymelaea	-	-	-	-	-	-	-	-	-	-	-	-	-	-	-
Valerianella coronata	-	-	-	-	1	-	-	-	-	-	-	-	-	-	-
Valerianella vesicaria	-	-	-	-	-	-	-	-	-	-	-	-	-	-	-
Valerianella	-	-	-	-	-	-	-	-	1	-	-	-	-	-	-
Peganum harmala	-	-	-	-	-	-	-	-	-	1	-	-	-	-	-
Zygophyllum	-	-	-	-	-	-	-	-	-	-	-	-	-	-	-
YH-unknown 9	-	-	-	-	-	-	-	-	-	-	-	-	-	-	-
YH-unknown 21	-	1	-	-	-	-	-	-	-	-	1	1	-	-	-
YH-unknown 23	-	-	1	-	-	-	-	-	-	-	-	-	-	-	-
YH-unknown 27	-	-	-	-	-	-	-	-	-	1	-	-	-	-	-
YH-unknown 29	-	-	-	-	-	-	-	-	-	-	-	-	-	-	-
YH-unknown 30	-	-	-	-	-	-	-	-	-	-	-	-	-	-	-
YH-unknown 35	-	-	-	1	-	-	-	-	-	-	-	-	-	-	-
Unknown	-	7	11	1	7	3	-	-	7	11	3	-	-	-	-
PLANT PARTS (pp)															
Hordeum internode (2-row)	-	2	1	-	2	-	-	-	-	-	-	-	-	-	-
Hordeum internode (6-row)	-	1	1	-	-	-	-	-	-	-	-	-	-	-	-
Hordeum internode (compact)	-	-	1	-	-	-	-	-	-	-	-	-	-	-	-
Triticum aestivum int.	-	-	-	-	-	-	-	-	-	-	-	-	-	-	-
T. aestivum/durum int.	-	3	3	-	-	2	-	-	-	-	-	3	-	-	-
T. aestivum/durum int. 1	-	-	-	-	-	-	-	-	-	-	-	-	-	-	-

Table F2, cont'd.: Columns 106–120

Column no.	106	107†	108	109	110	111	112	113	114*	115	116	117	118	119	120
YH no.	31439	30990	32303	32578	29541	31931	32692	30171	32598	32818	32590	32860	20127	20303	20144
T. aestivum/durum int. 2	-	-	-	-	-	-	-	-	-	-	-	-	-	-	-
T. durum internode	-	-	-	-	-	-	-	-	-	-	-	-	-	-	-
T. monococcum spikelet fork	-	-	1	-	-	-	-	-	-	-	-	-	-	-	-
T. dicoccum spikelet fork	-	-	-	-	-	-	-	-	-	-	-	-	-	-	-
T. monococcum/dicoccum spikelet fork	-	1	2	-	-	-	-	-	-	-	-	-	-	-	-
Triticum rachis base	-	-	-	-	-	-	-	-	-	-	-	-	-	-	-
Triticum, rachis fragment, square cross-section	-	-	-	-	-	-	-	-	-	-	-	-	-	-	-
Cereal culm node	-	1	3	-	1	-	-	-	-	-	-	-	-	-	-
Cereal culm root base?	-	-	-	-	-	-	-	-	-	1	-	-	-	-	-
cf. Oryza glume (charred)	-	-	-	-	-	-	-	-	-	-	-	-	-	-	-
Oryza glume (silicified phytolith)	-	-	-	-	-	-	-	-	-	-	-	-	-	-	-
Vitis peduncle	-	-	-	-	-	-	-	-	-	-	-	-	-	-	-
Onopordum achene collar	-	-	-	-	-	-	-	-	-	-	-	-	-	-	-
Asteraceae head, misc.	-	-	-	-	-	-	-	-	-	-	-	-	-	-	-
Euclidium silique	-	-	-	-	-	-	-	-	-	-	-	-	-	-	-
Capparis thorn	-	-	-	-	-	-	-	-	-	-	-	-	-	-	-
Atriplex bract (YH-pp 12)	-	-	-	-	-	-	-	-	-	-	-	-	-	-	-
Salsola inflor (YH-pp 7, YH-unk 8)	-	-	-	-	-	-	-	-	-	-	-	-	-	-	-
Cyperaceae stem fragment	-	+	-	-	-	-	-	-	-	-	-	-	-	-	-
Alhagi pod section	-	-	1	-	-	-	-	-	-	-	-	-	-	-	-
Alhagi leaf?	-	-	-	-	-	-	-	-	-	-	-	-	-	-	-
Aegilops glume base	-	-	-	-	-	-	-	-	-	-	-	-	-	-	-
Hordeum internode (wild)	-	-	-	-	-	-	-	-	-	-	-	-	-	-	-
Taeniatherum int. (YH-pp 8)	-	-	-	-	-	-	-	-	-	-	-	-	-	-	-
Poaceae, mystery 1 rachis frag.	-	-	-	-	-	-	-	-	-	-	-	-	-	-	-
YH-plant part 9	-	-	-	-	-	-	-	-	-	-	1	-	-	-	-
UNCHARRED-MINERALIZED															
Arnebia linearis	-	-	-	-	-	-	-	-	-	-	-	-	-	-	-
Arnebia/Lithospermum	4	-	1	-	2	+	-	-	4	5	2	-	-	-	-
Lithospermum tenuifolium	-	-	-	-	-	-	-	1	-	-	-	-	-	-	-
Lithospermum arvense	-	-	-	-	-	-	-	-	-	-	-	-	-	-	-
Lithospermum, misc.	-	-	-	-	-	-	-	-	-	-	-	-	-	-	-
Moltkia	-	-	2	-	-	1	-	-	-	-	-	1	1	-	-
BORAGINACEAE	-	-	-	-	-	-	-	-	-	-	1	-	-	-	-
Eleocharis (Cyperaceae-a)	-	-	-	-	-	-	-	-	-	12	2	-	-	-	-
Carex	-	-	-	-	-	-	-	-	-	-	-	-	-	-	-
Ficus	-	-	1	-	-	-	-	-	-	-	-	-	-	-	-
Glaucium	-	-	-	1	-	1	-	1	-	-	-	-	-	-	-
Papaver (white)	-	1	-	-	-	-	-	-	-	-	-	-	-	-	-

Table F2, cont'd.: Columns 121–135

Column no.	121	122	123	124	125	126	127	128	129	130†	131†	132†	133	134	135
YH no.	20221	20241	20249	20668	25869	21515	21516	25942	27260	26398	27461	27471	25945	28572	29312
Context type	debris	debris	debris	debris	pyro	mixed	mixed	surf	surf	pit	pit	pit	pit	pit	pit
Operation	6	6	6	6	8	5	5	9	9	9	9	9	9	9	9
Locus	6	11	13	13	3	27	27	5	16	12	12	12	7	20	25
Lot	9	13	14	18	13	34	34	27	49	39	59	60	29	84	98
Phase	660	660	660	660	660.01	700	700	705	705	705.01	705.01	705.01	705.02	705.03	705.04
Soil volume (liters)	8	7	8	7	2	6.5	4.5	15	11	13	1.2	10	2	15	6
Charcoal (>2 mm, g)	1.15	0.92	1.21	0.94	-	5.43	2.22	25.33	2.07	3.07	1.33	2.93	7.80	2.46	5.19
Seed (>2 mm, g)	0.15	0.14	0.12	0.03	-	0.47	0.79	0.12	0.28	0.21	0.80	0.33	0.27	0.50	0.74
Other (>2 mm, g)	-	0.03	+	-	-	0.05	0.17	-	+	-	0.23	-	0.47	0.01	0.01
Wild/weedy (#)	16	17	9	7	1	67	92	4	87	47	821	110	242	94	161
ECONOMIC PLANTS															
Hordeum vulgare (g)	0.05	0.06	0.01	-	-	0.11	0.42	0.03	0.05	0.07	0.50	0.09	0.14	0.13	0.17
H. vulgare var. nudum (g)	-	0.01	-	-	-	-	-	-	-	-	-	-	-	-	-
Triticum aestivum/durum (g)	0.04	0.02	0.01	-	-	0.12	0.13	0.01	0.18	0.04	0.16	0.19	0.02	0.14	0.20
Triticum monococcum (g)	-	0.01	+	-	-	0.02	0.02	0.01	0.02	0.02	0.02	-	-	0.02	0.03
Triticum dicoccum (g)	-	-	+	0.01	-	-	0.02	-	-	-	-	-	-	-	-
Triticum sp. (g)	-	-	-	0.01	-	0.04	-	+	-	-	0.02	-	0.02	0.10	0.18
Cereal, indet. (g)	0.07	0.05	0.04	0.02	-	0.15	0.30	0.07	0.10	0.06	0.35	0.16	0.09	0.18	0.36
Secale cereale (g)	-	-	-	-	-	-	-	-	-	-	-	-	-	-	-
Setaria italica (#)	-	-	-	-	-	2	-	-	-	-	-	1	-	-	-
Oryza (g)	-	-	-	-	-	-	-	-	-	-	-	-	-	-	-
Vicia ervilia (g)	+	+	-	-	-	0.03	-	-	+	0.01	+	-	0.02	0.01	-
Lens (g)	-	-	-	-	-	-	-	-	-	-	-	-	-	-	-
Pulse (g)	-	+	-	-	-	0.03	-	0.01	-	0.01	-	-	-	-	-
cf. Pistacia/nutshell (g)	-	-	-	-	-	-	-	-	-	-	-	-	-	-	-
cf. Quercus (g)	-	-	-	-	-	-	-	-	-	-	-	-	-	-	-
Ficus carica (#)	-	-	-	-	-	-	-	-	-	-	-	-	-	-	-
Vitis vinifera (# whole, total g)	-	-	-	-	-	-	+	-	-	-	-	-	-	-	-
WILD AND WEEDY (counts)															
Bupleurum	-	-	-	-	-	-	-	-	-	-	1	-	3	-	-
cf. Daucus	-	-	-	-	-	-	-	-	-	-	-	-	-	-	-
Torilis leptophylla?	-	-	-	-	-	-	-	-	-	-	2	-	-	-	-
YH-Apiaceae 2	-	-	-	-	-	-	-	-	-	-	-	-	1	-	1
YH-Apiaceae 4/8	-	-	-	-	-	-	-	-	-	-	-	-	-	-	-
YH-Apiaceae 6	-	-	-	-	-	-	-	-	-	-	-	-	-	-	-
YH-Apiaceae 7	-	-	-	-	-	-	-	-	-	-	-	-	-	-	-
YH-Apiaceae 10	-	-	-	-	-	-	-	-	-	-	-	-	-	-	-
YH-unknown 31 (Apiaceae)	-	-	-	-	-	-	-	-	-	-	-	-	-	-	-
APIACEAE	-	-	-	-	-	1	-	-	-	-	-	-	-	-	-
Anthemis/Matricaria	-	-	-	-	-	-	-	-	-	-	-	-	-	-	-
Artemisia	-	-	-	-	-	-	-	-	-	-	1	-	-	-	-
Carthamus	-	-	-	-	-	-	-	-	-	-	-	-	-	-	-
Centaurea cyanus-type	-	-	-	-	-	-	-	-	-	-	-	-	-	-	-
Centaurea	-	-	-	-	-	-	-	-	1	-	-	-	4	-	1
Onopordum	-	-	-	-	-	-	-	-	-	-	-	-	-	1	-
YH-Asteraceae 1	-	-	-	-	-	-	-	-	-	-	-	-	-	-	-

Table F2, cont'd.: Columns 121–135

Column no.	121	122	123	124	125	126	127	128	129	130†	131†	132†	133	134	135
YH no.	20221	20241	20249	20668	25869	21515	21516	25942	27260	26398	27461	27471	25945	28572	29312
YH-Asteraceae 2	-	-	-	-	-	-	-	-	2	-	-	-	-	-	-
YH-Asteraceae 5	-	-	-	-	-	-	-	-	-	-	2	-	-	-	-
YH-Asteraceae 9	-	-	-	-	-	-	-	-	-	-	-	-	-	-	-
ASTERACEAE	-	-	-	-	-	-	-	-	-	-	-	-	-	-	-
Arnebia/Lithospermum	-	-	-	-	-	-	-	-	-	-	-	-	-	-	-
Heliotropium	-	-	-	-	-	1	-	-	-	-	10	-	-	-	-
cf. Alyssum	-	-	-	-	-	-	-	-	-	-	-	-	-	-	-
cf. Camelina rumelica	-	-	-	-	-	-	-	-	-	-	-	-	-	-	2
cf. Camelina sativa	-	-	-	-	-	-	-	-	-	-	-	-	-	-	-
Conringia	-	-	-	-	-	2	-	-	-	-	-	-	4	2	-
cf. Lepidium	-	-	-	-	-	-	-	-	-	-	100	-	-	-	-
Sisymbrium altissimum-type	-	-	-	-	-	-	-	-	-	-	-	-	-	-	-
Thlaspi	-	-	-	-	-	-	-	-	-	-	-	-	-	-	-
YH-Brassicaceae 2	-	-	-	-	-	-	-	-	-	-	-	-	-	-	-
YH-Brassicaceae 3/5	-	-	-	-	-	1	-	-	-	-	-	1	1	-	-
YH-Brassicaceae 7	-	-	-	-	-	-	-	-	-	-	-	-	-	-	-
YH-Brassicaceae 10	-	-	-	-	-	1	-	-	-	-	2	-	-	-	-
YH-Brassicaceae 11	-	-	-	-	-	-	1	-	-	-	-	2	-	-	3
YH-Brassicaceae 12	-	-	-	-	-	-	-	-	1	-	2	1	-	-	-
BRASSICACEAE	-	-	-	-	-	3	-	-	-	1	-	-	-	3	-
Bufonia	-	-	-	-	-	-	-	-	-	-	-	-	-	-	-
Gypsophila	1	-	-	-	-	-	-	-	1	-	-	1	-	4	2
Silene	-	-	-	-	-	-	-	-	-	-	-	-	-	-	-
Vaccaria (YH-unknown 6)	-	-	-	-	-	-	-	-	-	-	-	1	-	-	-
CARYOPHYLLACEAE	-	-	-	-	-	-	1	-	-	-	-	1	-	-	-
YH-Caryophyllaceae 1	-	-	-	-	-	-	-	-	-	-	-	-	-	-	-
Atriplex	-	-	-	-	-	-	-	-	-	-	-	-	-	-	-
Chenopodium	-	-	-	-	-	-	-	-	5	-	-	1	-	-	-
Salsola kali-type	-	-	-	-	-	-	-	-	-	-	-	-	-	-	-
Salsola soda-type	-	-	-	-	-	-	-	-	-	-	-	5	-	-	-
Salsola	1	-	-	-	-	-	-	-	1	-	-	-	-	1	-
Suaeda	-	-	-	1	-	-	-	-	-	1	5	2	2	-	2
YH-Chenopodiaceae 2	-	-	-	-	-	-	-	-	-	-	-	-	-	-	-
CHENOPODIACEAE	-	-	-	-	-	-	-	-	-	-	9	-	-	-	-
Helianthemum	-	-	-	-	-	-	-	-	-	-	-	-	-	-	-
Carex	-	-	-	-	-	2	-	1	2	4	18	1	22	-	2
Carex 3	-	-	-	-	-	-	-	-	-	-	-	-	2	-	-
Eleocharis (YH-Cyperaceae a)	-	-	-	-	-	-	-	-	-	1	25	-	55	-	1
YH-Cyperaceae 1	-	-	1	-	-	12	-	-	4	2	32	2	15	3	-
YH-Cyperaceae 3	-	-	-	-	-	-	-	-	-	-	-	-	-	-	-
YH-Cyperaceae 4	-	-	-	-	-	-	-	-	1	-	-	-	1	-	-
YH-Cyperaceae 5	-	-	-	-	-	-	-	-	1	-	-	-	1	-	-
YH-Cyperaceae 7	-	-	-	-	-	-	-	-	-	-	-	-	-	-	1
YH-Cyperaceae 8	-	-	-	-	-	-	-	-	-	-	-	-	-	-	-
CYPERACEAE	-	-	-	1	-	2	-	-	-	4	53	3	5	1	1
Cephalaria	-	-	-	-	-	-	-	-	-	-	-	-	-	-	-
Scabiosa	-	-	-	-	-	-	-	-	-	-	-	-	-	-	-
Euphorbia	-	-	-	-	-	-	-	-	-	-	-	-	-	-	-

Table F2, cont'd.: Columns 121–135

Column no.	121	122	123	124	125	126	127	128	129	130†	131†	132†	133	134	135
YH no.	20221	20241	20249	20668	25869	21515	21516	25942	27260	26398	27461	27471	25945	28572	29312
Alhagi	-	-	-	-	-	-	-	-	-	-	-	-	-	-	-
Astragalus	-	-	-	-	-	-	-	-	1	1	-	5	-	1	-
Medicago	-	-	-	-	-	-	1	-	-	-	-	-	-	-	1
Onobrychis	-	-	-	-	-	-	-	-	-	-	-	-	-	-	-
Trifolium/Melilotus	1	-	-	-	-	2	-	-	1	-	9	-	1	2	1
Trigonella	1	2	-	1	-	3	-	-	16	2	6	15	-	18	8
Trigonella cf. astroites	-	-	-	-	-	-	-	-	11	-	3	1	-	1	-
YH-Fabaceae 2	-	-	-	-	-	-	-	-	-	-	-	-	-	-	-
FABACEAE	-	-	-	1	-	5	2	-	-	-	-	8	2	10	-
cf. Erodium	-	-	-	-	-	-	-	-	-	-	-	-	-	-	-
Teucrium	1	-	-	-	-	1	1	-	-	-	-	-	1	-	-
Ziziphora	-	-	-	1	-	-	-	-	-	1	-	4	-	-	-
YH-Lamiaceae 2	-	-	-	-	-	-	-	-	-	-	-	-	-	-	-
YH-Lamiac 3 (Nepeta?)	-	-	-	-	-	-	-	-	-	-	-	-	-	-	-
YH-Lamiaceae 5	-	-	-	-	-	-	-	-	-	-	-	-	-	1	-
LAMIACEAE	2	-	-	-	-	-	-	-	-	-	3	-	-	-	-
LILIACEAE	-	-	-	-	-	-	-	-	-	-	-	-	-	-	-
cf. Malva	-	1	-	-	-	-	-	-	-	-	-	-	-	-	-
Glaucium	-	1	-	-	-	-	-	-	-	-	-	-	-	-	-
Papaver	-	-	-	-	-	-	-	-	-	-	-	-	-	-	-
Fumaria	-	-	-	-	-	-	-	-	-	-	-	-	-	-	-
Plantago	-	-	-	-	-	-	-	-	-	-	-	-	3	-	-
Aegilops	-	-	-	-	-	1	-	-	-	-	-	-	-	-	-
Avena	-	-	-	-	-	-	-	-	-	1	-	-	-	-	-
Bromus japonicus-type	-	-	-	-	-	-	-	-	-	1	4	-	-	-	-
Bromus tectorum-type	-	-	-	-	-	-	-	-	1	-	10	-	2	-	-
Bromus	-	-	-	-	-	-	1	-	-	-	16	-	1	-	-
Eremopyrum	-	-	-	-	-	-	1	-	2	-	-	-	-	4	1
Hordeum cf. murinum	-	1	1	-	-	-	1	-	1	2	4	2	-	3	1
Hordeum spontaneum?	-	-	-	1	-	-	-	-	-	-	1	-	-	-	-
Hordeum misc.	-	-	-	-	-	-	-	-	-	-	-	-	-	-	-
cf. Lolium	-	-	-	-	-	-	-	-	-	-	-	-	-	-	-
cf. Phalaris	-	-	-	-	-	-	-	-	-	-	-	-	3	-	-
cf. Poa bulbosa	-	-	-	-	-	-	-	-	-	-	-	-	-	-	-
Setaria	-	-	-	-	-	-	-	-	-	-	-	-	-	-	-
Stipa	-	-	-	-	-	-	-	-	-	-	-	-	1	-	-
Taeniatherum	-	-	-	-	-	1	-	-	-	1	1	-	-	-	-
"Triticoid"	-	1	-	-	-	-	-	-	-	-	-	1	1	-	-
Triticum boeoticum	-	-	1	-	-	-	-	-	-	-	-	-	-	4	-
YH-Poaceae 1	2	4	1	-	-	3	-	-	3	1	16	9	1	7	6
YH-Poaceae 2	-	-	-	-	-	-	-	-	-	-	-	-	21	-	-
YH-Poaceae 3	-	-	-	-	-	-	-	-	-	1	1	1	-	-	1
YH-Poaceae 4	-	-	-	-	-	-	-	-	-	-	-	-	-	-	-
YH-Poaceae 5	-	-	1	-	-	-	-	-	-	-	-	-	-	-	-
YH-Poaceae 8	-	2	1	1	-	-	-	-	1	-	3	5	1	1	10
YH-Poaceae 10/15	1	-	-	-	-	1	-	-	-	-	-	-	3	-	-
YH-Poaceae 11	-	-	-	-	-	-	-	-	-	-	-	-	11	-	-
YH-Poaceae 13	-	-	-	-	-	-	-	-	-	-	-	1	2	-	-

Table F2, cont'd.: Columns 121–135

Column no.	121	122	123	124	125	126	127	128	129	130†	131†	132†	133	134	135
YH no.	20221	20241	20249	20668	25869	21515	21516	25942	27260	26398	27461	27471	25945	28572	29312
YH-Poaceae 14	-	-	-	-	-	-	-	-	-	-	-	-	-	-	-
YH-Poaceae 16	-	-	-	-	-	-	-	-	-	-	-	1	-	-	-
YH-Poaceae 17/18	-	-	-	-	-	-	-	-	-	-	-	1	-	-	-
YH-Poaceae 20 Aeg 1a	-	-	-	-	-	-	-	-	-	-	-	-	-	-	15
YH-Poaceae 21 Aeg 1b	-	-	-	-	-	-	-	-	-	-	-	-	-	-	20
POACEAE	4	2	2	-	-	5	7	-	15	5	279	10	16	9	42
Polygonum	-	-	-	-	-	-	1	-	-	-	-	-	-	-	-
Polygonum (was YH-Cyperaceae 2)	-	-	-	-	-	-	25	-	-	-	-	-	-	-	1
Polygonum (was YH-Cyperaceae 6)	-	-	-	-	-	-	-	-	2	-	-	-	-	-	-
Rumex	-	-	-	-	-	-	-	-	-	-	1	-	-	1	
Portulaca	-	-	-	-	-	-	-	-	-	-	-	-	-	-	-
Androsace	-	-	-	-	-	-	-	-	-	-	-	-	-	-	-
YH-Primulaceae 1	-	-	-	-	-	-	-	-	-	-	-	-	-	-	-
YH-Primulaceae 2	-	-	-	-	-	-	-	-	-	-	-	-	5	-	-
Aconitum	-	-	-	-	-	-	-	-	-	-	1	-	-	-	-
Adonis	-	-	-	-	-	-	-	-	-	1	4	-	-	-	-
Ceratocephalus (YH-pp 2)	-	-	-	-	-	-	-	-	-	-	-	-	-	-	-
Ranunculus (YH-unknown 4)	-	-	-	-	-	1	-	-	-	-	-	-	-	-	-
Ranunculus arvensis	-	-	-	-	-	-	-	-	-	-	-	-	-	-	-
Reseda	-	-	-	-	-	-	-	-	-	-	-	-	-	-	-
cf. Asperula (YH-Rubiaceae 2)	-	-	-	-	-	-	-	-	-	-	-	-	-	-	1
Galium	1	2	1	-	-	6	43	1	1	2	13	5	1	4	3
YH-Rubiaceae 1	-	-	-	-	-	1	-	-	-	1	-	-	5	2	5
YH-Scrophulariaceae 1	-	1	-	-	-	-	-	-	-	-	-	-	-	-	-
Hyoscyamus	1	-	-	-	-	-	1	-	-	-	-	-	-	1	-
SOLANACEAE	-	-	-	-	-	-	-	-	1	-	-	-	-	-	-
Thymelaea	-	-	-	-	-	-	-	-	-	-	1	-	13	-	1
Valerianella coronata	-	-	-	-	-	-	-	-	-	-	-	-	-	-	-
Valerianella vesicaria	-	-	-	-	-	-	-	-	-	-	-	-	-	-	-
Valerianella	-	-	-	-	-	-	-	-	-	-	-	-	-	-	-
Peganum harmala	-	-	-	-	-	-	-	-	-	-	-	-	-	-	-
Zygophyllum	-	-	-	-	-	-	-	-	-	-	-	-	-	-	-
YH-unknown 9	-	-	-	-	-	-	-	-	-	-	-	-	-	-	-
YH-unknown 21	-	-	-	-	-	-	-	-	-	-	3	-	14	1	-
YH-unknown 23	-	-	-	-	1	2	-	-	-	-	-	1	-	-	-
YH-unknown 27	-	-	-	-	-	-	-	-	-	-	-	-	-	-	-
YH-unknown 29	-	-	-	-	-	-	-	-	-	-	3	-	-	-	-
YH-unknown 30	-	-	-	-	-	-	-	-	-	-	-	-	-	-	-
YH-unknown 35	-	-	-	-	-	-	-	-	-	-	-	2	1	-	-
Unknown	3	3	-	-	-	10	6	2	12	9	182	13	20	10	28
PLANT PARTS (pp)															
Hordeum internode (2-row)	-	-	-	-	-	-	-	-	-	-	148	1	5	1	1
Hordeum internode (6-row)	-	-	-	-	-	-	-	-	-	-	7	-	2	-	-
Hordeum internode (compact)	-	-	-	-	-	-	-	-	1	-	1	-	-	-	-
Triticum aestivum int.	-	-	-	-	-	1	-	-	-	-	-	-	-	-	-
T. aestivum/durum int.	-	-	-	-	-	3	-	-	1	2	24	-	-	3	12
T. aestivum/durum int. 1	-	-	-	-	-	-	-	-	-	-	-	-	-	-	-

Table F2, cont'd.: Columns 121–135

Column no.	121	122	123	124	125	126	127	128	129	130†	131†	132†	133	134	135
YH no.	20221	20241	20249	20668	25869	21515	21516	25942	27260	26398	27461	27471	25945	28572	29312
T. aestivum/durum int. 2	-	-	-	-	-	-	-	-	-	-	-	-	-	-	-
T. durum internode	-	-	-	-	-	-	-	-	-	-	1	-	-	-	-
T. monococcum spikelet fork	-	-	-	-	-	-	-	-	1	-	3	-	-	-	15
T. dicoccum spikelet fork	-	-	-	-	-	-	-	-	-	-	-	-	-	-	-
T. monococcum/dicoccum spikelet fork	3	2	1	1	-	1	3	-	1	-	-	2	15	1	40
Triticum rachis base	-	-	-	-	-	1	-	-	-	-	-	-	-	-	-
Triticum, rachis fragment, square +cross-section	-	-	-	-	-	-	-	-	-	-	2	-	-	-	2
Cereal culm node	-	-	1	-	-	6	-	-	1	1	25	2	3	2	8
Cereal culm root base?	-	-	-	-	-	-	-	-	-	-	-	-	-	-	-
cf. Oryza glume (charred)	-	-	-	-	-	-	-	-	-	-	-	-	-	-	-
Oryza glume (silicified phytolith)	-	-	-	-	-	-	-	-	-	-	-	-	-	-	-
Vitis peduncle	-	-	-	-	-	-	-	-	-	-	1	1	-	-	-
Onopordum achene collar	-	-	-	-	-	-	-	-	-	-	-	-	-	-	-
Asteraceae head, misc.	-	-	-	-	-	-	-	-	-	-	1	-	-	-	-
Euclidium silique	-	-	-	-	-	-	-	-	-	-	-	-	-	-	-
Capparis thorn	-	-	-	-	-	-	-	-	-	-	-	-	-	-	-
Atriplex bract (YH-pp 12)	-	-	-	-	-	-	-	-	-	-	-	-	-	-	-
Salsola inflor (YH-pp 7, YH-unk 8)	-	-	-	-	-	-	-	-	-	-	-	-	-	-	-
Cyperaceae stem fragment	-	-	-	-	-	1	-	-	-	-	-	-	-	-	-
Alhagi pod section	-	-	-	-	-	-	-	-	-	-	-	-	-	-	-
Alhagi leaf?	-	-	-	-	-	-	1	-	-	-	-	-	-	-	-
Aegilops glume base	-	-	-	-	-	-	-	-	-	-	-	-	-	-	-
Hordeum internode (wild)	-	-	-	-	-	-	-	-	-	-	2	-	2	-	-
Taeniatherum int. (YH-pp 8)	-	-	-	-	-	-	-	-	-	-	-	-	-	-	-
Poaceae, mystery 1 rachis frag.	-	-	-	-	-	-	-	-	-	-	-	-	-	-	-
YH-plant part 9	-	-	-	-	-	-	-	-	-	-	-	-	-	-	-
UNCHARRED-MINERALIZED															
Arnebia linearis	-	-	-	-	-	-	-	-	-	-	-	-	-	-	-
Arnebia/Lithospermum	1	1	1	1	-	-	-	-	4	-	-	5	-	3	2
Lithospermum tenuifolium	-	-	-	-	-	-	-	-	-	-	-	-	-	-	-
Lithospermum arvense	-	-	-	-	-	-	6	-	-	-	-	-	-	-	-
Lithospermum, misc.	-	-	-	-	-	-	-	-	-	-	-	-	-	-	-
Moltkia	-	-	-	-	-	-	-	-	-	-	-	-	-	-	-
BORAGINACEAE	-	-	-	-	-	-	-	-	-	-	-	-	-	-	-
Eleocharis (Cyperaceae-a)	-	-	-	-	-	1	-	-	8	100s	37	11	200+	10	1
Carex	-	-	-	-	-	-	-	-	-	-	-	-	3	-	-
Ficus	-	-	-	-	-	-	-	-	-	-	-	1	-	-	-
Glaucium	-	-	-	-	-	1	-	-	-	-	-	-	-	1	-
Papaver (white)	-	-	-	-	-	-	-	-	-	-	-	-	-	-	-

Table F2, cont'd.: Columns 136–150

Column no.	136	137†	138	139†	140	141	142†	143	144†	145†	146	147†	148	149	150†
YH no.	30571	22706	27967	31666	28605	31113	31128	31753	26562	27981	27277	30494	26715	28130	28176
Context type	pit	pit	pit	pit	pit	pit	pit	pit	pit	pit	pyro	pit	pyro	coll	pit
Operation	10	6	8	11	8	10	10	10	8	8	9	11	9	9	14
Locus	4	26	9	28	9.7	10	11	14	7	9	18	31	13	21	25
Lot	7	47	49	124	77	20	25	32	27	55	54	119	41	75	49
Phase	705.05	705.07	705.10	705.11	705.12	705.13	705.14	705.15	705.16	705.16	705.18	705.23	705.25	720	730
Soil volume (liters)	6.5	8	11	7	1.8	5.5	11	6	1	3.5	0.6	1.3	8.5	6	7
Charcoal (>2 mm, g)	14.19	3.14	1.73	5.35	7.13	0.61	3.90	1.00	0.47	0.30	9.61	7.75	12.76	4.82	4.00
Seed (>2 mm, g)	0.02	1.35	0.61	1.90	0.11	0.16	0.31	0.24	1.22	1.77	-	0.28	0.11	0.16	0.66
Other (>2 mm, g)	-	0.03	+	0.24	0.08	+	-	+	0.02	-	-	0.07	+	-	0.18
Wild/weedy (#)	1	207	21	440	8	20	145	28	56	107	1	100	21	46	106
ECONOMIC PLANTS															
Hordeum vulgare (g)	-	0.57	0.27	0.64	0.06	0.07	0.10	0.08	0.51	0.61	-	0.12	0.02	0.02	0.23
H. vulgare var. nudum (g)	-	-	-	-	-	-	-	-	-	-	-	-	-	-	-
Triticum aestivum/durum (g)	0.01	-	0.12	0.51	0.03	0.04	0.03	0.10	0.27	0.49	-	0.04	0.08	0.04	0.21
Triticum monococcum (g)	-	0.16	-	-	-	-	+	0.01	0.01	-	-	0.04	0.01	-	-
Triticum dicoccum (g)	-	-	0.02	0.03	-	-	-	-	-	-	-	-	-	-	-
Triticum sp. (g)	-	0.20	-	0.26	-	-	0.04	0.03	0.13	0.30	-	-	-	0.06	0.03
Cereal, indet. (g)	-	0.39	0.19	0.64	0.08	0.08	0.19	0.12	0.65	1.02	-	0.07	+	0.05	0.10
Secale cereale (g)	-	-	-	-	-	-	-	-	-	-	-	-	-	-	-
Setaria italica (#)	-	1	-	1	-	-	-	-	-	-	-	-	-	-	-
Oryza (g)	-	-	-	-	-	-	-	-	-	-	-	-	-	-	-
Vicia ervilia (g)	-	0.02	0.03	0.03	-	+	0.02	-	-	-	-	-	-	-	0.05
Lens (g)	-	-	+	-	-	-	-	-	-	-	-	-	-	-	-
Pulse (g)	-	0.03	+	-	-	-	-	-	-	-	-	-	-	-	0.01
cf. Pistacia/nutshell (g)	-	-	-	+	-	-	+	-	-	-	-	-	-	-	-
cf. Quercus (g)	-	-	-	-	-	-	-	-	-	-	-	-	+	-	-
Ficus carica (#)	-	-	-	-	-	-	-	-	-	-	-	-	-	-	-
Vitis vinifera (# whole, total g)	-	-	-	1(0.02)	-	-	-	-	-	-	-	-	+	-	-
WILD AND WEEDY (counts)															
Bupleurum	-	-	-	-	-	-	-	-	-	-	-	-	-	-	-
cf. Daucus	-	-	-	-	-	-	-	-	-	-	-	-	-	-	-
Torilis leptophylla?	-	-	-	-	-	-	-	-	-	-	-	-	-	-	-
YH-Apiaceae 2	-	1	-	1	-	-	-	-	1	-	-	-	-	-	-
YH-Apiaceae 4/8	-	-	-	2	-	-	-	-	-	-	-	-	-	-	-
YH-Apiaceae 6	-	-	-	-	-	-	-	-	-	-	-	-	-	-	-
YH-Apiaceae 7	-	-	-	-	-	-	-	-	-	-	-	-	-	-	-
YH-Apiaceae 10	-	-	-	-	-	-	-	-	-	-	-	-	-	-	-
YH-unknown 31 (Apiaceae)	-	-	-	-	-	-	-	-	-	-	-	-	-	-	-
APIACEAE	-	-	-	4	-	-	-	-	-	-	-	-	-	-	-
Anthemis/Matricaria	-	-	-	-	-	-	-	-	-	-	-	-	-	-	-
Artemisia	-	-	-	-	-	-	-	-	-	-	-	-	-	-	-
Carthamus	-	-	-	-	-	-	-	-	-	-	-	-	-	-	-
Centaurea cyanus-type	-	-	-	-	-	-	-	-	-	-	-	-	-	-	-
Centaurea	-	1	-	16	-	-	-	-	5	1	-	1	-	-	-
Onopordum	-	-	-	1	-	-	-	-	-	-	-	-	1	-	-
YH-Asteraceae 1	-	-	-	-	-	-	-	-	-	-	-	1	-	-	-

Table F2, cont'd.: Columns 136–150

Column no.	136	137†	138	139†	140	141	142†	143	144†	145†	146	147†	148	149	150†
YH no.	30571	22706	27967	31666	28605	31113	31128	31753	26562	27981	27277	30494	26715	28130	28176
YH-Poaceae 14	-	-	-	-	-	-	-	-	-	-	-	-	-	-	-
YH-Poaceae 16	-	-	-	-	-	-	-	-	-	-	-	-	-	-	-
YH-Poaceae 17/18	-	-	-	-	-	-	1	-	-	-	-	-	1	-	-
YH-Poaceae 20 Aeg 1a	-	-	-	-	-	-	-	-	-	-	-	-	-	-	-
YH-Poaceae 21 Aeg 1b	-	-	-	-	-	-	-	-	-	-	-	-	-	-	-
POACEAE	-	47	4	42	-	3	17	5	6	11	-	8	2	7	16
Polygonum	-	-	-	-	-	-	-	-	-	-	-	1	-	-	-
Polygonum (was YH-Cyperaceae 2)	-	-	-	-	-	-	-	-	-	-	-	-	-	-	2
Polygonum (was YH-Cyperaceae 6)	-	-	-	2	-	-	-	-	-	1	-	-	-	-	-
Rumex	-	-	-	-	-	-	1	-	-	-	-	-	-	-	-
Portulaca	-	-	-	-	-	-	-	-	-	-	-	-	-	-	-
Androsace	-	-	-	2	-	-	2	-	-	-	-	-	-	-	-
YH-Primulaceae 1	-	1	-	-	-	-	-	-	-	-	-	-	-	-	-
YH-Primulaceae 2	-	-	-	-	-	-	-	-	-	-	-	-	-	-	-
Aconitum	-	-	-	-	-	-	-	-	-	-	-	-	-	-	-
Adonis	-	2	-	-	-	-	-	-	1	1	-	2	-	-	-
Ceratocephalus (YH-pp 2)	-	-	-	-	-	-	-	-	-	-	-	-	-	-	-
Ranunculus (YH-unknown 4)	-	-	-	5	-	-	-	-	-	-	-	-	-	-	-
Ranunculus arvensis	-	-	-	-	-	-	-	-	-	-	-	-	-	-	-
Reseda	-	-	-	-	-	-	-	-	-	-	-	-	-	-	-
cf. Asperula (YH-Rubiaceae 2)	-	-	-	-	-	-	-	-	-	-	-	-	-	-	-
Galium	-	14	1	22	1	4	-	3	-	-	-	15	1	2	10
YH-Rubiaceae 1	-	3	-	-	-	-	-	-	-	-	-	-	-	-	-
YH-Scrophulariaceae 1	-	-	-	-	-	-	-	-	-	-	-	-	-	-	-
Hyoscyamus	-	-	-	1	-	-	-	-	-	-	-	-	-	-	1
SOLANACEAE	-	-	-	-	-	-	-	-	-	-	-	-	-	-	-
Thymelaea	-	-	-	5	-	1	-	-	-	-	-	-	-	-	-
Valerianella coronata	-	-	-	1	-	-	-	-	-	-	-	1	-	-	-
Valerianella vesicaria	-	-	-	-	-	-	-	-	-	-	-	2	-	-	-
Valerianella	-	-	-	-	-	-	-	-	-	-	-	-	-	-	-
Peganum harmala	-	-	-	1	-	-	1	-	-	-	-	-	-	-	-
Zygophyllum	-	-	-	-	-	-	-	-	-	-	-	-	-	-	-
YH-unknown 9	-	-	-	-	-	-	-	-	-	-	-	-	-	-	-
YH-unknown 21	-	-	-	-	-	-	-	-	-	-	-	-	-	-	-
YH-unknown 23	-	1	-	1	-	-	1	-	-	-	-	-	-	-	-
YH-unknown 27	-	-	-	-	-	-	1	-	-	-	-	-	-	-	-
YH-unknown 29	-	1	-	-	-	-	-	-	-	-	-	-	-	-	-
YH-unknown 30	-	-	-	-	-	-	-	-	-	-	-	-	-	-	-
YH-unknown 35	-	-	-	-	-	-	-	-	-	-	-	-	-	-	-
Unknown	-	26	-	23	2	-	15	-	-	5	-	22	8	13	29
PLANT PARTS (pp)															
Hordeum internode (2-row)	-	65	2	36	-	2	3	1	28	28	-	-	-	-	5
Hordeum internode (6-row)	-	-	4	-	-	-	-	-	-	5	-	1	-	-	-
Hordeum internode (compact)	-	10	-	3	-	-	-	-	3	2	-	-	-	-	-
Triticum aestivum int.	-	-	-	48	-	-	-	-	-	-	-	-	-	-	-
T. aestivum/durum int.	-	36	1	74	1	2	-	-	2	4	-	9	-	1	2
T. aestivum/durum int. 1	-	-	-	-	-	-	-	-	-	-	-	-	-	-	-

Table F2, cont'd.: Columns 136–150

Column no.	136	137†	138	139†	140	141	142†	143	144†	145†	146	147†	148	149	150†
YH no.	30571	22706	27967	31666	28605	31113	31128	31753	26562	27981	27277	30494	26715	28130	28176
T. aestivum/durum int. 2	-	-	-	-	-	-	-	-	-	-	-	-	-	-	-
T. durum internode	-	-	-	-	-	-	-	-	-	-	-	-	-	-	-
T. monococcum spikelet fork	-	-	-	-	-	1	-	-	-	-	-	11	-	-	-
T. dicoccum spikelet fork	-	-	-	-	-	-	-	-	-	-	-	4	-	-	-
T. monococcum/dicoccum spikelet fork	-	12	1	6	-	-	1	2	1	-	-	22	-	2	2
Triticum rachis base	-	-	-	11	-	-	1	-	1	2	-	2	-	-	-
Triticum, rachis fragment, square cross-section	-	-	-	5	-	-	-	-	-	-	-	1	-	-	-
Cereal culm node	-	29	4	99	-	-	4	-	1	-	-	2	-	1	3
Cereal culm root base?	-	-	-	13	-	-	-	-	-	-	-	-	-	-	-
cf. Oryza glume (charred)	-	-	-	-	-	-	-	-	-	-	-	-	-	-	-
Oryza glume (silicified phytolith)	-	-	-	-	-	-	-	-	-	-	-	-	-	-	-
Vitis peduncle	-	-	-	-	-	-	-	-	-	-	-	-	-	-	-
Onopordum achene collar	-	-	-	-	-	-	-	-	-	-	-	-	-	-	-
Asteraceae head, misc.	-	-	-	1	-	-	-	-	-	-	-	-	-	-	-
Euclidium silique	-	-	-	1	-	-	-	-	-	-	-	-	-	-	-
Capparis thorn	-	-	-	-	-	-	-	-	-	-	-	-	-	-	-
Atriplex bract (YH-pp 12)	-	-	-	-	-	-	-	-	-	-	-	-	-	-	-
Salsola inflor (YH-pp 7, YH-unk 8)	-	1	-	3	-	-	-	-	-	-	-	-	-	-	-
Cyperaceae stem fragment	-	-	-	-	-	-	-	-	-	-	-	-	-	-	-
Alhagi pod section	-	-	-	-	-	-	-	-	-	-	-	-	-	-	-
Alhagi leaf?	-	-	-	-	-	-	-	-	-	-	-	-	-	-	-
Aegilops glume base	-	-	-	-	-	-	-	-	-	-	-	-	-	-	1
Hordeum internode (wild)	-	-	-	-	-	-	-	-	-	-	-	-	-	-	-
Taeniatherum int. (YH-pp 8)	-	-	-	-	-	-	-	-	-	-	-	-	-	-	-
Poaceae, mystery 1 rachis frag.	-	-	-	-	-	-	-	-	-	-	-	-	-	-	-
YH-plant part 9	-	-	-	-	-	-	-	-	-	-	-	-	-	-	-
UNCHARRED-MINERALIZED															
Arnebia linearis	-	-	-	-	-	-	-	-	-	-	-	-	-	-	-
Arnebia/Lithospermum	-	9	-	1	-	3	1	3	-	-	-	-	-	-	2
Lithospermum tenuifolium	-	5	-	1	-	-	1	-	1	-	-	-	-	-	-
Lithospermum arvense	-	-	-	-	-	-	-	-	-	-	-	-	-	-	-
Lithospermum, misc.	-	-	-	-	-	-	-	-	-	-	-	-	-	-	-
Moltkia	-	3	-	1	-	-	1	-	-	-	-	-	-	-	2
BORAGINACEAE	-	-	-	-	-	-	-	-	-	-	-	-	-	-	-
Eleocharis (Cyperaceae-a)	-	-	-	2	-	2	1	6	-	-	-	-	-	1	18
Carex	-	-	-	-	-	-	-	-	-	-	-	-	-	-	-
Ficus	-	-	-	-	-	-	-	-	-	-	-	-	-	-	-
Glaucium	-	-	-	-	-	-	-	-	-	-	-	-	-	-	-
Papaver (white)	-	-	-	-	-	-	-	-	-	-	-	-	-	-	-

Table F2, cont'd.: Columns 151–165

Column no.	151	152	153†	154	155†	156	157†	158*	159	160	161	162	163†	164*†	165†
YH no.	30349	28845	20987	21000	22281	22496	22192	28491	23237	23246	23571	22466	21301	21526	29473
Context type	pit	pit	pit	pit	pit	pit	pit	pit	pit	jar	pit	pit	pit	pyro	pit
Operation	14	14	5	5	5	3	5	14	5	5	5	3	5	3	9
Locus	27	36	22	22	37	45	35	25	39	40	40	42	24	17	26
Lot	97	73	26	26	63	81	55	65	73	74	82	74	27	48	107
Phase	730.01	730.02	730.02	730.02	730.02	730.02	730.03	730.04	730.04	730.04	730.04	730.06	730.07	735.01	740.01
Soil volume (liters)	9	6	9	7	12	15	8	7	0.7	8	4	7	8	18	12
Charcoal (>2 mm, g)	4.71	3.18	5.19	1.65	5.86	19.35	12.76	8.91	4.68	18.46	4.63	3.11	1.60	6.65	4.92
Seed (>2 mm, g)	0.35	0.34	0.28	0.23	2.21	0.76	1.38	0.89	0.07	0.56	0.31	0.25	0.45	1.41	1.40
Other (>2 mm, g)	0.01	-	0.13	0.01	0.06	0.04	0.04	0.02	0.02	+	+	-	+	-	0.03
Wild/weedy (#)	101	75	42	26	294	97	725	60	5	38	33	62	77	341	791
ECONOMIC PLANTS															
Hordeum vulgare (g)	0.13	0.07	0.10	0.05	0.77	0.46	0.52	0.16	0.02	0.10	0.02	0.06	0.14	0.62	0.91
H. vulgare var. nudum (g)	-	-	-	-	-	-	-	-	-	-	-	-	-	-	-
Triticum aestivum/durum (g)	0.10	0.16	0.07	0.09	0.80	0.17	-	0.11	0.05	0.08	0.09	0.06	0.14	0.21	0.12
Triticum monococcum (g)	0.01	0.02	-	+	0.04	+	-	0.40	0.01	0.26	0.10	0.10	0.01	0.06	-
Triticum dicoccum (g)	-	-	-	-	0.08	-	-	-	-	-	-	-	0.05	0.08	-
Triticum sp. (g)	0.03	-	0.05	-	0.12	-	0.14	0.01	-	0.04	0.01	0.02	0.08	0.17	0.08
Cereal, indet. (g)	0.20	0.18	0.07	0.05	0.87	0.23	0.35	0.39	0.02	0.21	0.09	0.09	0.16	0.33	0.41
Secale cereale (g)	-	-	-	-	-	-	-	-	-	-	-	-	-	-	-
Setaria italica (#)	2	-	-	-	-	-	-	-	-	-	-	-	-	-	-
Oryza (g)	-	-	-	-	-	-	-	-	-	-	-	-	-	-	-
Vicia ervilia (g)	+	-	-	0.01	-	-	0.52	0.01	-	-	0.02	-	-	0.01	0.03
Lens (g)	-	-	0.01	-	0.01	-	-	0.02	-	+	-	-	-	-	0.02
Pulse (g)	0.02	-	-	-	-	-	-	0.02	-	-	-	+	0.02	+	-
cf. Pistacia/nutshell (g)	-	-	-	-	-	-	+	0.04	-	-	-	-	-	-	-
cf. Quercus (g)	-	-	-	-	-	-	-	-	-	-	-	-	-	-	-
Ficus carica (#)	-	-	-	-	-	-	-	-	-	-	-	-	-	-	-
Vitis vinifera (# whole, total g)	-	-	-	-	-	-	-	-	-	-	-	-	-	-	-
WILD AND WEEDY (counts)															
Bupleurum	-	-	-	-	-	-	-	-	-	-	-	-	-	-	-
cf. Daucus	-	-	-	-	-	-	-	-	-	-	-	-	-	-	-
Torilis leptophylla?	-	-	-	-	-	-	-	-	-	-	-	-	-	1	-
YH-Apiaceae 2	-	-	-	-	-	-	-	-	-	-	-	-	-	-	-
YH-Apiaceae 4/8	-	-	-	-	-	-	-	-	-	-	-	-	-	-	-
YH-Apiaceae 6	-	-	-	-	-	1	-	-	-	-	-	-	-	2	-
YH-Apiaceae 7	-	-	-	-	-	-	-	-	-	-	-	-	-	-	-
YH-Apiaceae 10	-	-	-	-	-	-	-	-	-	-	-	-	-	-	-
YH-unknown 31 (Apiaceae)	-	-	-	-	-	-	-	-	-	-	-	-	-	-	-
APIACEAE	-	-	-	-	-	-	-	-	-	-	-	-	-	-	5
Anthemis/Matricaria	-	-	-	-	-	-	-	-	-	-	-	-	-	-	-
Artemisia	-	-	-	-	-	-	-	-	-	-	-	-	-	-	-
Carthamus	-	-	-	-	-	-	-	-	-	-	-	-	-	-	-
Centaurea cyanus type	-	-	-	-	-	-	-	-	-	-	-	-	-	-	-
Centaurea	-	2	1	-	7	-	2	1	-	-	1	5	-	5	3
Onopordum	-	-	-	-	1	-	-	-	-	-	-	-	-	1	-
YH-Asteraceae 1	-	-	-	-	1	1	-	-	-	-	-	-	-	-	-

Table F2, cont'd.: Columns 181–195

Column no.	181	182	183	184†	185	186	187	188*	189	190	191*†	192†	193	194	195*
YH no.	21024	21453	21463	21046	21047	21484	21489	21498	22708	28165	21542	22178	22270	22482	31053
YH-Asteraceae 2	-	-	-	-	-	-	-	-	-	-	-	-	3	-	2
YH-Asteraceae 5	-	-	-	-	-	-	-	-	-	-	-	-	2	-	-
YH-Asteraceae 9	-	-	-	-	-	-	-	-	-	-	-	-	-	-	-
ASTERACEAE	-	-	-	-	-	-	-	-	-	-	-	-	-	-	-
Arnebia/Lithospermum	-	-	-	-	-	-	-	-	-	-	-	-	-	-	-
Heliotropium	-	2	-	-	-	1	-	-	-	-	-	-	-	-	7
cf. Alyssum	-	-	-	-	-	-	-	-	-	-	-	-	-	-	-
cf. Camelina rumelica	-	-	-	-	-	-	-	-	-	-	-	-	-	-	-
cf. Camelina sativa	-	-	-	-	-	1	-	-	-	-	-	-	-	-	2
Conringia	-	-	-	-	-	-	-	-	-	-	-	-	-	-	2
cf. Lepidium	-	-	-	-	-	-	-	-	-	-	-	-	-	-	-
Sisymbrium altissimum-type	-	-	-	-	-	1	-	-	-	-	-	-	-	-	1
Thlaspi	-	-	-	-	-	-	-	-	-	-	-	-	-	-	-
YH-Brassicaceae 2	-	-	-	1	-	-	1	-	-	-	-	-	-	-	-
YH-Brassicaceae 3/5	-	-	-	-	-	-	-	-	-	-	-	-	7	-	5
YH-Brassicaceae 7	-	-	-	-	-	-	-	-	-	-	-	-	-	-	2
YH-Brassicaceae 10	-	-	-	-	-	-	-	-	-	-	-	1	-	-	-
YH-Brassicaceae 11	-	1	-	2	-	1	-	-	-	-	-	1	-	-	-
YH-Brassicaceae 12	-	1	-	-	-	1	-	-	-	-	2	-	-	-	1
BRASSICACEAE	-	-	-	-	2	2	-	-	-	7	3	-	6	-	11
Bufonia	-	-	-	-	-	1	-	-	-	-	-	-	-	-	1
Gypsophila	-	-	2	-	-	2	-	2	-	-	2	-	2	-	7
Silene	-	-	-	-	-	-	-	-	-	-	-	-	-	-	1
Vaccaria (YH-unknown 6)	-	-	-	1	-	-	-	-	-	-	-	1	-	-	1
CARYOPHYLLACEAE	-	-	-	-	-	-	-	-	-	-	-	-	3	-	2
YH-Caryophyllaceae 1	-	-	-	-	-	-	-	-	-	-	-	-	-	-	-
Atriplex	-	-	-	-	-	-	-	-	-	-	2	-	-	-	-
Chenopodium	-	2	1	-	-	3	2	1	-	-	3	3	14	-	3
Salsola kali-type	-	-	-	8	1	-	-	-	-	-	-	-	-	-	8
Salsola soda-type	-	-	-	-	-	-	-	-	-	-	-	1	2	-	-
Salsola	-	-	-	6	2	-	1	-	-	-	-	-	-	-	-
Suaeda	1	-	3	-	-	2	-	1	-	2	1	-	1	-	1
YH-Chenopodiaceae 2	1	-	-	-	2	11	1	-	-	-	-	-	-	-	-
CHENOPODIACEAE	-	-	-	30	2	-	-	-	-	-	-	-	3	-	3
Helianthemum	-	1	-	-	-	1	-	-	-	-	-	-	-	-	-
Carex	-	-	1	3	-	3	1	-	-	6	2	6	4	-	5
Carex 3	-	-	-	-	-	-	-	-	-	-	-	-	-	-	-
Eleocharis (YH-Cyperaceae a)	-	-	-	-	-	-	-	-	-	-	-	-	-	-	1
YH-Cyperaceae 1	-	-	1	-	1	1	1	1	-	1	2	1	1	-	5
YH-Cyperaceae 3	-	-	-	-	-	-	-	-	-	-	-	-	-	-	-
YH-Cyperaceae 4	-	-	-	-	-	-	-	-	-	-	-	-	-	-	-
YH-Cyperaceae 5	-	-	-	-	-	1	-	-	-	-	-	-	-	-	-
YH-Cyperaceae 7	-	-	-	-	-	-	-	-	-	-	-	-	-	-	2
YH-Cyperaceae 8	-	-	-	-	-	-	-	-	-	-	-	-	-	-	-
CYPERACEAE	-	-	-	1	1	-	-	-	-	-	-	4	1	-	-
Cephalaria	-	-	-	-	-	-	-	-	-	-	-	-	-	-	-
Scabiosa	-	-	-	-	-	-	-	-	-	-	-	-	-	-	-
Euphorbia	-	-	-	-	-	-	-	-	-	-	-	-	-	-	-

Table F2, cont'd.: Columns 196–210

Column no.	196†	197	198†	199	200	201	202	203†	204	205	206	207†	208	209†	210†
YH no.	22276	22491	22780	30623	33165	22775	22799	23208	23221	33187	23200	29965	32447	31603	31836
YH-Asteraceae 2	-	2	1	-	-	3	-	-	-	-	1	1	-	9	-
YH-Asteraceae 5	2	-	-	-	2	-	-	1	-	-	7	-	-	-	-
YH-Asteraceae 9	-	-	-	-	-	-	-	-	-	-	-	-	-	-	-
ASTERACEAE	1	-	-	-	4	-	-	-	-	-	1	1	-	2	-
Arnebia/Lithospermum	-	-	-	-	-	-	-	-	-	-	-	-	-	-	-
Heliotropium	1	2	-	-	3	3	-	-	-	1	1	1	-	-	-
cf. Alyssum	-	-	-	-	-	-	-	-	-	-	-	-	-	-	-
cf. Camelina rumelica	-	-	-	-	-	-	-	-	-	-	-	-	-	-	-
cf. Camelina sativa	-	-	-	-	-	-	-	-	-	-	-	-	-	-	-
Conringia	1	1	-	-	5	-	-	-	1	-	-	5	-	6	-
cf. Lepidium	-	-	-	-	1	-	-	-	-	-	-	-	-	1	-
Sisymbrium altissimum-type	-	-	-	-	1	1	-	-	-	-	-	2	-	1	-
Thlaspi	-	-	-	-	-	-	-	-	-	-	-	-	-	-	-
YH-Brassicaceae 2	1	-	-	-	-	-	1	-	-	-	1	-	-	-	-
YH-Brassicaceae 3/5	11	1	4	-	10	12	1	-	1	1	2	2	-	1	-
YH-Brassicaceae 7	-	-	-	-	-	-	-	-	-	-	-	-	1	-	-
YH-Brassicaceae 10	-	-	-	-	-	-	-	-	-	-	-	1	-	3	-
YH-Brassicaceae 11	7	-	-	-	-	-	4	-	-	-	-	-	-	6	-
YH-Brassicaceae 12	-	-	1	-	2	-	-	-	-	-	2	-	-	-	-
BRASSICACEAE	14	-	9	-	19	8	10	-	-	2	12	6	-	40	2
Bufonia	1	-	-	-	-	-	-	-	-	-	-	-	-	1	-
Gypsophila	5	11	3	-	5	6	2	-	6	1	2	16	1	21	-
Silene	-	1	1	-	-	-	-	-	-	-	-	-	-	-	-
Vaccaria (YH-unknown 6)	-	-	-	-	1	1	-	-	-	1	-	1	-	13	-
CARYOPHYLLACEAE	7	-	3	-	-	-	-	2	-	-	-	-	1	3	-
YH-Caryophyllaceae 1	-	1	-	-	-	-	-	-	-	-	-	-	-	-	-
Atriplex	-	-	1	-	-	-	-	-	-	-	-	-	-	-	-
Chenopodium	4	-	5	-	16	34	1	8	-	-	-	13	2	-	6
Salsola kali-type	-	-	3	-	-	-	-	-	-	-	-	-	-	-	-
Salsola soda-type	1	-	3	-	-	-	-	-	-	-	-	4	2	-	-
Salsola	3	2	9	-	2	4	2	-	-	-	-	-	-	-	-
Suaeda	17	11	-	-	4	5	-	2	3	2	2	7	-	8	1
YH-Chenopodiaceae 2	-	-	-	-	1	-	-	-	-	-	-	-	-	-	-
CHENOPODIACEAE	-	10	-	-	19	-	1	-	4	-	10	-	-	4	-
Helianthemum	-	-	1	-	-	-	-	-	-	-	-	-	1	-	-
Carex	5	8	27	-	5	4	4	2	2	1	2	2	1	1	3
Carex 3	-	-	-	-	-	-	-	-	-	-	-	-	-	-	-
Eleocharis (YH-Cyperaceae a)	-	-	2	-	2	1	-	-	-	1	-	-	1	1	-
YH-Cyperaceae 1	5	4	11	-	10	3	2	3	1	1	3	2	-	2	2
YH-Cyperaceae 3	-	-	-	-	-	-	-	-	-	-	-	-	-	-	-
YH-Cyperaceae 4	-	-	-	-	-	-	-	-	-	-	-	-	1	-	-
YH-Cyperaceae 5	-	-	-	-	1	-	-	-	1	-	-	-	-	1	-
YH-Cyperaceae 7	-	-	-	-	-	-	-	2	-	-	2	1	-	-	-
YH-Cyperaceae 8	-	-	-	-	-	-	-	-	-	-	-	-	-	-	-
CYPERACEAE	-	-	-	-	10	-	-	-	-	1	-	1	3	8	-
Cephalaria	-	-	-	-	1	1	-	-	-	-	-	1	-	-	-
Scabiosa	-	-	-	-	-	1	-	-	-	-	-	-	-	-	-
Euphorbia	-	1	-	-	-	-	-	-	-	-	-	-	-	-	-

Table F2, cont'd.: Columns 196–210

Column no.	196†	197	198†	199	200	201	202	203†	204	205	206	207†	208	209†	210†
YH no.	22276	22491	22780	30623	33165	22775	22799	23208	23221	33187	23200	29965	32447	31603	31836
Alhagi	-	-	-	-	-	-	-	-	-	-	-	-	-	1	-
Astragalus	3	6	-	-	-	2	-	-	-	-	1	2	-	-	-
Medicago	1	-	-	-	1	4	-	-	4	-	-	1	-	-	1
Onobrychis	-	-	-	-	-	-	-	-	-	-	-	-	-	-	-
Trifolium/Melilotus	2	21	5	-	3	3	2	3	6	2	3	9	-	-	-
Trigonella	82	61	76	7	37	52	2	13	32	11	64	16	1	29	18
Trigonella cf. astroites	9	34	28	-	5	9	4	1	-	-	5	10	-	-	4
YH-Fabaceae 2	-	-	-	-	-	-	-	-	-	-	-	-	-	-	-
FABACEAE	7	36	14	3	8	5	-	4	-	3	6	15	9	8	5
cf. Erodium	-	-	-	-	-	-	1	-	-	-	-	-	-	-	-
Teucrium	-	-	1	-	-	-	-	-	-	-	-	1	-	-	-
Ziziphora	4	3	1	41	4	-	-	2	3	-	1	4	-	5	2
YH-Lamiaceae 2	-	-	-	-	-	-	-	-	-	-	-	-	-	-	-
YH-Lamiac 3 (Nepeta?)	1	-	1	-	-	-	-	-	-	-	-	-	-	-	-
YH-Lamiaceae 5	-	-	-	-	-	-	-	-	2	-	-	1	-	1	-
LAMIACEAE	-	-	-	-	-	1	1	-	-	-	-	-	-	-	-
LILIACEAE	-	-	1	-	-	-	-	-	-	-	-	-	-	-	-
cf. Malva	-	-	-	-	-	-	-	-	-	-	-	4	-	-	1
Glaucium	-	1	1	-	-	3	-	-	-	-	-	-	-	-	-
Papaver	1	-	-	-	-	-	-	1	-	-	-	-	-	1	-
Fumaria	1	-	-	-	-	-	-	-	-	2	-	-	-	-	-
Plantago	-	-	-	-	-	-	-	-	-	-	-	-	-	-	-
Aegilops	-	-	-	-	-	-	1	-	-	-	-	1	-	-	-
Avena	-	-	-	-	1	-	-	-	-	-	1	-	-	-	1
Bromus japonicus-type	2	-	2	-	-	-	-	-	-	-	-	1	-	3	-
Bromus tectorum-type	-	1	-	-	4	-	-	2	-	-	-	1	-	9	-
Bromus	-	-	-	-	8	-	1	-	-	2	-	2	-	-	-
Eremopyrum	5	8	9	-	-	1	3	-	4	-	4	5	-	15	3
Hordeum cf. murinum	1	-	-	-	2	1	-	-	-	-	1	1	-	2	-
Hordeum spontaneum?	-	-	-	-	-	-	-	-	-	-	-	-	-	2	-
Hordeum misc.	-	1	-	-	-	-	1	-	1	1	4	-	1	-	-
cf. Lolium	-	-	-	-	-	-	-	-	-	-	-	-	-	1	-
cf. Phalaris	-	-	-	-	1	-	-	-	-	-	-	-	-	2	-
cf. Poa bulbosa	-	-	-	-	-	-	-	-	-	-	-	-	-	-	-
Setaria	-	-	-	-	-	-	-	-	-	-	1	-	-	-	-
Stipa	3	2	3	-	2	1	1	-	1	-	2	5	-	1	1
Taeniatherum	-	2	2	-	1	1	1	-	-	-	-	5	-	3	-
"Triticoid"	-	2	1	-	-	-	-	-	-	-	-	4	-	-	-
Triticum boeoticum	-	-	1	-	-	-	-	-	-	-	-	-	-	-	-
YH-Poaceae 1	6	2	-	-	3	-	-	1	1	-	-	-	-	6	-
YH-Poaceae 2	-	-	-	-	-	-	-	-	-	-	-	-	-	-	-
YH-Poaceae 3	-	-	8	-	2	-	-	-	1	1	-	-	1	4	1
YH-Poaceae 4	5	-	20	-	17	3	-	1	1	-	-	-	-	-	-
YH-Poaceae 5	-	-	-	-	-	-	-	-	-	-	-	-	-	-	-
YH-Poaceae 8	9	-	33	-	30	7	2	4	-	7	9	10	1	1	-
YH-Poaceae 10/15	2	-	-	-	25	2	-	4	-	1	-	-	-	2	-
YH-Poaceae 11	1	-	-	-	-	-	-	-	-	-	-	-	-	-	-
YH-Poaceae 13	-	1	1	-	4	-	-	-	1	-	-	-	-	-	-

Table F2, cont'd.: Columns 196–210

Column no.	196†	197	198†	199	200	201	202	203†	204	205	206	207†	208	209†	210†
YH no.	22276	22491	22780	30623	33165	22775	22799	23208	23221	33187	23200	29965	32447	31603	31836
YH-Poaceae 14	-	-	-	-	3	-	-	1	-	-	-	-	-	-	-
YH-Poaceae 16	-	-	-	-	-	-	-	-	-	-	-	-	-	-	-
YH-Poaceae 17/18	1	2	-	-	2	2	-	-	-	1	-	-	-	1	-
YH-Poaceae 20 Aeg 1a	-	-	-	-	-	-	-	-	-	-	-	-	-	-	-
YH-Poaceae 21 Aeg 1b	-	-	-	-	-	-	-	-	-	-	-	-	-	-	-
POACEAE	12	25	67	-	67	23	9	16	3	2	14	19	4	41	10
Polygonum	-	-	-	-	-	-	-	-	-	-	-	-	-	-	-
Polygonum (was YH-Cyperaceae 2)	-	3	3	-	2	2	-	-	-	-	-	3	-	1	-
Polygonum (was YH-Cyperaceae 6)	10	-	-	-	5	-	3	-	2	-	-	1	-	1	-
Rumex	1	3	-	-	-	-	-	1	-	-	1	1	-	-	-
Portulaca	-	-	-	-	-	-	-	-	-	-	-	-	-	-	-
Androsace	1	-	-	-	-	-	-	-	-	-	-	1	1	-	-
YH-Primulaceae 1	-	-	-	-	-	-	-	-	-	1	-	-	-	-	-
YH-Primulaceae 2	-	-	-	-	-	-	-	-	-	-	-	-	-	-	-
Aconitum	-	-	-	-	-	-	-	-	-	-	-	-	-	-	-
Adonis	-	-	-	-	2	1	-	-	-	-	-	-	-	1	-
Ceratocephalus (YH-pp 2)	3	-	1	-	1	-	1	-	-	-	-	-	-	-	-
Ranunculus (YH-unknown 4)	-	2	-	-	-	-	-	1	-	-	-	-	1	2	-
Ranunculus arvensis	-	-	-	-	-	-	-	-	-	-	-	-	-	1	-
Reseda	-	-	-	-	-	-	-	-	-	-	-	-	-	-	-
cf. Asperula (YH-Rubiaceae 2)	-	-	-	-	-	-	-	-	-	-	-	-	-	1	-
Galium	5	18	10	3	20	4	1	3	4	1	5	19	1	33	3
YH-Rubiaceae 1	6	1	4	-	1	1	-	-	1	-	2	15	-	6	-
YH-Scrophulariaceae 1	-	-	1	-	-	-	-	-	-	-	-	1	-	2	-
Hyoscyamus	1	1	2	-	-	-	1	-	-	-	-	-	-	3	1
SOLANACEAE	-	-	-	-	-	-	-	-	-	-	-	-	-	-	-
Thymelaea	2	-	1	-	1	1	-	2	-	-	-	-	-	1	-
Valerianella coronata	-	-	-	17	-	-	-	1	-	-	-	-	-	-	-
Valerianella vesicaria	-	-	-	-	-	-	-	-	-	-	-	-	-	-	-
Valerianella	-	-	-	-	-	-	-	-	-	-	-	-	1	-	-
Peganum harmala	4	-	3	-	-	-	-	2	-	1	-	-	6	-	-
Zygophyllum	-	-	-	-	-	-	-	-	-	-	-	-	-	-	-
YH-unknown 9	-	-	-	-	-	-	-	-	-	-	-	-	-	-	-
YH-unknown 21	-	-	-	-	1	1	-	-	-	-	1	-	1	-	-
YH-unknown 23	1	-	-	-	-	1	-	1	-	-	1	-	-	3	-
YH-unknown 27	-	-	-	-	-	1	-	-	-	-	-	-	-	-	-
YH-unknown 29	-	-	-	-	-	-	-	1	-	-	-	-	-	-	-
YH-unknown 30	3	-	-	-	-	1	-	-	-	-	-	-	-	-	-
YH-unknown 35	-	1	-	-	1	1	-	-	-	-	-	-	-	-	-
Unknown	43	18	72	9	38	22	1	14	3	8	31	40	-	39	7
PLANT PARTS (pp)															
Hordeum internode (2-row)	4	9	10	-	33	1	2	2	3	1	7	-	2	59	1
Hordeum internode (6-row)	2	-	3	-	6	-	-	2	-	-	-	-	-	5	-
Hordeum internode (compact)	-	2	1	-	6	1	2	1	-	-	1	2	-	10	-
Triticum aestivum int.	-	-	-	-	-	-	-	1	-	-	-	-	-	-	-
T. aestivum/durum int.	8	29	22	-	10	2	6	2	4	1	5	82	3	21	5
T. aestivum/durum int. 1	-	-	-	-	3	-	-	-	-	-	-	-	-	4	8

Table F2, cont'd.: Columns 196–210

Column no.	196†	197	198†	199	200	201	202	203†	204	205	206	207†	208	209†	210†
YH no.	22276	22491	22780	30623	33165	22775	22799	23208	23221	33187	23200	29965	32447	31603	31836
T. aestivum/durum int. 2	-	-	-	-	10	-	-	-	-	-	-	-	-	14	-
T. durum internode	-	-	-	-	-	-	-	-	-	-	-	-	-	-	-
T. monococcum spikelet fork	-	-	3	-	1	-	-	-	-	-	-	-	3	-	-
T. dicoccum spikelet fork	-	-	3	-	-	-	-	-	-	-	-	-	-	-	-
T. monococcum/dicoccum spikelet fork	1	2	6	-	-	2	-	-	-	-	-	2	6	-	1
Triticum rachis base	-	1	1	-	-	-	-	-	1	-	-	3	-	-	-
Triticum, rachis fragment, square cross-section	2	1	-	-	-	-	-	-	1	-	-	-	-	5	-
Cereal culm node	22	19	20	-	27	4	8	2	2	3	3	-	1	37	17
Cereal culm root base?	-	-	2	-	2	-	-	-	-	-	-	-	-	-	-
cf. Oryza glume (charred)	-	-	-	-	-	-	-	-	-	-	-	-	-	-	-
Oryza glume (silicified phytolith)	-	-	-	-	-	-	-	-	-	-	-	-	-	-	-
Vitis peduncle	-	-	1	-	-	-	-	-	-	-	-	-	-	-	-
Onopordum achene collar	-	-	-	-	-	-	-	-	-	-	-	-	-	-	-
Asteraceae head, misc.	-	-	-	-	-	-	-	-	-	-	-	-	-	-	-
Euclidium silique	-	-	-	-	-	-	1	-	-	-	-	-	-	-	-
Capparis thorn	-	-	-	-	-	-	-	-	-	-	-	1	-	-	-
Atriplex bract (YH-pp 12)	-	-	-	-	-	-	-	-	-	-	-	-	-	-	-
Salsola inflor (YH-pp 7, YH-unk 8)	-	-	-	-	-	-	2	-	-	-	1	-	-	1	-
Cyperaceae stem fragment	-	-	-	-	-	-	-	-	-	-	-	-	-	-	-
Alhagi pod section	-	-	-	-	-	-	-	-	-	-	-	-	-	-	-
Alhagi leaf?	-	-	-	-	-	-	-	-	-	-	-	-	-	-	-
Aegilops glume base	-	-	-	-	-	-	-	-	-	-	-	-	-	-	-
Hordeum internode (wild)	-	-	-	-	-	2	-	-	-	-	1	-	-	-	-
Taeniatherum int. (YH-pp 8)	-	-	-	-	-	-	-	-	-	-	-	2	-	-	-
Poaceae, mystery 1 rachis frag.	-	-	-	-	-	-	-	-	-	-	-	1	-	-	-
YH-plant part 9	-	-	-	-	-	-	-	-	-	-	-	-	-	-	-
UNCHARRED-MINERALIZED															
Arnebia linearis	-	-	-	-	-	-	-	-	-	-	-	-	-	-	-
Arnebia/Lithospermum	6	1	4	24	-	9	3	3	5	1	3	3	-	2	-
Lithospermum tenuifolium	1	3	-	-	-	-	-	-	-	-	-	-	-	-	-
Lithospermum arvense	-	-	-	-	-	-	-	-	-	-	-	-	-	-	-
Lithospermum, misc.	1	3	5	-	5	-	-	1	-	-	-	-	-	-	-
Moltkia	-	-	-	-	-	-	-	-	-	-	-	1	1	-	1
BORAGINACEAE	-	-	-	-	-	-	-	-	-	-	-	-	-	-	-
Eleocharis (Cyperaceae-a)	-	-	-	-	-	-	1	-	-	-	4	1	-	6	1
Carex	-	-	-	-	-	-	-	-	-	-	-	-	-	-	-
Ficus	-	-	-	-	-	-	-	-	-	-	-	-	-	-	-
Glaucium	-	-	-	-	-	1	-	1	-	-	2	-	-	1	1
Papaver (white)	-	-	-	-	-	-	-	-	5	-	-	-	-	-	-

Table F2, cont'd.: Columns 211–225

Column no.	211†	212	213†	214	215	216	217	218	219†	220	221	222	223	224†	225†
YH no.	31837	29943	31089	32407	32414	23168	26342	26682	28838	33122	33125	23236	23238	33280	33270
Context type	pit	surf	mixed	coll	coll	coll	debris	debris	debris	debris	debris	debris	debris	mixed	surf
Operation	14	9	14	14	14	5	14	14	14	14	14	5	5	14	14
Locus	52	24	17	60	60	38	10	10	10	60	60	38	38	61	61
Lot	132	115	117	147	148	68	21	29	72	165	164	72	72	173	171
Phase	870.04	900	900	950	950	950	970	970	970	970	970	970	970	1000	1030
Soil volume (liters)	8	10	11	6.5	3	15	14	13	9	8	6	8	8	6.5	5
Charcoal (>2 mm, g)	1.00	2.15	4.13	0.96	1.46	2.50	3.66	4.57	5.19	2.55	5.67	1.18	1.53	0.84	3.43
Seed (>2 mm, g)	0.55	0.32	0.56	0.14	0.22	0.23	0.18	0.26	0.48	0.22	0.53	0.45	0.51	0.32	0.13
Other (>2 mm, g)	0.02	-	+	-	+	-	+	+	0.01	0.03	0.01	0.01	+	+	-
Wild/weedy (#)	109	66	146	30	60	35	125	57	159	58	39	113	231	46	1203
ECONOMIC PLANTS															
Hordeum vulgare (g)	0.11	0.09	0.24	0.05	0.07	0.08	0.06	0.09	0.20	0.07	0.16	0.24	0.15	0.05	0.05
H. vulgare var. nudum (g)	-	-	-	-	-	-	-	-	-	-	-	-	-	-	-
Triticum aestivum/durum (g)	0.23	0.12	0.19	0.04	0.06	-	0.08	0.08	0.13	0.04	0.23	0.09	0.15	0.11	0.02
Triticum monococcum (g)	0.01	+	0.01	0.01	0.01	0.02	-	-	0.02	-	-	-	0.01	0.02	0.01
Triticum dicoccum (g)	-	-	-	-	-	-	-	0.01	-	-	-	-	-	0.01	-
Triticum sp. (g)	0.02	0.04	-	-	-	0.08	-	-	0.01	0.05	0.03	0.06	-	-	-
Cereal, indet. (g)	0.17	0.19	0.27	0.10	0.15	0.02	0.07	0.13	0.25	0.12	0.25	0.10	0.21	0.22	+
Secale cereale (g)	-	-	-	-	-	-	-	-	-	-	-	-	-	-	-
Setaria italica (#)	-	-	-	-	-	-	-	-	-	-	-	-	-	-	-
Oryza (g)	-	-	-	-	-	-	-	-	-	-	-	-	-	-	-
Vicia ervilia (g)	0.06	-	0.06	-	-	-	0.01	0.01	0.02	-	0.05	0.02	-	+	-
Lens (g)	0.02	-	-	-	-	-	-	-	-	-	-	-	0.01	-	-
Pulse (g)	0.01	-	-	-	-	0.02	-	-	0.02	0.01	-	-	0.02	-	-
cf. Pistacia/nutshell (g)	-	-	+	-	-	-	-	-	1 (0.02)	-	-	-	-	-	-
cf. Quercus (g)	-	-	-	-	-	-	-	-	-	-	-	-	-	-	-
Ficus carica (#)	-	-	-	-	-	-	-	-	-	-	-	-	-	1	-
Vitis vinifera (# whole, total g)	-	-	-	-	-	-	-	-	-	-	-	-	-	1 (0.01)	+
WILD AND WEEDY (counts)															
Bupleurum	-	-	-	-	-	-	-	-	-	-	-	-	-	-	-
cf. Daucus	-	-	2	-	-	-	-	-	-	-	-	-	-	-	-
Torilis leptophylla?	-	-	-	-	-	-	-	-	-	-	-	-	-	-	-
YH-Apiaceae 2	-	-	-	-	-	-	-	-	-	-	-	-	-	-	-
YH-Apiaceae 4/8	-	-	-	-	-	-	-	-	-	-	-	-	-	-	-
YH-Apiaceae 6	-	-	-	-	-	-	-	-	-	-	-	-	-	-	-
YH-Apiaceae 7	-	-	-	-	-	-	-	-	-	-	-	-	-	-	-
YH-Apiaceae 10	-	-	-	-	-	1	-	-	-	-	-	-	-	-	-
YH-unknown 31 (Apiaceae)	-	-	-	-	-	-	-	-	-	-	-	-	-	-	-
APIACEAE	-	-	-	-	-	-	-	-	-	-	-	-	-	-	-
Anthemis/Matricaria	-	-	-	-	-	-	-	-	1	-	-	-	-	-	-
Artemisia	-	1	-	-	-	-	-	-	-	-	-	-	-	-	-
Carthamus	-	-	-	-	-	-	-	-	-	-	-	-	-	-	-
Centaurea cyanus-type	-	-	-	-	-	-	-	-	-	-	-	-	-	-	-
Centaurea	1	1	4	-	-	-	-	-	1	-	1	-	1	-	-
Onopordum	1	-	-	-	-	-	-	-	-	-	-	-	-	-	-
YH-Asteraceae 1	1	-	-	-	1	-	-	-	-	-	-	-	4	1	1

Table F2, cont'd.: Columns 211–225

Column no.	211†	212	213†	214	215	216	217	218	219†	220	221	222	223	224†	225†
YH no.	31837	29943	31089	32407	32414	23168	26342	26682	28838	33122	33125	23236	23238	33280	33270
YH-Asteraceae 2	-	-	-	-	-	-	-	-	-	-	-	-	-	-	-
YH-Asteraceae 5	-	-	-	-	-	-	-	-	-	-	-	-	-	-	-
YH-Asteraceae 9	-	1	-	-	-	-	-	-	-	-	-	-	-	-	-
ASTERACEAE	1	-	-	-	-	-	-	-	1	-	-	3	-	-	-
Arnebia/Lithospermum	-	-	-	-	-	-	-	-	-	-	-	1	-	-	-
Heliotropium	-	-	-	1	1	-	-	-	-	-	-	3	-	-	1
cf. Alyssum	-	-	-	-	-	-	-	-	-	-	-	-	-	-	-
cf. Camelina rumelica	-	-	-	-	-	-	-	-	-	-	-	-	-	-	-
cf. Camelina sativa	-	-	-	-	-	-	-	-	-	-	-	-	-	-	-
Conringia	2	-	-	-	-	-	-	-	-	-	-	-	-	-	-
cf. Lepidium	-	-	-	-	-	-	-	-	-	-	-	-	-	-	-
Sisymbrium altissimum-type	-	-	-	-	-	-	-	-	-	-	-	-	-	-	-
Thlaspi	-	-	-	-	-	-	-	-	-	-	-	-	-	-	-
YH-Brassicaceae 2	-	-	-	-	-	-	-	-	-	-	-	-	-	-	-
YH-Brassicaceae 3/5	-	-	-	-	-	-	-	-	-	-	-	1	2	-	-
YH-Brassicaceae 7	-	-	-	-	-	-	-	-	-	-	-	-	-	-	-
YH-Brassicaceae 10	-	-	-	-	-	-	-	-	1	-	1	1	-	-	-
YH-Brassicaceae 11	-	-	2	-	-	-	-	-	-	-	-	1	6	-	-
YH-Brassicaceae 12	-	-	-	-	-	-	-	-	-	-	-	-	-	-	-
BRASSICACEAE	4	3	1	-	-	1	-	-	-	-	-	2	-	-	-
Bufonia	-	-	-	-	-	-	-	-	-	-	-	-	-	-	-
Gypsophila	-	1	5	-	-	-	-	-	-	-	-	-	1	-	-
Silene	-	-	-	-	-	-	-	-	-	-	-	1	-	-	-
Vaccaria (YH-unknown 6)	-	-	-	-	-	-	-	-	2	-	-	-	-	-	-
CARYOPHYLLACEAE	-	-	-	-	-	-	-	-	-	-	-	-	-	1	-
YH-Caryophyllaceae 1	-	-	-	-	-	-	-	-	-	-	-	-	-	-	-
Atriplex	-	-	-	-	-	-	-	-	2	-	-	-	-	-	-
Chenopodium	12	-	-	-	-	2	4	3	4	-	-	4	1	-	-
Salsola kali-type	-	-	-	-	-	-	-	-	-	-	-	-	-	-	-
Salsola soda-type	-	-	4	-	-	-	-	-	-	-	-	-	-	2	-
Salsola	-	-	-	-	-	1	-	-	-	-	-	-	-	-	-
Suaeda	-	1	3	-	-	-	3	6	1	2	1	1	-	-	-
YH-Chenopodiaceae 2	-	-	-	-	-	1	-	-	-	-	-	-	-	-	-
CHENOPODIACEAE	-	-	1	-	1	-	-	-	1	2	-	-	-	-	-
Helianthemum	-	-	-	-	-	-	-	-	-	-	-	-	-	-	-
Carex	5	3	3	-	1	2	3	1	11	2	1	7	22	-	2
Carex 3	-	-	-	-	-	-	-	-	-	-	1	-	-	-	-
Eleocharis (YH-Cyperaceae a)	-	-	-	-	-	-	-	-	-	-	-	-	-	-	-
YH-Cyperaceae 1	-	2	1	1	2	1	-	-	1	1	3	2	3	1	-
YH-Cyperaceae 3	-	-	-	-	-	-	-	-	-	-	-	-	5	-	-
YH-Cyperaceae 4	-	-	-	-	2	-	-	-	-	-	-	-	-	-	-
YH-Cyperaceae 5	-	-	-	-	-	-	-	3	1	-	1	1	1	-	-
YH-Cyperaceae 7	-	-	-	-	-	-	-	-	-	-	-	1	-	-	-
YH-Cyperaceae 8	-	-	-	-	-	-	-	-	-	-	-	-	-	-	-
CYPERACEAE	-	1	-	-	-	1	1	-	-	-	-	2	-	1	2
Cephalaria	-	-	-	-	-	-	-	-	-	-	-	-	-	-	-
Scabiosa	-	-	-	-	-	-	-	-	-	-	-	-	-	-	-
Euphorbia	-	-	-	-	-	-	-	-	-	-	-	-	-	-	-

Table F2, cont'd.: Columns 211–225

Column no.	211†	212	213†	214	215	216	217	218	219†	220	221	222	223	224†	225†
YH no.	31837	29943	31089	32407	32414	23168	26342	26682	28838	33122	33125	23236	23238	33280	33270
Alhagi	-	-	-	-	-	-	-	-	-	-	-	-	-	-	-
Astragalus	-	2	2	-	1	-	-	1	4	2	-	-	17	-	1
Medicago	-	1	-	-	-	-	2	2	2	-	-	1	7	-	-
Onobrychis	-	-	-	-	-	-	-	-	-	-	-	-	-	-	-
Trifolium/Melilotus	2	1	-	-	3	2	1	1	10	-	-	4	8	-	-
Trigonella	21	15	24	5	10	-	77	20	78	6	12	37	98	3	3
Trigonella cf. astroites	10	4	11	3	6	-	6	5	4	21	1	3	5	3	-
YH-Fabaceae 2	-	-	-	-	-	-	-	-	-	-	-	-	-	-	-
FABACEAE	6	-	2	4		-	13	-	-	-	2	10	17	-	1
cf. Erodium	-	-	-	-	-	-	-	-	-	-	-	-	-	-	-
Teucrium	1	-	-	-	-	-	-	-	-	-	-	-	2	-	-
Ziziphora	1	-	2	1	2	1	-	-	1	-	-	1	3	-	-
YH-Lamiaceae 2	-	-	-	-	-	-	-	-	-	-	-	-	-	-	-
YH-Lamiac 3 (Nepeta?)	-	-	-	-	-	-	-	-	-	-	-	-	-	-	-
YH-Lamiaceae 5	-	-	-	-	-	-	-	-	-	-	-	-	-	-	-
LAMIACEAE	-	-	-	-	-	-	-	1	-	-	-	-	-	-	-
LILIACEAE	-	-	-	-	1	-	-	-	-	-	-	-	-	-	-
cf. Malva	2	-	-	-	-	-	-	-	-	1	-	-	1	-	14
Glaucium	-	-	1	-	1	-	-	-	1	-	-	-	-	-	1
Papaver	-	-	-	-	-	-	-	-	-	-	-	-	-	-	2
Fumaria	-	-	1	-	-	-	-	-	-	-	-	-	-	-	-
Plantago	-	-	-	-	-	-	-	-	-	-	-	-	-	-	-
Aegilops	-	-	-	-	-	1	-	-	-	-	-	-	-	-	-
Avena	-	-	-	-	-	-	-	-	-	-	-	-	-	-	-
Bromus japonicus-type	-	-	-	-	-	-	-	-	2	3	-	1	1	-	-
Bromus tectorum-type	-	-	-	-	-	-	-	-	-	-	1	-	-	-	-
Bromus	-	-	2	-	-	-	2	-	-	1	-	1	1	-	-
Eremopyrum	2	4	4	-	-	1	-	-	-	-	-	3	3	-	-
Hordeum cf. murinum	6	-	1	1	1	4	-	-	2	2	1	-	-	-	1
Hordeum spontaneum?	-	-	-	-	1	-	-	-	-	-	-	-	-	-	-
Hordeum misc.	-	-	3	-	-	-	-	-	-	-	-	-	-	-	-
cf. Lolium	-	-	-	-	-	-	-	-	-	-	-	-	-	-	-
cf. Phalaris	-	-	-	-	-	-	-	-	-	-	-	-	-	-	-
cf. Poa bulbosa	-	-	-	-	-	-	-	-	-	-	-	-	-	-	-
Setaria	-	-	-	-	-	-	-	-	-	-	-	-	-	-	-
Stipa	2	-	3	-	2	-	-	-	-	-	-	-	-	1	-
Taeniatherum	-	-	2	-	-	-	-	2	-	1	-	-	-	-	-
"Triticoid"	-	-	-	-	-	-	-	-	-	-	-	-	-	-	-
Triticum boeoticum	-	-	-	-	-	-	-	-	-	-	-	-	-	-	-
YH-Poaceae 1	-	1	-	-	1	-	-	1	1	-	3	4	5	3	2
YH-Poaceae 2	-	-	-	-	-	-	-	-	-	-	-	-	-	-	-
YH-Poaceae 3	1	-	-	-	-	-	1	-	1	-	-	1	-	-	1
YH-Poaceae 4	-	-	-	-	-	-	-	-	-	-	-	-	-	-	-
YH-Poaceae 5	-	-	-	-	-	-	-	-	2	-	1	-	-	1	-
YH-Poaceae 8	3	2	5	-	7	-	1	1	-	-	-	2	1	-	1
YH-Poaceae 10/15	-	-	-	-	-	-	-	-	-	-	-	-	-	-	-
YH-Poaceae 11	-	-	-	-	-	-	-	-	-	-	-	-	-	-	-
YH-Poaceae 13	-	-	-	-	-	2	-	1	-	-	-	-	-	1	-

Table F2, cont'd.: Columns 211–225

Column no.	211†	212	213†	214	215	216	217	218	219†	220	221	222	223	224†	225†
YH no.	31837	29943	31089	32407	32414	23168	26342	26682	28838	33122	33125	23236	23238	33280	33270
YH-Poaceae 14	-	-	-	-	-	-	-	-	-	-	-	-	-	-	-
YH-Poaceae 16	-	-	-	-	-	-	-	-	-	-	-	-	-	-	-
YH-Poaceae 17/18	-	-	-	-	3	-	2	-	-	-	-	-	-	-	-
YH-Poaceae 20 Aeg 1a	-	-	-	-	-	-	-	-	-	-	-	-	-	-	-
YH-Poaceae 21 Aeg 1b	-	-	-	-	-	-	-	-	-	-	-	-	-	-	-
POACEAE	8	6	12	6	3	4	7	7	8	3	5	4	13	14	14
Polygonum	-	-	-	-	-	-	-	-	-	-	-	-	-	-	4
Polygonum (was YH-Cyperaceae 2)	1	-	4	-	-	-	-	-	-	-	-	-	-	-	-
Polygonum (was YH-Cyperaceae 6)	-	-	-	1	-	-	-	-	-	-	-	-	-	-	-
Rumex	-	-	-	-	-	2	-	-	-	-	-	-	-	-	-
Portulaca	-	-	-	-	-	-	-	-	-	-	-	-	-	-	-
Androsace	1	-	-	-	-	-	-	-	-	-	-	-	-	-	-
YH-Primulaceae 1	-	-	-	-	-	-	-	-	-	-	-	-	-	-	-
YH-Primulaceae 2	-	-	-	-	-	-	-	-	-	-	-	-	-	-	-
Aconitum	-	-	-	-	-	-	-	-	-	-	-	-	-	-	-
Adonis	-	-	-	-	-	-	-	-	-	-	-	-	-	2	1
Ceratocephalus (YH-pp 2)	-	-	-	-	-	-	-	-	-	-	-	-	-	-	-
Ranunculus (YH-unknown 4)	-	-	1	-	-	-	1	-	-	-	-	-	-	-	-
Ranunculus arvensis	-	-	-	-	-	-	-	-	-	-	-	-	-	-	-
Reseda	-	-	-	-	-	-	-	-	-	-	-	-	-	-	-
cf. Asperula (YH-Rubiaceae 2)	-	-	-	-	-	-	-	-	-	-	-	1	-	-	-
Galium	3	-	5	2	-	2	-	-	3	5	4	3	-	1	9
YH-Rubiaceae 1	-	-	-	-	1	-	-	-	-	-	-	1	-	-	-
YH-Scrophulariaceae 1	-	-	-	-	-	-	-	-	-	-	-	1	-	-	-
Hyoscyamus	-	-	-	-	-	-	-	-	-	-	-	-	-	-	6
SOLANACEAE	-	-	-	-	-	-	-	-	-	-	-	-	-	-	-
Thymelaea	-	1	-	-	-	-	-	-	1	-	-	-	1	-	-
Valerianella coronata	-	-	-	-	-	-	-	-	-	-	-	-	-	-	-
Valerianella vesicaria	-	-	-	-	-	-	-	-	-	-	-	-	-	-	-
Valerianella	-	-	-	-	-	-	-	-	-	-	-	-	-	-	-
Peganum harmala	1	-	-	-	-	-	-	-	-	-	-	-	-	-	-
Zygophyllum	-	-	-	-	-	-	-	-	-	-	-	-	-	-	-
YH-unknown 9	-	-	-	-	-	-	-	-	-	-	-	-	-	-	-
YH-unknown 21	-	-	-	-	-	-	-	-	-	-	-	-	-	-	-
YH-unknown 23	-	-	-	-	-	-	-	-	-	-	-	-	-	1	-
YH-unknown 27	-	-	-	-	-	-	-	-	-	-	-	-	-	-	-
YH-unknown 29	-	-	-	-	-	-	-	-	-	-	-	-	-	-	-
YH-unknown 30	-	-	-	-	-	-	-	-	-	-	-	-	-	-	-
YH-unknown 35	-	-	-	1	1	-	-	-	-	-	-	-	-	-	-
Unknown	11	15	35	4	8	7	-	2	12	6	-	-	4	11	12
PLANT PARTS (pp)															
Hordeum internode (2-row)	1	1	7	2	-	-	-	3	2	-	1	2	3	-	-
Hordeum internode (6-row)	-	-	-	-	-	-	-	-	-	-	-	-	-	-	-
Hordeum internode (compact)	2	-	5	-	-	1	1	2	-	-	-	2	1	-	-
Triticum aestivum int.	-	-	-	-	-	-	-	-	-	-	-	-	-	3	-
T. aestivum/durum int.	-	2	12	3	1	-	2	6	3	4	-	1	4	-	-
T. aestivum/durum int. 1	4	-	-	-	-	-	-	-	-	-	2	-	-	-	-

Table F2, cont'd.: Columns 211–225

Column no.	211†	212	213†	214	215	216	217	218	219†	220	221	222	223	224†	225†
YH no.	31837	29943	31089	32407	32414	23168	26342	26682	28838	33122	33125	23236	23238	33280	33270
T. aestivum/durum int. 2	20	-	-	-	-	-	-	-	-	-	-	-	-	-	-
T. durum internode	-	-	-	-	-	-	-	-	-	-	-	-	-	-	-
T. monococcum spikelet fork	-	1	-	-	-	-	-	-	-	-	-	-	-	3	-
T. dicoccum spikelet fork	-	-	-	-	-	-	-	-	-	-	-	-	-	2	-
T. monococcum/dicoccum spikelet fork	-	-	2	1	2	-	-	-	-	1	-	4	4	10	2
Triticum rachis base	-	-	1	-	-	-	-	-	-	-	-	-	2	-	-
Triticum, rachis fragment, square cross-section	-	-	-	-	-	-	-	-	-	-	-	-	1	-	-
Cereal culm node	16	2	7	-	-	2	-	3	6	3	5	4	5	2	-
Cereal culm root base?	-	-	-	-	-	-	-	-	-	-	-	1	-	-	-
cf. Oryza glume (charred)	-	-	-	-	-	-	-	-	-	-	-	-	-	-	-
Oryza glume (silicified phytolith)	-	-	-	-	-	-	-	-	-	-	-	-	-	-	-
Vitis peduncle	-	-	-	-	-	-	1	-	-	-	-	-	-	-	1
Onopordum achene collar	-	-	-	-	-	-	-	-	-	-	-	-	-	-	-
Asteraceae head, misc.	-	-	-	-	-	-	-	-	-	-	-	-	-	-	-
Euclidium silique	-	-	-	-	-	-	-	-	-	-	-	-	-	-	-
Capparis thorn	-	-	-	-	-	-	-	-	-	-	-	-	-	-	-
Atriplex bract (YH-pp 12)	-	-	-	-	-	-	-	-	-	-	-	-	-	-	-
Salsola inflor (YH-pp 7, YH-unk 8)	-	-	1	-	-	-	-	-	-	-	-	-	-	-	-
Cyperaceae stem fragment	-	-	-	-	-	-	-	-	-	-	-	-	-	-	-
Alhagi pod section	-	-	-	-	-	-	-	-	-	-	-	-	-	-	-
Alhagi leaf?	-	-	-	-	-	-	-	-	-	-	-	-	-	-	-
Aegilops glume base	-	-	-	-	-	-	-	-	-	-	-	-	-	-	-
Hordeum internode (wild)	-	-	1	-	-	-	-	-	-	-	-	-	-	-	-
Taeniatherum int. (YH-pp 8)	-	-	-	-	-	-	-	-	-	-	-	-	-	-	-
Poaceae, mystery 1 rachis frag.	-	-	-	-	-	-	-	-	-	-	-	-	-	-	-
YH-plant part 9	-	-	-	-	-	-	-	-	-	-	-	-	-	-	2
UNCHARRED-MINERALIZED															
Arnebia linearis	-	-	-	-	-	-	-	-	-	-	-	-	1	-	-
Arnebia/Lithospermum	2	-	10	80	-	-	2	-	8	1	-	-	-	1	-
Lithospermum tenuifolium	-	-	-	-	1	2	-	-	-	-	-	-	3	-	-
Lithospermum arvense	-	-	-	-	-	-	-	-	-	-	-	-	-	-	-
Lithospermum, misc.	1	-	-	-	-	-	-	-	-	-	-	4	11	-	-
Moltkia	1	1	1	-	-	-	-	-	-	-	-	-	1	-	-
BORAGINACEAE	-	-	-	-	-	-	1	-	-	-	-	-	2	-	-
Eleocharis (Cyperaceae-a)	1	-	-	1	-	-	-	1	-	-	2	1	1	-	-
Carex	-	-	-	-	-	-	-	-	-	-	-	-	-	-	-
Ficus	-	-	-	-	-	-	-	-	-	-	-	2	-	-	-
Glaucium	3	1	1	-	-	-	-	-	11	-	7	-	-	-	-
Papaver (white)	-	-	-	-	-	-	-	-	-	-	-	-	-	-	-

Table F2, cont'd.: Columns 226–240

Column no.	226*	227	228†	229*†	230†	231	232*†	233†	234	235*	236	237†	238	239	240
YH no.	32144	33234	33230	33243	33246	33554	33555	33573	33575	33580	33587	33590	33613	21068	21183
Context type	surf	jar	surf	jar	surf	jar	jar	surf	surf	jar	surf	jar	surf	surf	surf
Operation	1	1	1	1	1	1	1	1	1	1	1	1	1	2	2
Locus	100	100	100	100	100	100	100	100	100	100	100	100	100	21	21
Lot	193	203	203	205	205	205	205	205	205	205	205	205	208	42	61
Phase	620	620	620	620	620	620	620	620	620	620	620	620	620	350	350
Soil volume (liters)	7	n/a	n/a	150 ml	0.1	50 ml	80 ml	0.2	3	45 ml	50 ml	75 ml	0.2	2.3	6.4
Charcoal (>2 mm, g)	3.12	2.67	3.00	0.12	0.03	0.30	3.06	1.20	1.60	3.01	1.52	0.95	1.87	7.42	7.79
Seed (>2 mm, g)	0.05	0.03	15.18	16.16	10.21	6.04	0.13	10.62	17.29	3.61	5.74	-	9.43	0.01	0.05
Other (>2 mm, g)	-	-	-	-	-	-	-	-	+	-	-	0.21	-	4.62	1.07
Wild/weedy (#)	7		29	106	49	-	208	28	57	5	1	1	-	10	-
ECONOMIC PLANTS															
Hordeum vulgare (g)	0.01	-	10.97	0.04	-	4.61	-	10.40	0.03	0.09	8.96	7.81	7.15	-	0.01
H. vulgare var. nudum (g)	-	-	-	-	-	-	-	-	-	-	-	-	-	-	-
Triticum aestivum/durum (g)	0.02	0.03	0.03	0.04	6.75	-	-	-	0.04	3.39	-	0.01	0.08	0.01	0.04
Triticum monococcum (g)	0.01	-	-	0.01	-	-	-	-	0.01	-	-	-	-	-	-
Triticum dicoccum (g)	-	-	0.31	0.02	-	-	-	-	-	-	-	-	-	-	-
Triticum sp. (g)	-	-	-	0.01	5.61	0.01	-	0.07	-	-	0.04	-	-	-	-
Cereal, indet. (g)	0.01	-	3.67	-	-	2.34	-	6.47	-	0.09	4.78	0.52	4.62	-	0.01
Secale cereale (g)	-	-	-	-	-	-	-	-	-	-	-	-	-	-	-
Setaria italica (#)	3	-	-	-	-	-	-	-	-	-	-	-	-	-	-
Oryza (g)	-	-	-	-	-	-	-	-	-	-	-	-	-	-	-
Vicia ervilia (g)	-	-	-	0.14	-	-	-	-	0.19	-	-	-	-	-	-
Lens (g)	-	+	+	15.7	-	-	0.01	0.04	19.71	-	-	-	-	-	-
Pulse (g)	-	-	-	0.10	-	-	0.06	-	0.01	-	-	-	-	-	0.01
cf. Pistacia/nutshell (g)	-	-	-	-	-	-	-	-	-	-	-	-	-	-	-
cf. Quercus (g)	-	-	-	-	-	-	-	-	-	-	-	-	-	-	-
Ficus carica (#)	-	-	-	-	-	-	-	-	-	-	-	-	-	-	-
Vitis vinifera (# whole, total g)	-	-	-	-	-	-	-	-	-	-	-	-	-	+	-
WILD AND WEEDY (counts)															
Bupleurum	-	-	-	-	-	-	-	-	-	-	-	-	-	-	-
cf. Daucus	-	-	-	-	-	-	-	-	-	-	-	-	-	-	-
Torilis leptophylla?	-	-	-	-	-	-	-	-	-	-	-	-	-	-	-
YH-Apiaceae 2	-	-	-	-	1	-	-	-	-	-	-	-	-	-	-
YH-Apiaceae 4/8	-	-	-	-	-	-	-	-	-	-	-	-	-	-	-
YH-Apiaceae 6	-	-	-	-	-	-	-	-	-	-	-	-	-	-	-
YH-Apiaceae 7	-	-	-	-	-	-	-	-	-	-	-	-	-	-	-
YH-Apiaceae 10	-	-	-	-	-	-	-	-	-	-	-	-	-	-	-
YH-unknown 31 (Apiaceae)	-	-	-	-	-	-	-	-	-	-	-	-	-	-	-
APIACEAE	-	-	-	-	-	-	-	1	1	-	-	-	-	-	-
Anthemis/Matricaria	-	-	-	-	-	-	-	-	-	-	-	-	-	-	-
Artemisia	-	-	-	-	-	-	-	-	-	-	-	-	-	-	-
Carthamus	-	-	-	-	-	-	-	-	-	-	-	-	-	-	-
Centaurea cyanus-type	-	-	-	-	-	-	-	-	-	-	-	-	-	-	-
Centaurea	-	-	-	-	-	-	-	-	-	-	-	-	-	-	-
Onopordum	-	-	-	-	-	-	-	-	-	-	-	-	-	-	-
YH-Asteraceae 1	-	-	-	-	-	-	-	-	-	-	-	-	-	-	-

Table F2, cont'd.: Columns 226–240

Column no.	226*	227	228†	229*†	230†	231	232*†	233†	234	235*	236	237†	238	239	240
YH no.	32144	33234	33230	33243	33246	33554	33555	33573	33575	33580	33587	33590	33613	21068	21183
YH-Asteraceae 2	-	-	-	-	-	-	-	-	-	-	-	-	-	-	-
YH-Asteraceae 5	-	-	-	-	-	-	-	-	-	-	-	-	-	-	-
YH-Asteraceae 9	-	-	-	-	-	-	-	-	-	-	-	-	-	-	-
ASTERACEAE	-	-	-	-	-	-	-	-	-	-	-	-	-	-	-
Arnebia/Lithospermum	-	-	-	-	-	-	3	-	-	-	-	-	-	-	-
Heliotropium	-	-	-	-	-	-	-	-	-	-	-	-	-	-	-
cf. Alyssum	-	-	-	-	-	-	-	-	-	-	-	-	-	-	-
cf. Camelina rumelica	-	-	-	-	-	-	-	-	-	-	-	-	-	-	-
cf. Camelina sativa	-	-	-	-	-	-	-	-	-	-	-	-	-	-	-
Conringia	-	-	-	-	-	-	-	-	-	-	-	-	-	-	-
cf. Lepidium	-	-	-	-	-	-	-	-	-	-	-	-	-	-	-
Sisymbrium altissimum-type	-	-	-	-	-	-	-	-	-	-	-	-	-	-	-
Thlaspi	-	-	-	-	-	-	33	-	-	-	-	-	-	-	-
YH-Brassicaceae 2	-	-	-	-	-	-	-	-	-	-	-	-	-	-	-
YH-Brassicaceae 3/5	-	-	-	-	-	-	-	-	-	-	-	-	-	-	-
YH-Brassicaceae 7	-	-	-	-	-	-	-	-	-	-	-	-	-	-	-
YH-Brassicaceae 10	-	-	-	-	-	-	-	-	-	-	-	-	-	-	-
YH-Brassicaceae 11	-	-	-	-	-	-	-	-	-	-	-	-	-	-	-
YH-Brassicaceae 12	-	-	-	-	-	-	-	-	-	-	-	-	-	-	-
BRASSICACEAE	-	-	-	-	-	-	139	-	-	-	-	-	-	-	-
Bufonia	-	-	-	-	-	-	-	-	-	-	-	-	-	-	-
Gypsophila	-	-	-	-	-	-	-	-	-	-	-	-	-	-	-
Silene	-	-	-	-	-	-	2	-	-	-	-	-	-	-	-
Vaccaria (YH-unknown 6)	-	-	-	-	-	-	-	-	-	-	-	-	-	-	-
CARYOPHYLLACEAE	-	-	-	-	-	-	-	-	-	-	-	-	-	-	-
YH-Caryophyllaceae 1	-	-	-	-	-	-	-	-	-	-	-	-	-	-	-
Atriplex	-	-	-	-	-	-	-	-	-	-	-	-	-	-	-
Chenopodium	-	-	-	-	-	-	-	-	-	-	-	-	-	-	-
Salsola kali-type	-	-	-	-	-	-	-	-	-	-	-	-	-	-	-
Salsola soda-type	-	-	-	-	-	-	-	-	-	-	-	-	-	-	-
Salsola	-	-	-	-	-	-	-	-	-	-	-	-	-	1	-
Suaeda	-	-	-	-	-	-	-	-	-	-	-	-	-	-	-
YH-Chenopodiaceae 2	-	-	-	-	-	-	-	-	-	-	-	-	-	-	-
CHENOPODIACEAE	2	-	-	-	-	-	-	-	-	-	-	-	-	-	-
Helianthemum	-	-	-	-	-	-	-	-	-	-	-	-	-	-	-
Carex	-	-	-	-	-	-	-	-	-	-	-	-	-	2	-
Carex 3	-	-	-	-	-	-	-	-	-	-	-	-	-	-	-
Eleocharis (YH-Cyperaceae a)	-	-	-	-	-	-	-	-	-	-	-	-	-	-	-
YH-Cyperaceae 1	-	-	-	-	-	-	-	-	-	-	-	-	-	2	-
YH-Cyperaceae 3	-	-	-	-	-	-	-	-	-	-	-	-	-	-	-
YH-Cyperaceae 4	-	-	-	-	-	-	-	-	-	-	-	-	-	-	-
YH-Cyperaceae 5	-	-	-	1	-	-	-	-	5	-	-	-	-	-	-
YH-Cyperaceae 7	-	-	-	-	-	-	-	-	-	-	-	-	-	-	-
YH-Cyperaceae 8	-	-	-	-	-	-	-	-	-	-	-	-	-	-	-
CYPERACEAE	-	-	-	-	-	-	22	-	-	2	-	-	-	-	-
Cephalaria	-	-	-	-	-	-	-	-	-	-	-	-	-	-	-
Scabiosa	-	-	-	-	-	-	-	-	-	-	-	-	-	-	-
Euphorbia	-	-	-	-	-	-	-	-	-	-	-	-	-	-	-

Table F2, cont'd.: Columns 226–240

Column no.	226*	227	228†	229*†	230†	231	232*†	233†	234	235*	236	237†	238	239	240
YH no.	32144	33234	33230	33243	33246	33554	33555	33573	33575	33580	33587	33590	33613	21068	21183
Alhagi	–	–	–	–	–	–	–	–	–	–	–	–	–	–	–
Astragalus	–	–	–	–	–	–	–	–	–	–	–	–	–	–	–
Medicago	–	–	–	–	–	–	–	–	–	–	–	–	–	–	–
Onobrychis	–	–	–	–	–	–	–	–	–	–	–	–	–	–	–
Trifolium/Melilotus	–	–	–	–	–	–	–	–	–	–	–	–	–	–	–
Trigonella	–	–	–	–	–	–	–	–	–	–	–	–	–	–	–
Trigonella cf. astroites	1	–	–	–	–	–	–	–	–	–	–	–	–	–	–
YH-Fabaceae 2	–	–	–	–	–	–	–	–	–	–	–	–	–	–	–
FABACEAE	1	–	–	–	–	–	–	–	–	–	–	–	–	3	–
cf. Erodium	–	–	–	–	–	–	–	–	–	–	–	–	–	–	–
Teucrium	–	–	–	–	–	–	–	–	–	–	–	–	–	–	–
Ziziphora	–	–	–	–	–	–	–	–	–	–	–	–	–	–	–
YH-Lamiaceae 2	–	–	–	–	–	–	–	–	–	–	–	–	–	–	–
YH-Lamiac 3 (Nepeta?)	–	–	–	–	–	–	–	–	–	–	–	–	–	–	–
YH-Lamiaceae 5	–	–	–	–	–	–	–	–	–	–	–	–	–	–	–
LAMIACEAE	–	–	–	–	–	–	–	–	1	–	–	–	–	–	–
LILIACEAE	–	–	–	–	–	–	–	–	–	–	–	–	–	–	–
cf. Malva	–	–	–	–	–	–	–	–	–	–	–	–	–	–	–
Glaucium	–	–	–	–	–	–	–	–	–	–	–	–	–	–	–
Papaver	–	–	–	–	–	–	–	–	–	–	–	–	–	1	–
Fumaria	–	–	–	–	–	–	1	–	–	–	–	–	–	–	–
Plantago	–	–	–	–	–	–	–	–	–	–	–	–	–	–	–
Aegilops	–	–	–	–	–	–	–	–	–	–	–	–	–	–	–
Avena	–	–	–	–	–	–	–	1	–	–	–	1	–	–	–
Bromus japonicus-type	–	–	–	–	–	–	–	–	–	–	–	–	–	–	–
Bromus tectorum-type	–	–	–	–	–	–	–	–	–	–	–	–	–	–	–
Bromus	–	–	–	1	–	–	–	–	–	–	–	1	–	–	–
Eremopyrum	–	–	–	–	–	–	–	–	–	–	–	–	–	–	–
Hordeum cf. murinum	–	–	–	–	–	–	–	18	–	–	–	–	–	–	–
Hordeum spontaneum?	–	–	–	–	–	–	–	–	–	–	–	–	–	–	–
Hordeum misc.	–	–	–	–	–	–	–	4	–	–	–	–	–	–	–
cf. Lolium	–	–	–	2	–	–	5	–	–	–	–	–	–	–	–
cf. Phalaris	–	–	–	–	–	–	–	–	–	–	–	–	–	–	–
cf. Poa bulbosa	–	–	–	–	–	–	–	–	–	–	–	–	–	–	–
Setaria	–	–	–	2	–	–	–	–	–	1	–	–	–	–	–
Stipa	–	–	–	–	1	–	–	–	–	–	–	–	–	–	–
Taeniatherum	–	–	–	–	–	–	–	–	–	–	–	–	–	–	–
"Triticoid"	–	–	–	–	–	–	–	–	–	–	–	–	–	–	–
Triticum boeoticum	–	–	–	–	–	–	–	–	–	–	–	+	–	–	–
YH-Poaceae 1	1	–	–	–	–	–	–	–	–	–	–	–	–	–	–
YH-Poaceae 2	–	–	–	–	–	–	–	–	–	–	–	–	–	–	–
YH-Poaceae 3	–	–	–	–	–	–	–	–	–	–	–	–	–	–	–
YH-Poaceae 4	–	–	–	–	–	–	–	–	–	–	–	–	–	–	–
YH-Poaceae 5	–	–	–	–	–	–	–	–	–	–	–	–	–	–	–
YH-Poaceae 8	–	–	–	–	–	–	–	–	–	–	–	–	–	–	–
YH-Poaceae 10/15	–	–	–	–	–	–	–	–	–	–	–	–	–	–	–
YH-Poaceae 11	–	–	–	–	–	–	–	–	–	–	–	–	–	–	–
YH-Poaceae 13	–	–	–	–	–	–	–	–	–	–	–	–	–	–	–

Table F2, cont'd.: Columns 226–240

Column no.	226*	227	228†	229*†	230†	231	232*†	233†	234	235*	236	237†	238	239	240
YH no.	32144	33234	33230	33243	33246	33554	33555	33573	33575	33580	33587	33590	33613	21068	21183
YH-Poaceae 14	-	-	-	-	-	-	-	-	-	-	-	-	-	-	-
YH-Poaceae 16	-	-	-	-	-	-	-	-	-	-	-	-	-	-	-
YH-Poaceae 17/18	-	-	-	-	-	-	-	-	-	-	-	-	-	-	-
YH-Poaceae 20 Aeg 1a	-	-	-	-	-	-	-	-	-	-	-	-	-	-	-
YH-Poaceae 21 Aeg 1b	-	-	-	-	-	-	-	-	-	-	-	-	-	-	-
POACEAE	2	-	26	-	-	-	1	3	3	-	-	-	-	1	-
Polygonum	-	-	-	-	-	-	-	-	-	-	-	-	-	-	-
Polygonum (was YH-Cyperaceae 2)	-	-	-	1	-	-	-	-	-	2	-	-	-	-	-
Polygonum (was YH-Cyperaceae 6)	-	-	-	-	45	-	-	-	-	-	-	-	-	-	-
Rumex	-	-	-	-	-	-	1	-	-	-	-	-	-	-	-
Portulaca	-	-	-	-	-	-	-	-	-	-	-	-	-	-	-
Androsace	-	-	-	-	-	-	-	-	-	-	-	-	-	-	-
YH-Primulaceae 1	-	-	-	-	-	-	-	-	-	-	-	-	-	-	-
YH-Primulaceae 2	-	-	-	-	-	-	-	-	-	-	-	-	-	-	-
Aconitum	-	-	-	-	-	-	-	-	-	-	-	-	-	-	-
Adonis	-	-	-	-	-	-	-	-	-	-	-	-	-	-	-
Ceratocephalus (YH-pp 2)	-	-	-	-	-	-	-	-	-	-	-	-	-	-	-
Ranunculus (YH-unknown 4)	-	-	-	-	-	-	-	-	-	-	-	-	-	-	-
Ranunculus arvensis	-	-	-	-	-	-	-	-	-	-	-	-	-	-	-
Reseda	-	-	-	-	-	-	-	-	-	-	-	-	-	-	-
cf. Asperula (YH-Rubiaceae 2)	-	-	-	8	-	-	-	-	23	-	-	-	-	-	-
Galium	-	-	3	91	1	-	1	1	24	-	-	-	-	-	-
YH-Rubiaceae 1	-	-	-	-	-	-	-	-	-	-	-	-	-	-	-
YH-Scrophulariaceae 1	-	-	-	-	-	-	-	-	-	-	-	-	-	-	-
Hyoscyamus	-	-	-	-	-	-	-	-	-	-	-	-	-	-	-
SOLANACEAE	-	-	-	-	-	-	-	-	-	-	-	-	-	-	-
Thymelaea	-	-	-	-	-	-	-	-	-	-	-	-	-	-	-
Valerianella coronata	-	-	-	-	-	-	-	-	-	-	-	-	-	-	-
Valerianella vesicaria	-	-	-	-	-	-	-	-	-	-	-	-	-	-	-
Valerianella	-	-	-	-	-	-	-	-	-	-	-	-	-	-	-
Peganum harmala	-	-	-	-	-	-	-	-	-	-	-	-	-	-	-
Zygophyllum	-	-	-	-	-	-	-	-	-	-	-	-	-	-	-
YH-unknown 9	-	-	-	-	-	-	-	-	-	-	-	-	-	-	-
YH-unknown 21	-	-	-	-	-	-	-	-	-	-	-	-	-	-	-
YH-unknown 23	-	-	-	-	-	-	-	-	-	-	-	-	-	-	-
YH-unknown 27	-	-	-	-	1	-	-	-	-	-	-	-	-	-	-
YH-unknown 29	-	-	-	-	-	-	-	-	-	-	-	-	-	-	-
YH-unknown 30	-	-	-	-	-	-	-	-	-	-	-	-	-	-	-
YH-unknown 35	-	-	-	-	-	-	-	-	-	-	-	-	-	-	-
Unknown	10	-	-	-	3	-	27	-	7	-	-	-	1	1	-
PLANT PARTS (pp)															
Hordeum internode (2-row)	-	-	1	-	-	-	-	-	-	-	-	-	1	-	1
Hordeum internode (6-row)	-	-	-	-	-	-	-	-	-	-	-	-	-	-	-
Hordeum internode (compact)	-	-	-	-	-	-	-	-	-	-	-	-	-	-	-
Triticum aestivum int.	-	-	-	-	-	-	-	-	-	-	-	-	-	-	-
T. aestivum/durum int.	-	-	-	-	1	-	-	-	-	-	-	-	-	-	1
T. aestivum/durum int. 1	-	-	-	-	-	-	-	-	-	-	-	-	-	-	-

Table F2, cont'd.: Columns 226–240

Column no.	226*	227	228†	229*†	230†	231	232*†	233†	234	235*	236	237†	238	239	240
YH no.	32144	33234	33230	33243	33246	33554	33555	33573	33575	33580	33587	33590	33613	21068	21183
T. aestivum/durum int. 2	-	-	-	-	-	-	-	-	-	-	-	-	-	-	-
T. durum internode	-	-	-	-	-	-	-	-	-	-	-	-	-	-	-
T. monococcum spikelet fork	-	-	-	2	-	-	-	-	-	-	-	-	-	-	-
T. dicoccum spikelet fork	-	-	-	2	-	-	-	-	-	-	-	-	-	-	-
T. monococcum/dicoccum spikelet fork	-	-	-	-	1	-	-	-	-	-	-	-	-	-	-
Triticum rachis base	-	-	-	-	-	-	-	-	-	-	-	-	-	-	-
Triticum, rachis fragment, square cross-section	-	-	-	-	-	-	-	-	-	-	-	-	-	-	-
Cereal culm node	-	-	-	-	-	-	-	•	1	1	-	3	-	-	•
Cereal culm root base?	-	-	-	-	-	-	-	-	-	-	-	-	-	-	1
cf. Oryza glume (charred)	-	-	-	-	-	-	-	-	-	-	-	-	-	-	-
Oryza glume (silicified phytolith)	-	-	-	-	-	-	-	-	-	-	-	-	-	-	-
Vitis peduncle	-	-	-	-	-	-	-	-	-	-	-	-	-	-	-
Onopordum achene collar	-	-	-	-	-	-	-	-	-	-	-	-	-	-	-
Asteraceae head, misc.	-	-	-	-	-	-	-	-	-	-	-	-	-	-	-
Euclidium silique	-	-	-	-	-	-	-	-	-	-	-	-	-	-	-
Capparis thorn	-	-	-	-	-	-	-	-	-	-	-	-	-	-	-
Atriplex bract (YH-pp 12)	-	-	-	-	-	-	-	-	-	-	-	-	-	-	-
Salsola inflor (YH-pp 7, YH-unk 8)	-	-	-	-	-	-	-	-	-	-	-	-	-	-	-
Cyperaceae stem fragment	-	-	-	-	-	-	-	-	-	-	-	-	-	-	-
Alhagi pod section	-	-	-	-	-	-	-	-	-	-	-	-	-	-	-
Alhagi leaf?	-	-	-	-	-	-	-	-	-	-	-	-	-	-	-
Aegilops glume base	-	-	-	-	-	-	-	-	-	-	-	-	-	-	-
Hordeum internode (wild)	-	-	-	-	-	-	-	-	-	-	-	-	-	-	-
Taeniatherum int. (YH-pp 8)	-	-	-	-	-	-	-	-	-	-	-	-	-	-	-
Poaceae, mystery 1 rachis frag.	-	-	-	-	-	-	-	-	-	-	-	-	-	-	-
YH-plant part 9	-	-	-	-	-	-	-	-	-	-	-	-	-	-	-
UNCHARRED-MINERALIZED															
Arnebia linearis	-	-	-	-	-	-	-	-	-	-	-	-	-	-	-
Arnebia/Lithospermum	-	-	-	-	-	-	-	1	-	-	-	-	40	-	13
Lithospermum tenuifolium	-	-	-	-	-	-	-	-	-	-	-	-	-	-	-
Lithospermum arvense	-	-	-	-	-	-	-	-	-	-	-	-	-	-	-
Lithospermum, misc.	-	-	-	-	-	-	-	-	-	-	-	-	-	-	-
Moltkia	-	-	-	-	-	-	-	-	-	-	-	-	-	-	-
BORAGINACEAE	-	-	-	-	-	-	-	-	-	-	-	-	-	-	-
Eleocharis (Cyperaceae-a)	-	-	-	-	-	-	-	-	-	-	-	-	-	-	-
Carex	-	-	-	-	-	-	-	-	-	-	-	-	-	-	-
Ficus	-	-	-	-	-	-	-	-	-	-	-	-	-	-	-
Glaucium	-	-	-	-	-	-	-	-	-	-	-	-	-	-	-
Papaver (white)	-	-	-	-	-	-	-	-	-	-	-	-	-	-	-

Table F2, cont'd.: Columns 241–252

Column no.	241	242	243	244*†	245	246	247	248	249	250	251	252
YH no.	29483	29924	30416	33335	33368	33379	33382	33402	33414	33422	29923	33394
Context type	surf	surf	surf	surf	surf	surf	surf	surf	pyro	pyro	coll	pit
Operation	9	9	9	10	10	10	10	10	10	10	9	10
Locus	23	23	23	25	25	25	25	25	38	38	23	37
Lot	106	111	116	80	87	87	87	87	94	94	112	90
Phase	725	725	725	725	725	725	725	725	725.04	725.04	725.05	725.08
Soil volume (liters)	6	45 ml	70 ml	0.2	11	75 ml	1	0.2	6	9	12	2
Charcoal (>2 mm, g)	13.57	4.81	1.21	0.14	7.68	5.83	0.83	0.08	0.88	2.36	4.48	0.69
Seed (>2 mm, g)	0.56	0.03	3.97	5.78	9.11	4.14	3.59	13.19	0.87	1.30	0.80	9.74
Other (>2 mm, g)	0.20	-	-	0.29	0.06	-	-	-	-	0.01	0.64	-
Wild/weedy (#)	32	-	15	164	52	-	31	3	28	110	56	84
ECONOMIC PLANTS												
Hordeum vulgare (g)	0.13	0.01	3.80	0.30	6.88	0.36	0.19	0.06	0.08	0.69	0.55	0.25
H. vulgare var. nudum (g)	-	-	-	-	-	-	-	-	-	-	-	-
Triticum aestivum/durum (g)	0.17	-	+	0.02	0.72	-	2.41	13.66	0.40	0.29	0.10	4.93
Triticum monococcum (g)	0.07	0.02	-	-	0.05	-	-	+	0.05	0.11	0.02	0.11
Triticum dicoccum (g)	-	-	-	-	0.11	-	-	0.03	0.01	-	-	-
Triticum sp. (g)	0.06	-	+	0.01	0.22	4.37	-	-	0.12	0.18	0.03	4.65
Cereal, indet. (g)	0.20	+	2.99	0.10	1.60	2.11	2.77	1.13	0.19	0.64	0.18	1.30
Secale cereale (g)	-	-	-	-	-	-	-	-	- 3 (0.01)	-	-	-
Setaria italica (#)	-	-	-	-	-	-	1	-	-	9	-	5
Oryza (g)	-	-	-	-	-	-	-	-	-	-	-	-
Vicia ervilia (g)	0.01	-	-	17.85	0.01	-	0.02	-	-	0.05	+	0.02
Lens (g)	-	-	-	0.01	-	-	-	-	-	+	-	-
Pulse (g)	0.01	-	-	-	-	-	-	-	-	-	-	-
cf. Pistacia/nutshell (g)	-	-	-	-	-	-	-	-	-	-	-	-
cf. Quercus (g)	-	-	-	-	-	-	-	-	-	-	-	-
Ficus carica (#)	-	-	-	-	-	-	-	-	-	-	-	-
Vitis vinifera (# whole, total g)	-	-	-	-	-	-	-	-	-	-	-	-
WILD AND WEEDY (counts)												
Bupleurum	-	-	-	-	-	-	-	-	-	-	-	-
cf. Daucus	-	-	-	-	-	-	-	-	-	-	-	-
Torilis leptophylla?	-	-	-	-	-	-	-	-	-	-	-	-
YH-Apiaceae 2	-	-	1	-	-	-	-	-	-	-	-	-
YH-Apiaceae 4/8	-	-	-	-	-	-	-	-	-	-	-	-
YH-Apiaceae 6	-	-	-	-	-	-	-	-	-	-	-	-
YH-Apiaceae 7	-	-	-	-	-	-	-	-	-	-	-	-
YH-Apiaceae 10	-	-	-	1	-	-	-	-	-	-	-	2
YH-unknown 31 (Apiaceae)	-	-	-	-	-	-	-	-	-	-	-	-
APIACEAE	-	-	2	5	-	-	-	-	-	-	-	-
Anthemis/Matricaria	-	-	-	-	-	-	-	-	-	-	-	-
Artemisia	-	-	-	-	-	-	-	-	-	-	-	-
Carthamus	-	-	-	-	-	-	-	-	-	-	-	-
Centaurea cyanus-type	-	-	-	-	-	-	-	-	-	-	-	-
Centaurea	-	-	-	-	-	-	-	-	-	-	-	-
Onopordum	-	-	-	-	-	-	-	-	-	-	-	-
YH-Asteraceae 1	-	-	-	-	-	-	-	-	-	-	-	-

Table F2, cont'd.: Columns 241–252

Column no.	241	242	243	244*†	245	246	247	248	249	250	251	252
YH no.	29483	29924	30416	33335	33368	33379	33382	33402	33414	33422	29923	33394
YH-Asteraceae 2	-	-	-	-	-	-	-	-	-	-	-	-
YH-Asteraceae 5	2	-	-	-	-	-	-	-	-	-	-	-
YH-Asteraceae 9	-	-	-	-	-	-	-	-	-	-	-	-
ASTERACEAE	-	-	-	-	-	-	-	-	-	-	-	-
Arnebia/Lithospermum	-	-	-	-	-	-	-	-	-	-	-	-
Heliotropium	1	-	-	1	1	-	3	-	2	-	1	3
cf. Alyssum	-	-	-	-	-	-	-	-	-	-	-	-
cf. Camelina rumelica	-	-	-	-	-	-	-	-	-	-	-	-
cf. Camelina sativa	-	-	-	-	-	-	-	-	-	-	-	-
Conringia	-	-	-	2	-	-	-	-	-	-	-	-
cf. Lepidium	-	-	-	-	-	-	-	-	-	-	-	-
Sisymbrium altissimum-type	-	-	-	-	-	-	-	-	-	-	-	-
Thlaspi	-	-	-	-	-	-	-	-	-	-	-	-
YH-Brassicaceae 2	-	-	-	1	-	-	-	-	-	-	-	-
YH-Brassicaceae 3/5	-	-	-	-	-	-	-	-	-	-	-	-
YH-Brassicaceae 7	-	-	-	-	-	-	-	-	-	-	-	-
YH-Brassicaceae 10	-	-	-	-	-	-	-	-	-	-	-	-
YH-Brassicaceae 11	-	-	-	-	-	-	-	-	-	-	-	-
YH-Brassicaceae 12	-	-	-	-	-	-	-	-	-	-	-	1
BRASSICACEAE	-	-	-	-	-	-	-	-	-	-	2	-
Bufonia	-	-	-	-	-	-	-	-	-	-	-	-
Gypsophila	-	-	-	-	-	-	-	-	-	-	-	-
Silene	-	-	-	-	-	-	-	-	-	1	-	-
Vaccaria (YH-unknown 6)	-	-	-	2	-	-	-	-	-	-	-	1
CARYOPHYLLACEAE	1	-	-	-	-	-	-	-	-	-	-	2
YH-Caryophyllaceae 1	-	-	-	-	-	-	-	-	-	-	-	-
Atriplex	-	-	-	-	-	-	-	-	-	-	-	-
Chenopodium	-	-	-	-	-	-	-	-	-	-	-	-
Salsola kali-type	-	-	-	-	-	-	-	-	-	-	-	-
Salsola soda-type	-	-	-	-	-	-	-	-	-	-	-	-
Salsola	-	-	-	-	-	-	-	-	-	-	-	-
Suaeda	2	-	-	2	12	-	21	1	12	7	9	32
YH-Chenopodiaceae 2	-	-	-	-	-	-	-	-	-	-	-	-
CHENOPODIACEAE	-	-	-	-	-	-	-	-	-	3	-	-
Helianthemum	-	-	-	-	-	-	-	-	-	-	-	-
Carex	-	-	-	-	-	-	-	-	1	1	1	-
Carex 3	-	-	-	-	-	-	-	-	-	-	-	-
Eleocharis (YH-Cyperaceae a)	-	-	-	-	-	-	-	-	-	1	2	-
YH-Cyperaceae 1	-	-	-	-	2	-	-	-	-	2	2	3
YH-Cyperaceae 3	-	-	-	-	-	-	-	-	-	-	-	-
YH-Cyperaceae 4	-	-	-	-	-	-	-	-	-	-	-	-
YH-Cyperaceae 5	-	-	-	-	-	-	-	-	-	1	2	-
YH-Cyperaceae 7	-	-	-	-	-	-	-	-	-	-	-	-
YH-Cyperaceae 8	-	-	-	-	-	-	-	-	-	-	-	-
CYPERACEAE	-	-	-	3	-	-	-	-	1	-	6	3
Cephalaria	-	-	-	-	-	-	-	-	-	-	-	-
Scabiosa	-	-	-	-	-	-	-	-	-	-	-	-
Euphorbia	-	-	-	-	-	-	-	-	-	-	-	-

Table F2, cont'd.: Columns 241–252

Column no.	241	242	243	244*†	245	246	247	248	249	250	251	252
YH no.	29483	29924	30416	33335	33368	33379	33382	33402	33414	33422	29923	33394
Alhagi	-	-	-	-	-	-	-	-	-	-	-	-
Astragalus	1	-	-	-	-	-	-	-	-	-	-	-
Medicago	-	-	-	-	-	-	-	-	-	-	-	-
Onobrychis	-	-	-	-	-	-	-	-	-	-	-	-
Trifolium/Melilotus	1	-	-	-	-	-	-	-	-	1	1	-
Trigonella	4	-	-	-	3	-	-	-	3	-	4	5
Trigonella cf. astroites	-	-	-	-	-	-	1	-	-	2	-	-
YH-Fabaceae 2	-	-	-	-	-	-	-	-	-	-	-	-
FABACEAE	1	-	-	-	-	-	-	-	-	2	2	-
cf. Erodium	-	-	-	-	-	-	-	-	-	-	-	-
Teucrium	1	-	-	-	1	-	-	-	-	-	-	-
Ziziphora	-	-	-	-	-	-	-	-	-	1	-	-
YH-Lamiaceae 2	-	-	-	-	-	-	-	-	-	-	-	-
YH-Lamiac 3 (Nepeta?)	-	-	-	-	-	-	-	-	-	-	-	-
YH-Lamiaceae 5	-	-	-	-	-	-	-	-	-	-	-	-
LAMIACEAE	-	-	-	1	-	-	-	-	-	-	-	-
LILIACEAE	-	-	-	-	-	-	-	-	-	-	-	-
cf. Malva	-	-	-	-	-	-	-	-	-	-	-	-
Glaucium	-	-	-	-	-	-	-	-	-	-	-	-
Papaver	-	-	-	-	-	-	-	-	-	-	-	-
Fumaria	-	-	-	-	-	-	-	-	-	-	-	-
Plantago	-	-	-	-	-	-	-	-	-	-	-	-
Aegilops	-	-	-	-	-	-	-	-	1	-	-	-
Avena	-	-	-	-	-	-	-	-	-	-	-	-
Bromus japonicus-type	-	-	-	-	-	-	-	-	-	-	-	-
Bromus tectorum-type	-	-	-	-	-	-	-	-	-	-	-	-
Bromus	-	-	-	-	-	-	-	-	-	1	-	-
Eremopyrum	-	-	-	-	-	-	1	-	-	-	-	-
Hordeum cf. murinum	2	-	-	-	-	-	-	-	-	-	-	1
Hordeum spontaneum?	-	-	-	-	3	-	-	-	-	1	-	-
Hordeum misc.	-	-	-	-	-	-	-	-	1	-	-	-
cf. Lolium	-	-	-	-	-	-	-	-	-	-	-	-
cf. Phalaris	-	-	-	-	-	-	-	-	-	-	-	-
cf. Poa bulbosa	-	-	-	-	-	-	-	-	-	-	-	1
Setaria	-	-	-	-	-	-	-	-	-	8	-	-
Stipa	-	-	-	-	-	-	-	-	-	2	-	-
Taeniatherum	-	-	-	-	-	-	-	-	-	-	-	1
"Triticoid"	-	-	-	-	-	-	-	-	-	-	1	3
Triticum boeoticum	2	-	-	3	-	-	-	-	1	4	2	-
YH-Poaceae 1	-	-	-	-	-	-	-	-	-	-	1	-
YH-Poaceae 2	-	-	-	-	-	-	-	-	-	-	6	-
YH-Poaceae 3	-	-	-	-	-	-	-	-	-	-	-	-
YH-Poaceae 4	-	-	-	-	-	-	-	-	-	-	-	-
YH-Poaceae 5	-	-	-	-	-	-	-	-	-	-	-	-
YH-Poaceae 8	2	-	1	-	3	-	1	-	1	2	1	-
YH-Poaceae 10/15	-	-	-	-	-	-	-	-	-	-	-	-
YH-Poaceae 11	-	-	-	-	-	-	-	-	-	-	-	-
YH-Poaceae 13	-	-	-	-	-	-	-	-	-	-	-	-

Table F2, cont'd.: Columns 241–252

Column no.	241	242	243	244*†	245	246	247	248	249	250	251	252
YH no.	29483	29924	30416	33335	33368	33379	33382	33402	33414	33422	29923	33394
YH-Poaceae 14	-	-	-	-	-	-	-	-	-	-	-	-
YH-Poaceae 16	-	-	-	-	-	-	-	-	-	-	-	-
YH-Poaceae 17/18	-	-	-	-	-	-	-	-	-	-	-	-
YH-Poaceae 20 Aeg 1a	-	-	-	-	-	-	-	-	-	-	-	-
YH-Poaceae 21 Aeg 1b	-	-	-	-	-	-	-	-	-	-	-	-
POACEAE	5	-	7	3	10	-	-	-	2	18	9	6
Polygonum	1	-	-	-	-	-	-	-	-	-	-	-
Polygonum (was YH-Cyperaceae 2)	-	-	-	-	-	-	-	-	-	-	-	-
Polygonum (was YH-Cyperaceae 6)	-	-	-	-	1	-	-	1	1	-	1	-
Rumex	-	-	-	-	-	-	-	-	-	-	-	-
Portulaca	-	-	-	-	1	-	1	-	-	3	-	-
Androsace	-	-	-	-	-	-	-	-	-	2	-	-
YH-Primulaceae 1	-	-	-	-	-	-	-	-	-	-	-	-
YH-Primulaceae 2	-	-	-	-	-	-	-	-	-	-	-	-
Aconitum	-	-	-	-	-	-	-	-	-	-	-	-
Adonis	-	-	-	-	-	-	-	-	1	-	2	-
Ceratocephalus (YH-pp 2)	-	-	-	-	-	-	-	-	-	-	-	-
Ranunculus (YH-unknown 4)	-	-	-	-	-	-	-	-	-	-	-	-
Ranunculus arvensis	-	-	-	-	-	-	-	-	-	-	-	-
Reseda	-	-	-	-	-	-	-	-	-	-	-	-
cf. Asperula (YH-Rubiaceae 2)	-	-	-	1	-	-	1	-	-	-	-	-
Galium	-	-	1	131	5	-	-	-	-	1	1	1
YH-Rubiaceae 1	-	-	-	-	-	-	-	-	-	1	-	-
YH-Scrophulariaceae 1	-	-	-	-	-	-	-	-	-	-	-	-
Hyoscyamus	-	-	-	-	-	-	-	-	-	-	-	-
SOLANACEAE	-	-	-	-	-	-	-	-	-	-	-	-
Thymelaea	-	-	-	-	-	-	-	-	-	-	-	2
Valerianella coronata	-	-	-	-	-	-	-	-	-	-	-	-
Valerianella vesicaria	-	-	-	-	-	-	-	-	-	-	-	-
Valerianella	-	-	-	-	-	-	-	-	-	-	-	-
Peganum harmala	-	-	-	-	-	-	-	-	-	-	-	-
Zygophyllum	-	-	-	-	-	-	-	-	-	-	-	-
YH-unknown 9	-	-	-	-	-	-	-	-	-	-	-	-
YH-unknown 21	1	-	-	-	3	-	-	-	-	8	-	-
YH-unknown 23	-	-	-	-	-	-	-	-	-	-	-	-
YH-unknown 27	-	-	-	-	-	-	-	-	-	-	-	-
YH-unknown 29	-	-	-	-	-	-	-	-	-	-	-	-
YH-unknown 30	-	-	-	-	-	-	-	-	-	-	-	-
YH-unknown 35	-	-	-	-	-	-	-	-	-	-	-	-
Unknown	5	-	3	8	7	-	3	-	1	37	-	17
PLANT PARTS (pp)												
Hordeum internode (2-row)	22	-	1	-	3	-	-	1	-	-	4	2
Hordeum internode (6-row)	-	-	-	-	-	-	-	-	-	-	-	-
Hordeum internode (compact)	4	-	-	1	-	-	-	1	-	-	-	-
Triticum aestivum int.	-	-	-	-	-	-	-	-	-	-	-	-
T. aestivum/durum int.	3	-	-	-	2	-	-	-	-	-	-	1
T. aestivum/durum int. 1	-	-	-	-	-	-	-	-	-	-	-	-

Table F2, cont'd.: Columns 241–252

Column no.	241	242	243	244*†	245	246	247	248	249	250	251	252
YH no.	29483	29924	30416	33335	33368	33379	33382	33402	33414	33422	29923	33394
T. aestivum/durum int. 2	-	-	-	-	-	-	-	-	-	-	-	-
T. durum internode	-	-	-	-	-	-	-	-	-	-	-	-
T. monococcum spikelet fork	-	-	-	-	-	-	-	-	-	-	-	-
T. dicoccum spikelet fork	-	-	-	-	2	-	-	-	-	-	-	-
T. monococcum/dicoccum spikelet fork	8	1	1	1	2	-	6	-	-	-	2	43
Triticum rachis base	-	-	-	-	-	-	-	-	-	-	-	-
Triticum, rachis fragment, square cross-section	-	-	-	-	-	-	-	-	-	-	-	-
Cereal culm node	13	-	1	-	3	-	-	-	-	-	-	1
Cereal culm root base?	-	-	-	-	-	-	-	-	-	-	-	-
cf. Oryza glume (charred)	-	-	-	-	-	-	-	-	-	-	-	-
Oryza glume (silicified phytolith)	-	-	-	-	-	-	-	-	-	-	-	-
Vitis peduncle	-	-	-	-	-	-	-	-	-	-	-	-
Onopordum achene collar	-	-	-	-	-	-	-	-	-	-	-	-
Asteraceae head, misc.	-	-	-	-	-	-	-	-	-	-	-	-
Euclidium silique	-	-	-	-	-	-	-	-	-	-	-	-
Capparis thorn	-	-	-	-	-	-	-	-	-	-	-	-
Atriplex bract (YH-pp 12)	-	-	-	-	-	-	-	-	-	-	-	-
Salsola inflor (YH-pp 7, YH-unk 8)	-	-	-	-	-	-	-	-	-	-	-	-
Cyperaceae stem fragment	-	-	-	-	-	-	-	-	-	-	-	-
Alhagi pod section	-	-	-	-	-	-	-	-	-	-	-	-
Alhagi leaf?	-	-	-	-	-	-	-	-	-	-	-	-
Aegilops glume base	-	-	-	-	-	-	-	-	-	-	-	-
Hordeum internode (wild)	-	-	-	-	-	-	-	-	-	-	-	-
Taeniatherum int. (YH-pp 8)	-	-	-	-	-	-	-	-	-	-	-	-
Poaceae, mystery 1 rachis frag.	-	-	-	1	-	-	-	-	-	-	-	-
YH-plant part 9	-	-	-	-	-	-	-	-	-	-	-	-
UNCHARRED-MINERALIZED												
Arnebia linearis	-	-	-	-	-	-	-	-	-	-	-	-
Arnebia/Lithospermum	-	-	-	-	-	-	-	-	-	2	-	-
Lithospermum tenuifolium	-	-	-	-	-	-	-	-	-	-	-	-
Lithospermum arvense	-	-	-	-	-	-	-	-	-	-	-	-
Lithospermum, misc.	-	-	-	-	-	-	-	-	-	-	-	-
Moltkia	-	-	-	-	-	-	-	-	-	-	-	1
BORAGINACEAE	-	-	-	-	-	-	-	-	1	-	-	-
Eleocharis (Cyperaceae-a)	-	-	-	-	1	-	-	-	-	4	-	-
Carex	-	-	-	-	-	-	-	-	-	-	-	-
Ficus	-	-	-	-	1	-	-	-	-	-	-	-
Glaucium	-	-	-	-	-	-	-	-	-	2	-	-
Papaver (white)	-	-	-	-	-	-	-	-	-	2	-	-

* Economic Plants

Col.	YH no.	Additional contents
5	20605	1 Gossypium
12	22075	5 Gossypium
14	20480	1 Trigonella foenum-graecum
37	31277	1 (0.04 g) Cicer
45	21207	0.04 g Pisum
52	22084	1 Coriandrum
62	30039	1 (+ g) Pisum/Vicia
73	23307	1 cf. Lathyrus
77	26789	1 (0.04 g) cf. Lathyrus
81	27239	1 (0.03 g) cf. Pisum
94	26944	3 (0.03 g) Pisum/Vicia
100	30191	1 (+ g) cf. Lathyrus
101	31560	+ g Prunus (almond)
102	29540	1 (0.12 g) Prunus (cherry)

Col.	YH no.	Additional contents
103	31203	0.01g Prunus (almond)
114	32598	1.5 (0.02 g) cf. Lathyrus
158	28491	0.01 g Prunus (almond)
164	21526	0.02 g Prunus (almond)
168	32550	+ g Prunus (almond)
188	21498	0.03 g cf. Pisum
191	21542	0.03 g cf. Pisum
195	31053	0.05 g Pisum/Vicia
226	32144	0.01 g Prunus (almond)
229	33243	0.13 g Cicer arietinum
232	33555	0.06 g Cicer arietinum, 11760 Linum usitatis-simum
235	33580	0.15 g Cicer arietinum
244	33335	1 (0.03 g) cf. Lathyrus

†Wild and Weedy Plants

Col.	YH no.	Additional contents
2	26472	1 YH-unk 14
5	20605	1 YH-Asteraceae 12
9	22343	1 YH-Lamiaceae 6, 1 YH-Poaceae 7.1, 3 YH-unk 14. PP: 1 Beta inflorescence, was YH-pp16, 1 Vitis, dried
10	21728	1 YH-Asteraceae 7, 10 Fimbristylis
11	22074	1 Hypericum, 4 cf. Potentilla
14	20480	1 YH-unk 16
21	20825	PP: 0.48 g Phragmites stem fragments
25	21369	1 YH-unk 14
27	23634	5 Convolvulus, 1 Coronilla
29	23633	1 YH-unk 16. PP: 1 cf. Neslia silique
30	23637	1 Veronica, 2 YH-unk 32
31	23393	1 YH-Lamiaceae 6
32	25713	min'l: 1 YH-unknown 24
36	28338	1 Anthriscus, 1 Medicago radiata, 25 Trigonella capitata, 3 YH-Fabaceae 3, 38 cf. Geranium, 7 cf. Phleum pratense, 59 YH-Poaceae 7.1, 2 YH-unk 28. PP: 0.01 g Vitis fruit, 9 Carduae phyllary tips, 1 Boreava orientalis silique, 18 Medicago constricta pod fragments, Min'l 1 Nonea
37	31277	1 Vicia, 3 YH-unk 36
46	21245	1 Medicago radiata
47	21585	1 YH-Solanaceae 2
48	21719	25 Salsola (=YH-unk 10)
49	22669	1 Vicia. PP: + Trigonella pod frag.
50	27718	4 YH-Asteraceae 13. PP: 136 Carduae phyllaries, 142 Carduae phillary tips, 1 Centaurea head, 4 fragments Onopordum capitulum
51	22085	1 Bifora
53	22096	1 YH-Lamiaceae 6
54	22109	5 Eryngium, 1 Vicia, 14 YH-Fabaceae 4, 5 Lamium cf. amplexicaule, 1 YH-unk 12. PP: 1 cf. Cardaria draba silique, 14 Trigonella pod frags., >10 Lamiaceae stem fragments (=0.03 g), 1 cf. Peganum harmala fruit case; 1 YH-pp 14

Col.	YH no.	Additional contents
58	22337	1 Taraxacum, 1 cf. Cerastium
59	29295	1 Verbena officinalis. PP: 1 YH-unknown 25
61	25652	1 cf. Turgenia, 1 cf. Senecio 1 YH-Asteraceae 10
63	28856	3 cf. Cerastium, 4 Trigonella capitata
68	23009	6 cf. Senecio, 2 Valerianella dentata
69	23053	1 YH-Lamiaceae 1
73	23307	1 Artedia, 2 cf. Turgenia, 2 YH-Asteraceae 12. PP: Asteraceae head includes type 1 and type 2 (one each); 3 Hypecoum fruit pod segments, 1 Papaver disk fragment
77	26789	1 Mentha
82	26899	1 cf. Senecio, 1 Convolvulus
84	26870	1 Mentha, 1 YH-Poaceae 12, 1 Valerianella dentata
89	25248	1 YH-Unk 17 YH-Unk 32
90	28774	1 cf. Senecio
92	30664	1 YH-Asteraceae 11, 1 Convolvulus; 4 YH-Lamiaceae 6, 1 Linum. PP: 1 Brassicaceae silique 3, 3 Atriplex bract YH-pp 11
94	26944	1 Bifora, 1 YH-unk 32
98	28522	2 Convolvulus
100	30191	1 YH-Asteraceae 10, 1 Convolvulus, 1 Coronilla, 1 cf. Geranium, 5 YH-Lamiaceae 7, 1 Verbena officinalis
102	29540	min'l: 1 Hordeum
107	30990	1 Cirsium, 1 YH-Poaceae 6
115	32818	1 Y-Asteraceae 7
130	26398	3 YH-Lamiaceae 4
131	27461	102 YH-Lamiaceae 4
132	27471	1 YH-Asteraceae 10, 4 YH-unk 38
137	22706	2 YH-Lamiaceae 1, 4 YH-unk 3
139	31666	1 Koelpinia
142	31128	1 YH-unk 14
144	26562	1 YH-unk 14. PP: 5 YH-pp 15
145	27981	1 YH-Unk 40
147	30494	PP: + Medicago pod fragment
150	28176	16 YH-Fabaceae 3

†Wild and Weedy Plants cont'd.

Col.	YH no.	Additional contents
153	20987	1 YH-Asteraceae 13
155	22281	1 YH-unk 18
157	22192	2 YH-unk 36
163	21302	1 YH-Lamiaceae 1, 1 YH-Poaceae 7.1
164	21526	3 Verbascum, 1 Solanum, 3 YH-unk 12, 7 YH-unk 17. PP: 2 YH-pp 18
165	29473	1 YH-Lamiaceae 1
167	33319	1 YH-unk 18
177	28188	1 YH-Poaceae 7.1; 1 YH-unk 17
184	21046	1 YH-Lamiaceae 1, 1 Linum
191	21542	1 YH-unk 14
192	22178	2 Veronica
193	22270	PP: 1 silique 4
195	31053	1 YH-Poaceae 19
196	22276	1 YH-unk 36
198	22780	1 YH-unk 7, 1 YH-unk 36

Col.	YH no.	Additional contents
203	23208	1 YH-Asteraceae 13
207	29965	2 Koelpinia
209	31603	2 YH-Apiaceae 9
210	31836	1 Ajuga chamaepitys
211	31837	1 Medicago radiata. PP: 1 silique 4
213	31089	1 YH-Apiaceae 3
219	28838	PP: 1 YH-pp 6
224	33280	1 YH-unk 18
225	33270	1 Salsola (=YH-unk 10), 4 YH-unk 19, 1 YH-unk 37
228	33230	1 Raphanus. PP: 0.12 g glumes
229	33243	1 cf. Caucalis
230	33246	1 Anchusa cf. azurea
232	33555	33 Thlaspi
233	33573	1 cf. Turgenia
237	33590	PP: 0.21 g straw
244	33335	PP: 4 pods (Fabaceae); 4 Juncus capsules

Anderson, Seona. 1994/1995. Fuel, Fodder and Faeces: An Ethnographic and Botanical Study of Dung Fuel Use in Central Anatolia. M.Sc. thesis, Dept. of Archaeology and Prehistory, University of Sheffield.

Aytuğ, Burhan. 1970. Arkeolojik araştırmaların ışığı altında iç Anadolu stebi. *İstanbul Orman Fakültesi Dergisi,* Seri A 20:127–43.

_______ 1988. Le mobilier funéraire du roi Midas I. In *Wood and Archaeology,* ed. T. Hackens, A.V. Munaut, and C. Till, pp. 357–68. PACT. 22. Strasbourg: Council of Europe.

Aytuğ, Burhan, and Ertuğrul Gorcelioğlu. 1988. Gordion'daki 'P' Tümülüsünde bulunmuş mobilya'lar. *Arkeometri Sonuçları Toplantısı* 4:49–59.

Aytuğ, Burhan, and Sevil Pehlivan. 1989. Gordion'daki 'P' Tümülüsünden bulunmuş mobilyalar II ve bazı odun eşyalar. In *TÜBITAK, Aksay Ünitesi* I, pp. 123–48. Ankara.

Barth, Hans-Jörg, and Horst Strunk. The Die-back Phenomenon of *Juniperus procera* at the Al-Soudah Family Park. http://www.uni-regensburg.de/Fakultaeten/phil_Fak_III/Geographie/phygeo/downloads/barthjuniperus.pdf (verified February 19, 2007).

Bellinger, L. 1962. Textiles from Gordion. *The Bulletin of the Needle and Bobbin Club* 46:4–33.

Blanchette, Robert A., and Elizabeth Simpson. 1992. Soft Rot and Wood Pseudomorphs in an Ancient Coffin (700 BC) from Tumulus MM at Gordion, Turkey. *IAWA Bulletin* n.s. 13:201–13.

Bogach, V.S. 1985. *Wood as Fuel, Energy for Developing Countries.* New York: Praeger.

Bottema, S., and H. Woldring. 1984. Late Quaternary Vegetation and Climate of Southwestern Turkey, Part II. *Palaeohistoria* 26:123–49.

_______ 1990. Anthropogenic Indicators in the Pollen Record of the Eastern Mediterranean. In *The Impact of Ancient Man on the Landscape of the Eastern Mediterranean Region and the Near East,* ed. S. Bottema, G. Entjes-Nieborg, and W. van Zeist, pp. 231–65. Rotterdam: A. A. Balkema.

Bottema, S., H. Woldring, and B. Aytuğ. 1993/1994. Late Quaternary Vegetation History of Northern Turkey. *Palaeohistoria* 35/36:13–72.

Braadbaart, F., and P. F. van Bergen. 2005. Digital Imaging Analysis of Size and Shape of Wheat and Pea upon Heating under Anoxic Conditions as a Function of the Temperature. *Vegetation History and Archaeobotany* 14:67–75.

Braudel, Fernand. 1979. *The Structures of Everyday Life.* New York: Harper & Row.

Burke, Brendan. 2005. Textile Production at Gordion and the Phrygian Economy. In *The Archaeology of Midas and the Phrygians, Recent Work at Gordion,* ed. L. Kealhofer, pp. 69–81. Philadelphia: University of Pennsylvania Museum of Archaeology and Anthropology.

Dandoy, Jeremiah R., Page Selinsky, and Mary M. Voigt. 2002. Celtic Sacrifice. *Archaeology* 55(1): 44–49.

Davis, Peter. 1965–1988. *Flora of Turkey.* 10 vols. Edinburgh: The University Press.

DeVries, Keith. 2000. Gordion. *Expedition* 42(1): 18–20.

_______ 2005. Greek Pottery and Gordion Chronology. In *The Archaeology of Midas and the Phrygians, Recent Work at Gordion,* ed. L. Kealhofer, pp. 36–55. Philadelphia: University of Pennsylvania Museum of Archaeology and Anthropology.

DeVries, K., P. I. Kuniholm, G. K. Sams, and M. M. Voigt. 2003. New Dates for Iron Age Gordion. *Antiquity* 77:296.

Enneking, Dirk. 1995. *The Toxicity of* Vicia *Species*

and Their Utilisation as Grain Legumes. 2nd ed. Centre for Legumes in Mediterranean Agriculture (CLIMA) Occasional Publication No. 6. Nedlands W.A.: University of Western Australia.

Ertuğ, Füsun. 2000. An Ethnobotanical Study in Central Anatolia (Turkey). *Economic Botany* 54: 155–82.

Fahn, A., E. Werker, and P. Baas. 1986. *Wood Anatomy and Identification of Trees and Shrubs from Israel and Adjacent Regions.* Jerusalem: Israel Academy of Sciences and Humanities.

Forest Research Institute and Colleges. 1972. *Indian Forest Utilization.* 2 vols. Dehra Dun: Forest Research Institute and Colleges.

French, David H. 1971. An Experiment in Water-sieving. *Anatolian Studies* 21:59–64.

Fritts, H. C. 1976. *Tree Rings and Climate.* New York: Academic Press.

Goldman, Andrew. 2005. Reconstructing the Roman-period Town at Gordion. In *The Archaeology of Midas and the Phrygians, Recent Work at Gordion,* ed. L. Kealhofer, pp. 56–67. Philadelphia: University of Pennsylvania Museum of Archaeology and Anthropology.

Graves, H. S. 1919. *The Use of Wood for Fuel.* USDA Bulletin No. 753. Washington, D.C.

Gürsan-Salzmann, Ayşe. 2005. Ethnographic Lessons for Past Agro-Pastoral Systems in the Sakarya-Porsuk Valleys. In *The Archaeology of Midas and the Phrygians, Recent Work at Gordion,* ed. L. Kealhofer, pp. 172–89. Philadelphia: University of Pennsylvania Museum of Archaeology and Anthropology.

Hall, R. T. 1942. Wood Fuel in Wartime. *Farmers' Bulletin* No. 1912. Washington, D.C.: United States Department of Agriculture.

Helbaek, Hans. 1961. Late Bronze Age and Byzantine Crops at Beycesultan in Anatolia. *Anatolian Studies* 11:77–97.

______ 1959. Notes on the Evolution and History of Linum. *Kuml,* pp. 103–29.

______ 1969. Plant Collecting, Dry-Farming, and Irrigation Agriculture in Prehistoric Deh Luran. In *Prehistory and Human Ecology of the Deh Luran Plain,* by F. Hole, K. V. Flannery, and J. A. Neely, pp. 383–426. University of Michigan Museum of Anthropology Memoir 1. Ann Arbor.

Henrickson, Robert C. 1993. Politics, Economics, and Ceramic Continuity at Gordion in the Late Second and First Millennia B.C. In *The Social and Cultural Contexts of New Ceramic Technologies,* ed. W. D. Kingery, pp. 89–176. Westerville, OH: American Ceramic Society.

Heun, M., R. Schäfer-Pregl, D. Klawan, R. Castagna, M. Accerbi, B. Borghi, and F. Salamini. 1997. Site of Einkorn Wheat Domestication Identified by DNA Fingerprinting. *Science* 278:1312–14.

Hillman, Gordon C. 1984. Interpretation of Archaeological Plant Remains: The Application of Ethnographic Models from Turkey. In *Plants and Ancient Man,* ed. W. van Zeist and W. A. Casparie, pp. 1–41. Rotterdam: A. A. Balkema.

Hoffner, Harry A. 1974. *Alimenta Hethaeorum, Food Production in Hittite Asia Minor.* New Haven, CT: American Oriental Society.

Hubbard, R. N. L. B. 1975. Assessing the Botanical Component of Human Paleo-economies. *Bulletin of the Institute of Archaeology* 12:197–205.

______ 1976. Crops and Climate Change in Prehistoric Europe. *World Archaeology* 8(2): 159–68.

Kayacık, Hayrettin, and Burhan Aytuğ. 1968. Gordion kral mezari'nin ağaç malzemesi üzerinde ormancılık yönünden araştırmalar (Recherches au point de vue forestier sur les matériaux en bois du tombeau royal de Gordion). *İstanbul Üniversitesi Orman Fakültesi Dergisi* Seri A,18:37–54.

Kealhofer, Lisa. 2005. The Gordion Regional Survey. Settlement and Land Use. In *The Archaeology of Midas and the Phrygians, Recent Work at Gordion,* ed. L. Kealhofer, pp. 137–48. Philadelphia: University of Pennsylvania Museum of Archaeology and Anthropology.

Körte, Gustav, and Alfred Körte. 1904. *Gordion Ergebnisse der Ausgrabung im Jahre 1900.* Berlin: Druck und Verlag von Georg Reimer. On line: http://efts.lib.uchicago.edu/cgibin/eos/eos_title.pl?callnum=CC27.D481_vol5

Kroll, Helmut. 1991. Südosteuropa. In *Progress in Old World Palaeoethnobotany,* ed. W. van Zeist, K. Wasylikowa, and K.-E. Behre, pp. 161–77. Rotterdam: A. A. Balkema.

Kuniholm, Peter I. 1977. Dendrochronology at Gordion and on the Anatolian Plateau. Ph.D. diss., University of Pennsylvania. Philadelphia.

______ 1990. Aegean Dendrochronology Project: 1989–1990 Results/Ege'deki Dendrokronoloji Projesi:

1989–1990 Sonuçlari. *Arkeometri Sonuçlari Toplantisi* 6:138.

Kuniholm, Peter I., and Susan L. Tarter. 1989. Gordion: Midas Mound Tumulus, City Mound, TB2A, Tumuli B, P, Z, Koerte III, and Mama Derisi. Unpublished report. Gordion archive, University of Pennsylvania Museum of Archaeology and Anthropology. Philadelphia.

McGovern, Patrick E., Donald L. Glusker, Robert A. Moreau, Alberto Nuñez, Curt W. Beck, Elizabeth Simpson, Eric D. Butrym, Lawrence J. Exner, and Edith C. Stout. 1999. A Funerary Feast Fit for King Midas. *Nature* 402:863–64.

Mallowan, M. E. L. 1953. The Excavations at Nimrud (Kalhu), 1952. *Iraq* 15:1–42.

Marsh, Ben. 1993. Report of Geomorphic Investigations—Gordion & Environs. Unpublished report. Gordion archive, University of Pennsylvania Museum of Archaeology and Anthropology. Philadelphia.

______ 2000. Geomorphology of the Gordion Regional Survey. Unpublished report dated September 12, 2000. Gordion archive, University of Pennsylvania Museum of Archaeology and Anthropology. Philadelphia.

______ 2005. Physical Geography, Land Use, and Human Impact at Gordion. In *The Archaeology of Midas and the Phrygians, Recent Work at Gordion,* ed. L. Kealhofer, pp. 161–71. Philadelphia: University of Pennsylvania Museum of Archaeology and Anthropology.

Marston, John M. 2003. Plant Remains from Middle and Late Phrygian Domestic Contexts at Gordion, Turkey (1993–1994). MASCA Ethnobotanical Laboratory Report 34, on file, Gordion archive, University of Pennsylvania Museum of Archaeology and Anthropology. Philadelphia.

______ 2009. Modeling Wood Acquisition Strategies from Archaeological Charcoal Remains. *Journal of Archaeological Science* 36:2192–2200.

______ 2010. The Behavioral Ecology and Environmental Archaeology of Gordion, a City in the Central Anatolian Steppe. Ph.D. diss., University of California, Los Angeles.

______ n.d. Reconstructing the Functional Use of Wood at Phrygian Gordion through Charcoal Analysis. Ms. prepared for publication in *The Archaeology of Phrygian Gordion,* ed. Brian Rose. Philadelphia:

University of Pennsylvania Museum of Archaeology and Anthropology. Forthcoming.

Meteoroloji Işleri Genel Müdürlüğü. 1974. *Ortalama ve ekstrem kiymetler meteroloji bülteni.* Ankara: Devlet Işleri Genel Müdürlüğü. Başbakanlık Basımevi.

Miller, Naomi F. 1982. Economy and Environment of Malyan, a Third Millennium B.C. Urban Center in Southern Iran. Ph.D. diss., University of Michigan, Ann Arbor.

______ 1984. The Use of Dung as Fuel: An Ethnographic Model and an Archaeological Example. *Paléorient* 10(2): 71–79.

______ 1985. Paleoethnobotanical Evidence for Deforestation in Ancient Iran: A Case Study of Ancient Malyan. *Journal of Ethnobiology* 5:1–19.

______ 1988. Ratios in Paleoethnobotanical Analysis. In *Current Paleoethnobotany,* ed. C. Hastorf and V. Popper, pp. 72–85. Chicago: University of Chicago Press.

______ 1990. Clearing Land for Farmland and Fuel in the Ancient Near East. *Economy and Settlement in the Near East: Analyses of Ancient Sites and Materials.* MASCA Research Papers in Science and Archaeology, supplement to vol. 7, pp. 70–78. Philadelphia: MASCA, University Museum of Archaeology and Anthropology, University of Pennsylvania.

______ 1991. The Near East. In *Progress in Old World Palaeoethnobotany,* ed. W. van Zeist, K. Wasylikowa, and K.-E. Behre, pp. 133–60. Rotterdam: A.A. Balkema.

______ 1997a. The Macrobotanical Evidence for Vegetation in the Near East, c. 18000/16000 BC to 4000 BC. *Paléorient* 23(2): 197–207.

______ 1997b. Farming and Herding along the Euphrates: Environmental Constraint and Cultural Choice (Fourth to Second Millennia B.C.). In *Subsistence and Settlement in a Marginal Environment: Tell es-Sweyhat, 1989–1995 Preliminary Report,* by R. L. Zettler, J. A. Armstrong, A. Bell, M. Braithwaite, M. D. Danti, N. F. Miller, P. N. Peregrine, and J. A. Weber, pp. 123–32. MASCA Research Papers in Science and Archaeology 14. Philadelphia: MASCA, University of Pennsylvania Museum of Archaeology and Anthropology.

______ 1998. Patterns of Agriculture and Land Use at Medieval Gritille. In *The Archaeology of the*

Frontier in the Medieval Near East: Excavations at Gritille, Turkey, by Scott Redford, pp. 211–52. Archaeological Institute of America Monograph n.s. 3. Philadelphia: University of Pennsylvania Museum of Archaeology and Anthropology.

_____ 1999. Erosion, Biodiversity, and Archaeology: Preserving the Midas Tumulus at Gordion/Erozyon, bioçeşitlilik ve arkeoloji, Gordion'daki Midas Höyüğü'nun korunması. *Arkeoloji ve Sanat* 93:13–19 + plate.

_____ 2007a. Roman and Medieval Charcoal from the 2004 Excavation at Gordion, Operations 52, 53, 54, and 55. MASCA Ethnobotanical Laboratory Report 42, on file, Gordion Archive, University of Pennsylvania Museum of Archaeology and Anthropology. Philadelphia.

_____ 2007b. Roman Flotation Samples from the 2004 and 2005 Excavation at Gordion, Operations 44, 52, 53, 54, and 55. MASCA Ethnobotanical Laboratory Report 45, on file, Gordion Archive, University of Pennsylvania Museum of Archaeology and Anthropology. Philadelphia.

Miller, Naomi F., and Kurt Bluemel. 1999. Plants and Mudbrick: Preserving the Midas Tumulus at Gordion, Turkey. *Conservation and Management of Archaeological Sites* 3:225–37.

Miller, Naomi F., Kimberly E. Leaman, and Julie Unruh. 2006. Serendipity. Secrets of the Mudballs. *Expedition* 48(3): 40–41.

Miller, Naomi F., and Tristine Lee Smart. 1984. Intentional Burning of Dung as Fuel: A Mechanism for the Incorporation of Charred Seeds into the Archeological Record. *Journal of Ethnobiology* 4:15–28.

Miller, Naomi F., Melinda A. Zeder, and Susan R. Arter. 2009. From Food and Fuel to Farms and Flocks: The Integration of Plant and Animal Remains in the Study of Ancient Agropastoral Economies at Gordion, Turkey. *Current Anthropology* 50:915–24.

Minnis, Paul E. 1981. Seeds in Archaeological Sites: Sources and Some Interpretive Problems. *American Antiquity* 46:143–52.

Mohlenbrock, Robert H. 1991. Buffalo Beats, Ohio. *Natural History* (Dec.):18–20.

Morris, Ian. 2004. Economic Growth in Ancient Greece. *Journal of Institutional and Theoretical Economics* 160:709–42.

Nesbitt, M., and G. D. Summers. 1988. Some Recent Discoveries of Millet (*Panicum miliaceum* and *Setaria italica*) at Excavations in Turkey and Iran. *Anatolian Studies* 38:85–97.

Pabot, H. 1960. The Native Vegetation and Its Ecology in the Khuzistan River Basins. Ahwaz, Iran: Khuzistan Development Service. Mimeo.

Panshin, A. J., and C. de Zeeuw. 1970. *Textbook of Wood Technology*, Vol. 1. 3rd ed. New York: McGraw-Hill.

Popper, V. S. 1988. Selecting Quantitative Measurements in Paleoethnobotany. In *Current Paleoethnobotany,* ed. C. A. Hastorf and V. S. Popper, pp. 53–71. Chicago: University of Chicago Press.

Reynolds, R. V., and A. H. Pierson. 1942. Fuel Wood Used in the United States 1630–1930. USDA Circular no. 641. Washington, D.C.

Riehl, Simone. 1999. *Bronze Age Environment and Economy in the Troad. The Archaeobotany of Kumtepe and Troy.* BioArchaeologica 2. Tübingen: Mo Vince Verlag.

Sams, G. Kenneth. 1977. Beer in the City of Midas. *Archaeology* 30:108–15.

_____ 1988. The Early Phrygian Period at Gordion: Toward a Cultural Identity. *Source* 7(3/4): 9–15.

_____ 2005. Gordion. Explorations over a Century. In *The Archaeology of Midas and the Phrygians, Recent Work at Gordion,* ed. L. Kealhofer, pp. 10–21. Philadelphia: University of Pennsylvania Museum of Archaeology and Anthropology.

Samuel, Delwen. 2001. Archaeobotanical Evidence and Analysis. In *Peuplement rural et aménagements hydroacgricoles dans la moyenne vallée de l'Euphrate fin VIIe–XIXe siècle, région de Deir ez-Zor-Abu Kemal (Syrie),* ed. S. Berthier, pp. 347–481. Institut Français de Damas, Damascus.

Schoch, Werner H., Barbara Pawlik, and Fritz H. Schweingruber. 1988. *Botanische Makroreste/Botanical Macro-remains/Macrorestes botaniques.* Bern: Verlag Paul Haupt.

Schwarz, P., and R. Horsley. n.d. A Comparison of North American Two-row and Six-row Malting Barley. The Brewers' Market Guide, on-line at http://www.brewingtechniques.com/bmg/schwarz.html (verified October 28, 2008).

Schweingruber, F. H. 1990. *Anatomie europäischer Hölzer/Anatomy of European Wood.* Bern: Verlag Paul Haupt.

Smart, T. L., and E. S. Hoffman. 1988. Environmental Interpretation of Archaeological Charcoal. In *Current Paleoethnobotany,* ed. C. A. Hastorf and V. S. Popper, pp. 167–205. Chicago: University of Chicago Press.

Stein, Gil, Reinhard Bernbeck, Cheryl Coursey, Augusta McMahon, Naomi F. Miller, Adnan Misir, Jeffrey Nicola, Holly Pittman, Susan Pollock, and Henry Wright. 1966. Uruk Colonies and Anatolian Communities: An Interim Report on the 1992–1993 Excavations at Hacınebi, Turkey. *American Journal of Archaeology* 100:205–60.

Townsend, C. C., and E. Guest. 1985. *Monocotyledones. Flora of Iraq*, Vol. 8. Ministry of Agriculture and Agrarian Reform, Baghdad.

van Geel, B., N. A. Bokovenko, N. D. Burova, K. V. Chugunov, V. A. Dergachev, V. G. Dirksen, M. Kulkova, A. Nagler, H. Parzinger, J. van der Plicht, S. S. Vasiliev, and G. I. Zaitseva. 2004. Climate Change and the Expansion of the Scythian Culture after 850 BC: A Hypothesis. *Journal of Archaeological Science* 31:1735–42.

van Zeist, W., and J. A. H. Bakker-Heeres. 1982 [1985]. Archaeobotanical Studies in the Levant 1. Neolithic Sites in the Damascus Basin: Aswad, Ghoraifé, Ramad. *Palaeohistoria* 24:165–256.

______ 1984[1986]. Archaeobotanical Studies in the Levant 2. Neolithic and Halaf Levels at Ras Shamra Report. *Palaeohistoria* 26:151–70.

______ 1985 [1988]. Archaeobotanical Studies in the Levant 4. Bronze Age Sites on the North Syrian Euphrates. *Palaeohistoria* 27:247–316.

van Zeist, W., and S. Bottema. 1991. Late Quaternary Vegetation of the Near East. *Beihefte zum Tübinger Atlas des voerderen Orients,* Reihe A, nr. 18. Wiesbaden: Dr. Ludwig Reichert Verlag.

van Zeist, W., H. Woldring, and D. Stapert. 1975. Late Quaternary Vegetation and Climate of Southwestern Turkey. *Palaeohistoria* 17:53–143.

Voigt, Mary M. 1994. Excavations at Gordion 1988–89: The Yassıhöyük Stratigraphic Sequence. In *Anatolian Iron Ages 3: The Proceedings of the Third Anatolian Iron Ages Colloquium Held at Van, 6–12 August 1990 /Anadolu Demir Çağları: III. Anadolu Demir Çağları Sempozyumu Bildirileri,* ed. A. Çilingiroğlu and D. H. French. London: British Institute of Archaeology at Ankara.

______ 1996. Provenience Key: Gordion 1988–89. Report on file, University of Pennsylvania Museum, Philadelphia.

______ 2003. Celts at Gordion. *Expedition* 45(1): 14–19.

______ 2005. Old Problems and New Solutions. Recent Excavations at Gordion. In *The Archaeology of Midas and the Phrygians, Recent Work at Gordion*, ed. L. Kealhofer, pp. 22–35. Philadelphia: University of Pennsylvania Museum of Archaeology and Anthropology.

______ 2007. The Middle Phrygian Occupation at Gordion. In *Anatolian Iron Ages 6*, ed. A. Çilingiroğlu and A. Sagona, pp. 311–33. Leuven: Peeters.

Voigt, Mary M., and Robert C. Henrickson. 2000a. Formation of the Phrygian State: The Early Iron Age at Gordion. *Anatolian Studies* 50:37–54.

______ 2000b. The Early Iron Age at Gordion: The Evidence from the Yassıhöyük Stratigraphic Sequence. In *The Sea Peoples and Their World: A Reassessment,* ed. E. D. Oren, pp. 327–60. Philadelphia: University of Pennsylvania Museum.

Voigt, Mary M., and T. Cuyler Young, Jr. 1999. From Phrygian Capital to Achaemenid Entrepot: Middle and Late Phrygian Gordion. *Iranica Antiqua* 34:191–241.

Walter, Heinrich. 1956. Das Problem der Zentralanatolischen Steppe. *Die Naturwissenschaften* 5:97–102.

Wilding, L. P., and L. R. Drees. 1968. Biogenic Opal in Soils as an Index of Vegetative History in the Prairie Peninsula. In *The Quaternary of Illinois,* ed. R. E. Bergstrom, pp. 96–103. University of Illinois College of Agriculture Special Publication 14.

Young, R. S. 1960. Gordion: Phrygian Construction and Architecture. *Expedition* 2(2): 2–9.

______ 1981. *Three Great Early Tumuli*. Philadelphia: University Museum, University of Pennsylvania.

Zeder, M. A., and S. Arter. 1994. Changing Patterns of Animal Utilization at Ancient Gordion. *Paléorient* 20(2): 105–18.

Zohary, D., and M. Hopf. 1994. *Domestication of Plants in the Old World.* 2nd ed. Oxford: Clarendon Press.

Zohary, M. 1973. *Geobotanical Foundations of the Middle East.* 2 vols. Stuttgart: Fischer Verlag.

Index

(Page numbers in italics indicate illustrations)

Abandoned Village (YHSS 3A) *3*, 5, 21, 31, 61
agricultural year 15–16
agriculture. *See* crops, irrigation
agropastoral continuum 8, 64, 67–69
Alhagi (camelthorn) 16, 34, 51, *58*, *125–126*
Alnus viridis (alder) 28, 33, 77
animal husbandry 16, 64
    ancient 7, 54, 66–69
animals, hunted *67–68*

Burnt Reed House–BRH (YHSS 7) *3*, 4, 21, 31, 33, 42,
    47, 59

catchment 7, 18, 32, 66, 69
cereal yield 15
cereals. *See* crops
charcoal
    identification 22, App. B
    analysis, quantification 22–23, *24–27*
    data App. E
chronology 2, 4
climate 6, 9, 11–12
    ancient 17, 18, 34, 65
crop yield, cereal 15
crops 13, 15, 52
crops, ancient 40–48, 64
    concentrations 42, 47, 59, 61
    barley (*Hordeum vulgare*) 13, 43–44, *46*, *47*
    bitter vetch (*Vicia ervilia*) 13, *46*, *47*
    chickpeas (*Cicer arietinum*) 13, 47
    cotton (*Gossypium*) 48, 66
    cumin (*Cuminum*) 13
    flax (*Linum usitatissimum*) 13, 48, *49*
    grape (*Vitis vinifera*) 48
    lentil (*Lens culinaris*) 13, *46*, *47*
    millet (Italian) (*Setaria italica*) 44, *47*, 66

    rice (*Oryza sativa*) 44, 47, 66
    rye (*Secale cereale*) 13
    wheat (*Triticum*, various) 13, 40–43, *45*, *46*, 69
cultural affiliation 6, 8, 69–70

Destruction Level (YHSS 6A) *3*, 5, 18. *See also* Early
    Phrygian, Terrace Building 2A
dung 64. *See also* wild:cereal
    fuel 17, 30, 40
    Turkish words for 17

Early Iron Age (YHSS 7) 4, 71. *See also* Burnt Reed
    House
Early Phrygian period (YHSS 6) 4, 5, 44, 46, 71. *See also*
    Terrace Building 2A, Destruction Level
erosion *34*, 64, 65
Euphrates sites 54, 56, 67
    Gritille 56
    Hacınebi 56
    Kurban Höyük 67
    Sweyhat 56

flotation analysis 37–39, 51, App. A. *See also*
    quantification, sampling, seed identification
    lab procedures 23–24, App. A
    data App. F
fodder 13, 16, 39, 43, 47, 64, 67. *See also* wild:cereal
fruit (grape, *Vitis vinifera*; cherry, *Prunus* sp.; hackberry,
    Celtis sp.) 48, 64
fuel 16–17, 63, 69
    dung 17, 30, 40, 51
    wood 22, 25, 27, 30, 32, 40

Galatian (European Celts) 5, 6
Gordion 1
    in history 1

pre-1987 archaeological investigation 1–2, 5, 18, 44, 47–48

Hellenistic period (YHSS 3) 4, 6, 71. *See also* Abandoned Village

insects *46, 47*
irrigation 7, 12, 16, 56, 65–66. *See also* cotton, millet, rice
Late Bronze Age (YHSS 8/9) *2, 5, 65,* 71
Late Phrygian period (YHSS 4) 4, 71

Medieval period (YHSS 1) 5, 6, 71
Middle Bronze (YHSS 10) 2, 5, 70
Middle Phrygian (YHSS 5) 4, 5–6, 65, 66, 67, 71
migration 8
    Celtic 70
    Phrygian 5, 69, 70

nuts (wild almond, *Prunus* sp.; wild pistachio, *Pistacia* sp.) 48

Peganum harmala (wild rue) 16, 34, 56, *58*
percentages 55–57

quantification
    percentages 55–57
    seed:charcoal 51, 53, *59*
    ubiquity 27, 51
    wild:cereal 54–55, 56, *58, 60, 61,* 64, 68
    wild:charcoal 54, *59*
    wood charcoal 25, 27

regional settlement 1, 5
Roman period (YHSS 2) 4

Sakarya valley 7, 9, *13,* 16, 63
sampling 21–22, 37–38

seed identification criteria
    economic plants 40–49
    wild and weedy plants App. D
seed:charcoal 51, 53, *59,* 66
soils 10, 34
stratigraphy. *See* YHSS

Terrace Building 2A (YHSS 6A) *3,* 31, 33–34, 42–43, 59
trees *14,* 16, 27–28
Trigonella 53, 56, *61,* 64
tumuli 6, 18–19
    Tumulus MM, plant products 48
    Tumulus MM, vegetation survey 49, App. C
    Tumulus MM, wood 18, 76

ubiquity 27, 51

vegetation 6, 7, 9, 12, *13–14,* 27
    change 31, 63
    survey 12, 49, 50, App. C

wild:cereal 54–55, 56, *58, 60, 61,* 64, 68
wild:charcoal 54, *59*
wood use, ancient 18–19, 33–34

YHSS (Yassıhöyük Stratigraphic Sequence) 2, 4
    YHSS 1. *See* Medieval period
    YHSS 2. *See* Roman period
    YHSS 3. *See* Hellenistic period, Abandoned Village
    YHSS 4. *See* Late Phrygian period
    YHSS 5. *See* Middle Phrygian period
    YHSS 6. *See* Early Phrygian period, Destruction Level
    YHSS 7. *See* Early Iron Age
    YHSS 8/9. *See* Late Bronze Age
    YHSS 10. *See* Middle Bronze Age
Young, Rodney S. excavation 1, 18

Author Note

NAOMI F. MILLER is an archaeobotanist with extensive excavation experience in Turkey (Gordion, Hacınebi, Kurban Höyük) and neighboring countries (Iran and Syria). Her special area of interest is long-term human impact on the environment and ancient vegetation and land use, particularly in west Asia. Her other publications include *The Archaeology of Garden and Field* (co-edited with Kathryn Gleason; 1994) and *Yeki bud, yeki nabud, Essays on the Archaeology of Iran in Honor of William M. Sumner* (co-edited with Kamyar Abdi; 2003).